Contents of Volume 2

London Mathematical Society Lecture Note Series. 305

Groups St Andrews 2001 in Oxford

Volume II

Edited by

C.M. Campbell
University of St Andrews

E.F. Robertson
University of St Andrews

G.C. Smith
University of Bath

PUBLISHED BY THE PRESS SYNDICATE OF THE UNIVERSITY OF CAMBRIDGE
The Pitt Building, Trumpington Street, Cambridge, United Kingdom

CAMBRIDGE UNIVERSITY PRESS
The Edinburgh Building, Cambridge CB2 2RU, UK
40 West 20th Street, New York, NY 10011–4211, USA
477 Williamstown Road, Port Melbourne, VIC 3207, Australia
Ruiz de Alarcón 13, 28014 Madrid, Spain
Dock House, The Waterfront, Cape Town 8001, South Africa

http://www.cambridge.org

First published 2003

Printed in the United Kingdom at the University Press, Cambridge

Typeface Computer Modem 11/13.5 pt. *System* LaTeX [TB]

A catalogue record for this book is available from the British Library

Library of Congress Cataloguing in Publication data

ISBN 0 521 53740 1 hardback

Contents of Volume 1

INTRODUCTION

Groups St Andrews 2001 in Oxford was another highly successful conference in the continuing series. The conference, held from 5 August 2001 to 18 August 2001, was attended by 230 mathematicians from 35 countries. The lectures and talks were given in the Mathematical Institute of the University of Oxford. We thank Lady Margaret Hall, St Anne's College, The Queen's College and Merton College for providing accommodation. We also acknowledge with gratitude the financial support of the Edinburgh and London Mathematical Societies. The Mathematical Institute of the University of Oxford generously made lecture rooms available, and supported the conference in various other important ways.

The organizing committee consisted of C M Campbell (University of St Andrews), D P Groves (Merton), R P Martineau (Wadham), P M Neumann (The Queen's College), E F Robertson (University of St Andrews), G C Smith (University of Bath), W B Stewart (Exeter) and G A Stoy (LMH). Administrative support was provided by Jan Campbell and Maureen White.

The main speakers, who were invited to give a series of talks, were Marston D E Conder (Auckland), Persi Diaconis (Stanford), Peter P Pálfy (Eötvös Loránd, Budapest), Marcus du Sautoy (Cambridge), and Michael R Vaughan-Lee (Oxford). As has become the tradition, all these invited speakers have written substantial articles for these Proceedings. All papers have been subjected to a formal refereeing process comparable to that of a major international journal. Publishing constraints have forced the editors to exclude some very worthwhile papers, and this is of course a matter of regret.

The Theory of Groups continues to move forward on many fronts, and twenty years on from the announcement of the classification of the finite simple groups, it prospers perhaps surprisingly well (rather like Mark Twain). It is a measure of the success of this conference series and this subject that mathematical libraries around the world are collecting the series of St Andrews Conference Proceedings.

As LaTeX has become the almost universal typesetting language for mathematics, the nuts and bolts of the editorial job are perhaps getting a little easier. We hope that in future even more contributors will stick faithfully to the typesetting standard prescribed by the editors (so that these hacks will have more time to do research in group theory, lie on beaches and so on). Fran Burstall of the University of Bath provided typesetting counselling. It is hoped that the next conference in this series will be held in 2005 and will be Groups St Andrews in St Andrews (revisiting the scene of the original crime), and it is also hoped that in 2009 we will return once more to Bath.

As these proceedings go to press, news has just arrived that one of the grandfathers of the subject, B H Neumann, has died. He attended the first Groups St Andrews conference in 1981 and some later ones in the series. Many of our community knew him with great affection – his warm smile and shoestring tie will linger in the memory almost as long as his important contributions to mathematics.

CMC, EFR, GCS

GRACEFULNESS, GROUP SEQUENCINGS AND GRAPH FACTORIZATIONS

GIL KAPLAN, ARIEH LEV and YEHUDA RODITTY

Department of Computer Sciences
The Academic College of Tel-Aviv-Yaffo
4 Antokolsky st., Tel-Aviv, Israel 64044
and
School of Mathematical Sciences
Tel-Aviv-University
Tel-Aviv, Israel

1 Introduction

In this article we survey recent results on the connection between decompositions of complete graphs which admit given automorphism groups, and some combinatorial problems related to these groups. Although these problems are of combinatorial nature, group theorists may find interest in these problems as well.

Most of the results surveyed here are found in [10], [11], [12] and [14] (we do not give proofs in this article, and the interested reader is referred to the above papers). Some of these results extend and generalize previous results on graph labeling (see [5] for a survey) into results related to any finite groups, and some of the results apply known results on group sequencings. Another result applies the Z^*-Theorem of Glaubermann.

We begin with basic definitions. By K_n (respectively, K_n^*) we denote the complete undirected (respectively, directed) graph on n vertices. (Each pair of different vertices in K_n^* is joined by two arcs (directed edges) with opposite directions.)

Definition 1.1 Let n be a natural number and let F be a directed (res., undirected) graph on m vertices, where $m \leq n$. The problem of determining whether or not the edge set of K_n^* (res., K_n) can be partitioned into arc (res., edge) disjoint subgraphs isomorphic to F will be denoted by $P_n^*(F)$ (res., $P_n(F)$). If a solution to this problem exists, we shall say that there exists a *decomposition* of K_n^* (res., K_n) into arc (res., edge) disjoint subgraphs isomorphic to F, and we shall denote this fact $F \mid K_n^*$ (res., $F \mid K_n$). If $m = n$, i.e. each subgraph in the decomposition is a spanning subgraph of K_n^* (res., K_n), we shall say that $F \mid K_n^*$ (res., $F \mid K_n$) is a *factorization* of K_n^* (res., K_n) into factors isomorphic to F. In this case (i.e., when $m = n$) we shall write briefly $P^*(F)$ (res., $P(F)$) instead of $P_n^*(F)$ (res., $P_n(F)$).

Notice that a decomposition problem can always be handled as a factorization problem, by considering a corresponding factor with added isolated vertices.

Definition 1.2 Let F be a directed (res., undirected) graph on n vertices, and let G be a permutation group on the vertices of K_n^* (K_n). We say that K_n^* (K_n) admits a *G-transitive factorization*, into factors isomorphic to F, if there exists a factorization of K_n^* (K_n) such that G acts transitively on the factors in the factorization. If the action of G on these factors is regular, we shall say that the factorization is *G-regular.* The problem of determining whether or not K_n^* (K_n) admits a G-regular factorization into factors isomorphic to F is denoted by $RP^*(G;F)$ ($RP(G;F)$). The problem of determining whether or not K_n^* (K_n) admits *any* regular factorization into factors isomorphic to F is denoted by $RP^*(F)$ ($RP(F)$).

Notice that when F is a directed (undirected) factor of K_n^* (K_n) and $RP^*(G;F)$ ($RP(G;F)$) has a solution, then the order of G must be $n(n-1)/|E(F)|$ (res., $\frac{1}{2}n(n-1)/|E(F)|$), where $E(F)$ denotes the arc (edge) set of F.

This article is organized as follows: In Section 2 we deal with the connection between the problems $RP^*(G;F)$, $RP(G;F)$ and a certain type of labeling of the vertices of F by the elements of G. These labelings generalize and extend well known labeling problems by natural numbers.

In Section 3 we deal with the problem $RP^*(G;F)$, where F is a disjoint union of cycles. In order to treat these problems, we apply and extend the well known concept of *group sequencing.* In the particular case when the factor F consists of a single cycle, known results on group sequencing are applied to obtain corresponding classification theorems.

Section 4 deals with the connection between transitive large sets of disjoint decompositions and group sequencings (see Section 4 for the corresponding definitions). In order to obtain classification results on such large sets, we apply the known concept of R-sequencing of groups. We define also the notion of a Frobenius set, which is a set of permutations having some properties similar to Frobenius groups. We show that the existence of such sets is equivalent to the existence of corresponding large sets.

In Section 5 we study a special case of decompositions, the *inner-transitive Hering decompositions.* We show that the classification of such decompositions is equivalent to the classification of pairs (τ, N), where N is a group and τ is an automorphism of a special type of N, called an (n,t) automorphism (see Section 5 for the corresponding definitions). In order to obtain this equivalence, we use Glaubermann's Z^* Theorem [7] and a generalization of this theorem (proved by Artemovich [3]).

For a directed (or undirected) graph H, we shall use the standard notation $V(H)$ and $E(H)$ for the vertex set and arc (or edge) set of H, respectively. *All groups in this article are finite.* We use the standard notation of group theory.

2 Regular factorization and labeling of the vertices of a graph by group elements

When dealing with regular factorizations of the complete graph K_n^* (Definition 1.2) two fundamental problems arise:

1. For which factors F of K_n^* does there exist a solution to $RP^*(F)$;
2. Classify all pairs (G,F), where G is a group and F is a factor of K_n^*, for which $RP^*(G;F)$ has a solution.

It emerges that the above problems are closely related to a problem of a certain labeling of the vertices of F by the elements of G, the *G-graceful labeling.* The definition is as follows.

Definition 2.1 Let G be a group of order n and let H be a (not necessarily connected) directed graph with n vertices and $n-1$ arcs. We say that H is *G-graceful* if there exists a bijection $f : V(H) \to G$ such that $G - \{1\} = \{f(v)f(u)^{-1} \mid (u, v) \in E(H)\}$.

Notice that in Definition 2.1 all the elements of the multiset $\{f(v)f(u)^{-1} \mid (u, v) \in E(H)\}$ are distinct. Notice also that F may have isolated vertices.

Example 2.2 The following are three simple examples of G-graceful graphs. The corresponding G-graceful labelings are specified by the notation of the vertices.

1. Let G be the additive group $\mathbf{Z}_4 = \{0, 1, 2, 3\}$ and let H be the following directed graph: $V(H) = \{0, 1, 2, 3\}$, $E(H) = \{(1, 0), (1, 2), (1, 3)\}$.

2. Let G be the additive group $\mathbf{Z}_5$ and let H be the following directed graph: $V(H) = \{0, 1, 2, 3, 4\}$, $E(H) = \{(1, 2), (2, 4), (4, 3), (3, 1)\}$ (i.e., H has a unique isolated vertex).

3. Let G be the additive group $\mathbf{Z}_6$ and let H be the following directed graph: $V(H) = \{0, 1, 2, 3, 4, 5\}$, $E(H) = \{(0, 1), (0, 2), (0, 5), (2, 5), (1, 5)\}$ (i.e., H has two isolated vertices).

The following result ([10], Theorem 3.2) clarifies the connection between regular factorization problems and the notion of G-graceful graphs.

Theorem 2.3 *Let G be a group of order n and let F be a directed graph with n vertices and $n - 1$ arcs. Then $RP^*(G; F)$ has a solution if and only if F is G-graceful.*

Remark 2.4 When a bijection $f : V(F) \to G$ is given for a G-graceful graph F (like in Definition 2.1), we shall identify each vertex v of F with its image in G, i.e., with $f(v)$. Thus the edges of F are some ordered pairs of elements of G. Moreover, the distinct copies of F in the factorization $F \mid K_n^*$ (whose existence is ensured by Theorem 2.3) are given by group multiplication on the right. More precisely, the distinct copies of F are F^g, $g \in G$, where (x, y) is an arc of F if and only if (xg, yg) is an arc of F^g (here $x, y \in G$).

Theorem 2.3 extends a result of Rosa [16] on *β-valuations* and *ρ-valuations.* These notions are defined as follows.

Definition 2.5 Let H be an undirected graph.

1. An injection $f : V(H) \to \{0, 1, \ldots, |E(H)|\}$ is called a *β-valuation* (*graceful labeling*) if $|f(x) - f(y)|$ are distinct for all $\{x, y\} \in E(H)$. A graph which admits a β-valuation is called a *graceful* graph.

2. An injection $f : V(H) \to \{0, 1, \ldots, 2|E(H)|\}$ is called a *ρ-valuation* if $\{|f(x) - f(y)| \mid \{x, y\} \in E(H)\} = \{a_1, a_2, \ldots, a_{|E(H)|}\}$, where $a_i = i$ or $a_i = 2|E(H)| + 1 - i$.

Rosa introduced the above notions as a tool for decomposing the undirected complete graph K_{2n+1} into isomorphic subgraphs on $n+1$ vertices. In particular, these notions were a tool for attacking the well known conjecture of Ringel [15] which says:

Conjecture. Let T be any tree on $n+1$ vertices. Then $T \mid K_{2n+1}$.

Rosa proved that a tree T on $n+1$ vertices has a ρ-valuation if and only if Ringel's conjecture holds for T. Therefore, if T has a β-valuation then $T \mid K_{2n+1}$. Since Rosa published his results [16], problems on graph valuations attracted a lot of attention. An updated survey on this subject, including many references, is found in [5].

It can be shown that the above results of Rosa are, in a sense, particular cases of Theorem 2.3. For the result on β-valuation, the reader is referred to [10], Corollary 4.3. In order to see how the result on ρ-valuation follows from Theorem 2.3, we shall confine ourselves with the following example.

Example 2.6 Let T be the following undirected tree:

$$V(T) = \{v_0, v_1, v_2, v_3, v_4\},$$

$$E(T) = \{\{v_0, v_1\}, \{v_0, v_2\}, \{v_2, v_3\}, \{v_2, v_4\}\}.$$

Notice that T has 4 edges. We shall obtain by Theorem 2.3 that $T \mid K_9$ if and only if T has a ρ-valuation. Let $G = \mathbf{Z}_9$, and let T^* be the directed factor of K_9^* obtained from T by adding 4 isolated vertices and considering the directed graph obtained when each edge of T is replaced by two opposite arcs. Notice that T^* has 9 vertices and 8 arcs. Consider the following $\mathbf{Z}_9$-graceful labeling of T^*:

$$V(T^*) = \{0, 1, 2, \ldots, 8\},$$

$$E(T^*) = \{(0,1), (1,0), (0,2), (2,0), (2,5), (5,2), (2,6), (6,2)\}.$$

By Theorem 2.3 $T^* \mid K_9^*$, and so it easily follows that $T \mid K_9$. Furthermore, one easily verifies that the given $\mathbf{Z}_9$-graceful labeling of T^* yields a corresponding ρ-valuation of T. More generally, as in the current example, Theorem 2.3 implies that for a tree T with $n+1$ vertices, $T \mid K_{2n+1}$ if and only if T has a ρ-valuation.

We note that Theorem 2.3 extends also another well known problem, namely the using of difference sets for the construction of block designs. The reader is referred to [10], Section 4.

We conclude this section with the following theorem, which enables us to treat factorizations $F^* \mid K_n^*$ (or $F \mid K_n$), when F is an arc (edge) disjoint union of cycles. Such factorizations and their connection to group sequencing problems, will be treated in the next section.

Theorem 2.7 *Let F be a directed graph such that $|V(F)| = |E(F)| = n$, and let G be a group of order $n-1$. Then $RP^*(G; F)$ has a solution if and only if there exists a vertex $v_0 \in V(F)$ of outdegree 1 and indegree 1, such that $F - \{v_0\}$ is G-graceful.*

3 Regular Oberwolfach problems and group sequencings

We begin with the following definitions.

Definition 3.1 An *Oberwolfach factorization* of K_n^* (res., K_n) is a factorization of K_n^* (K_n) into copies of a graph F, where F is an arc (edge) disjoint union of directed (undirected) cycles.

Definition 3.2 Let G be a permutation group on $V(K_n^*)$ ($V(K_n)$). The problem of determining whether there exists a G-regular Oberwolfach factorization of K_n^* (K_n) into copies isomorphic to F will be denoted $ROP^*(G;F)$ ($ROP(G;F)$). (Recall the definition of regular factorizations given in Definition 1.2.)

The Oberwolfach factorization problem of K_n^* (or K_n) for a given F is far from being settled. See [1] for a survey on this subject. When F consists of a single cycle, then F is a hamiltonian cycle, and the corresponding factorization is called a *hamiltonian factorization.* The existence of a hamiltonian factorization for the undirected case (where n must be odd) is well known. The existence of a hamiltonian factorization for the directed case, for every n, was established by Tilson [17].

The notion of G-regular hamiltonian factorizations is closely related to the well known notion of group sequencing (see Theorem 3.4 below). We give first the corresponding definition.

Definition 3.3 Let G be a group of order n. A *sequencing* of G is a sequence $a_1, a_2, \ldots, a_n$ of all the distinct elements of G, such that all the partial products $a_1, a_1a_2, \ldots, a_1a_2\cdots a_n$ are distinct (note that necessarily $a_1 = 1$). A group having a sequencing is called a *sequenceable* group.

The classification of all the sequenceable groups is a well known problem, which is still not settled. However, various infinite families of sequenceable groups are known. It is conjectured that all the non-abelian groups of order larger than 8 are sequenceable (see [13], Section 5.4 for further details).

Applying Theorem 2.7, the following result can be proved (see [11], Corollary 3.1).

Theorem 3.4 *Let $n \geq 3$ and let G be a group of order $n-1$. Then K_n^* admits a G-regular hamiltonian factorization if and only if G is sequenceable.*

Using known results on group sequencing, we obtain (see [11], Corollary 3.2):

Corollary 3.5 *Let $n \geq 3$ and let G be a group of order $n-1$. Then K_n^* admits a G-regular hamiltonian factorization if one of the following cases holds:*

1. G is a solvable group with a unique element of order 2, except the case when G is the quaternion group Q_4;

2. G is a dihedral group, except the cases D_3 and D_4;

3. G is a dicyclic group, except the case when G is the quaternion group Q_4;

4. G is a non-abelian group of order pq, where $p < q$ are odd primes such that p has 2 as a primitive root.

We turn now to G-regular hamiltonian factorizations of the undirected graph K_n. Notice first that in this case n must be odd. Applying known results on group sequencing, we obtain a full classification, as specified in the following theorem.

Theorem 3.6 *Let $n \geq 3$ be an odd integer and let G be a group of order $(n-1)/2$. Then K_n admits a G-regular hamiltonian factorization if and only if $n \equiv 3 \pmod 4$.*

For the proof of this theorem we applied Proposition 3.8 below. We include first the definition of the known concept of *symmetric* sequencing.

Definition 3.7 Let G be a group of order $2m$ with a unique element of order 2, say z. A *symmetric sequencing* of G is a sequencing $a_0, a_1, \dots, a_{2m-1}$ of G such that $a_m = z$ and $a_{m+i} = a_{m-i}^{-1}$ for $1 \leq i \leq m-1$ (notice that the first element is denoted by a_0 instead of a_1).

Proposition 3.8 *Let $n \geq 3$ be an odd integer and let G be a group of order $(n-1)/2$. Let $G_1 = \langle z \rangle \times G$, where z is an element of order 2, and suppose that G_1 has a symmetric sequencing. Then K_n admits a G-regular hamiltonian factorization.*

Proposition 3.8 follows by Lemma 4.4 and Theorem 5.2 in [11]. For the proof of Theorem 3.6 we applied, besides Proposition 3.8, a result of Anderson and Ihrig [2], which asserts that every finite solvable group which has a unique element of order 2, except the quaternion group Q_4, has a symmetric sequencing. For the detailed proof (which uses also the theorem of Feit and Thompson) the reader is referred to [11], Sections 4 and 5.

The general case of G-regular Oberwolfach factorizations was also treated in [11]. It was shown there that the existence of G-regular Oberwolfach factorizations is equivalent to the existence of some sequencings (called $(l_1, \dots, l_t)$-sequencings) of the group G, which are extensions of the original notion of sequencing. These sequencings may be used to obtain new G-regular factorizations. We shall not discuss further these results, and the reader is referred to [11] for details.

4 Transitive large sets of disjoint decompositions and group sequencings

The notions of *large sets* and *transitive large sets* are defined as follows.

Definition 4.1 A *large set of disjoint decompositions* of K_n^* (K_n) into cycles of length k, denoted by k-LSD, is a partition of the set of all cycles of length k in K_n^* (K_n) into disjoint decompositions of K_n^* (K_n) (i.e., no two decompositions have a common cycle). We shall say that such a k-LSD is *transitive*, or *H-transitive*, if there exists a permutation group H on the vertices of K_n^* (K_n) such that H is transitive on the decompositions in the k-LSD. A transitive k-LSD will be denoted by k-TLSD.

Notice that an n-LSD for K_n^* (or K_n) is a large set of disjoint *hamiltonian factorizations.* This subject was treated by Bryant in [4]. In particular, it was proved there that for each odd $n \geq 3$ there exists a large set of disjoint hamiltonian factorizations of K_n.

Theorem 3.4 on G-regular hamiltonian factorizations enables us to obtain a result on an n-TLSD for K_n^* (Theorem 4.3 below). However, we need first the following definition.

Definition 4.2 Let G be a group. If one (and so, up to group isomorphism, all) of the decompositions in a k-TLSD is G-regular, we shall say that the k-TLSD is *G-regular.*

It emerges that when a regular hamiltonian factorization of K_n^* is given, or, equivalently (see Theorem 3.4), when a sequenceable group of order $n-1$ is given, we can construct a corresponding n-TLSD for K_n^*. Thus we have (see [12], Theorem 1):

Theorem 4.3 *Let G be a group of order $n-1$, where $n \geq 3$. Then there exists a G-regular n-TLSD for K_n^* if and only if G is sequenceable.*

Applying known results on group sequencing, we obtain (see [12], Corollary 1.1):

Corollary 4.4 *There exists a regular n-TLSD for the following values of n:*
1. All odd n, $n \geq 3$;
2. All n such that $n = pq+1$, where $p < q$ are odd primes and p has 2 as a primitive root.

For the undirected case, we have (see [12], Theorem 2):

Theorem 4.5 *Let $n \geq 3$ be an odd integer and let G be a group of order $(n-1)/2$. Then there exists a G-regular n-TLSD for K_n if and only if $n \equiv 3 \pmod 4$.*

For each odd n, $n \geq 3$, there exists a group G of order $n-1$ having a symmetric sequencing (see Definition 3.7 and [2]; we can choose, for instance, a cyclic group). By using this we can construct an n-TLSD for the undirected graph K_n. Thus we obtain the following (see [12], Theorem 3).

Theorem 4.6 *For each odd n, $n \geq 3$, there exists an n-TLSD for K_n.*

In [12] we have studied also $n-1$-LSDs, i.e. LSDs of decompositions into cycles of size $n-1$ of K_n^* or K_n. Such cycles were called *almost hamiltonian cycles,* and the corresponding decompositions were called *almost hamiltonian decompositions.* For introducing the corresponding results, we include first the known concept of R-sequencing.

Definition 4.7 Let G be a group of order n. An *R-sequencing* of G is a sequence $a_0 = 1, a_1, \ldots, a_{n-1}$ of all the distinct elements of G, such that $a_0 a_1 \cdots a_{n-1} = 1$ and such that all the partial products $a_0, a_0 a_1, \ldots, a_0 a_1 \cdots a_{n-2}$ are distinct. (Exactly one element of G does not occur among these partial products.) A group having an R-sequencing is called an *R-sequenceable group.*

It emerges that the connection between R-sequenceable groups and regular almost hamiltonian decompositions is similar to the connection between sequenceable groups and regular hamiltonian decompositions. We have (see [12], Theorem 4):

Theorem 4.8 *Let $n \geq 4$ be an integer and let G be a group of order n. Then K_n^* has a G-regular almost hamiltonian decomposition if and only if G is R-sequenceable.*

As for almost hamiltonian regular large sets, we have the following result ([12], Theorem 5).

Theorem 4.9 *Let n be a prime power. Then there exists a regular $(n-1)$-TLSD for K_n^*.*

The proof of Theorem 4.9 is based on a construction which is induced by a Frobenius group of order $n(n-1)$, where n is a prime power. It emerges that the concept of a *Frobenius set* defined below is closely related to regular $(n-1)$-TLSDs of K_n^*.

Definition 4.10 Let G be a regular subgroup of the symmetric group S_n and let H be a subgroup of S_n of order $n-1$. We shall say that the set $HG = \{hg \mid h \in H, g \in G\}$ is a *Frobenius set* if every $x \in HG - \{1\}$ fixes at most one letter from $\{1, 2, \ldots, n\}$. We shall call the groups G, H a *kernel* and a *complement* of the Frobenius set, respectively.

Notice that every Frobenius group of order $n(n-1)$, where n is a prime power, is a Frobenius set.

We define the following.

Definition 4.11 An $(n-1)$-LSD of K_n^* is *strongly S_{n-2}-transitive* if for every pair of different vertices $v_1, v_2 \in V = V(K_n^*)$, the subgroup of $Sym(V)$ of all permutations which fix both v_1 and v_2 acts transitively (and so regularly) on the $(n-2)!$ decompositions in the LSD.

The following result makes clear the connection between the last two definitions. (See Lemma 5.1 in [12].)

Proposition 4.12 *Let G be a group of order n. Then K_n^* admits a G-regular strongly S_{n-2}-transitive $(n-1)$-LSD if and only if there exists a G-regular almost hamiltonian decomposition of K_n^* such that $\langle h \rangle G$ is a Frobenius set, where h is the permutation induced by one of the cycles in the decomposition.*

We conclude this section with the following two conjectures.

Conjecture. A Frobenius set with a kernel G of order n and a cyclic complement H of order $n-1$ exists if and only if n is a prime power.

Clearly, the validity of this conjecture will imply:

Conjecture. There exists a G-regular strongly S_{n-2}-transitive $(n-1)$-LSD of K_n^* if and only if n is a prime power.

5 Inner transitive Hering configurations

In this section we deal with inner transitive factorizations of K_n^* [14].

Definition 5.1 Let $F \mid K_n^*$ be a factorization of K_n^*, where F is a disjoint union of directed cycles and (possibly) some isolated vertices. Denote by $F_1, F_2, \ldots, F_m$ all the (arc disjoint) factors in the factorization. For each factor F_i, $1 \leq i \leq m$, let $\sigma_i \in S_n$ be the permutation defined by $\sigma_i(k) = l$ if and only if (k, l) is an arc of F_i. Let G be the group generated by $\sigma_1, \ldots, \sigma_m$. Then the factorization is *inner transitive* if G permutes the factors $F_1, \ldots, F_m$ transitively.

In many cases the group generated by the σ_is will be S_n or A_n, and the factorization will not be inner transitive. It may be interesting, and certainly not easy, to classify the inner transitive factorizations.

In this section we classify the inner transitive Hering configurations. A Hering configuration is a special type of factorization of K_n^* defined as follows.

Definition 5.2 A factorization $F \mid K_n^*$ is a *Hering configuration* of type t and order n if the following conditiond hold:

1. The factor F is a disjoint union of $(n-1)/t$ directed cycles of size t and one isolated vertex;

2. any two factors in the factorization have exactly one edge (undirected arc) in common. That is, if the common edge is $\{i, j\}$, then the arc (i, j) lies in one factor, and the arc (j, i) lies in the second factor.

The Hering configuration was introduced by Hering [9]. Since then, various papers have been published (see [14], p. 380), which deal mainly with the problem of classifying the pairs (n, t) for which a Hering configuration of type t and order n exists, and with the classification of such configurations. In [14] a classification was given for all the pairs (n, t) for which an *inner transitive Hering configuration* of type t and order n exists. This classification used a deep result in group theory - Glaubermann's Z^*-theorem [6] and its generalization proved by Artemovich [3]. We note that the result of Artemovich uses the classification of the finite simple groups.

A key notion in the classification of inner transitive Hering configurations is the notion of (n, t)-automorphisms, defined as follows.

Definition 5.3 Let N be a finite group. An automorphism τ of N is an (n, t)-automorphism of N if the following conditions hold:

1. $o(\tau) = t$;

2. $[N : C_N(\tau^i)] = n$ for every $1 \le i \le t-1$;
3. $(t, |N|) = 1$;
4. Denote $G = \langle \tau \rangle N$, and let K be the conjugacy class of τ in G. Then for any distinct $x, y \in K$, xy^{-1} does not commute with any element of K.

It emerges that every (n, t)-automorphism induces an inner transitive Hering configuration of type t and order n. We have ([14], Theorem 1):

Theorem 5.4 *Let τ be an (n,t)-automorphism of a group N, where $t \ge 3$, and let $K = \{\tau_1, \tau_2, \ldots, \tau_n\}$ be the conjugacy class of τ in the semidirect product $G = \langle \tau \rangle N$. For each i, $1 \le i \le n$, let F_i be the factor of K_n^* defined as follows: for $v_k, v_l \in V(K_n^*)$, $(v_k, v_l) \in E(K_n^*)$ if and only if $\tau_i^{-1}\tau_k\tau_i = \tau_l$. Then the F_is provide an inner transitive Hering configuration of type t and order n (with corresponding transitive group $G = \langle \tau \rangle N$).*

Notice that Theorem 5.4 enables us to construct various Hering configurations. Furthermore, using the Z^*-theorem (and its generalization) we can show that *all* the inner transitive Hering configurations can be constructed in the way described in 5.4. More precisely, we have the following result ([14], Theorem 3).

Theorem 5.5 *Let $t \ge 3$ be a natural number. Then any inner transitive Hering configuration of type t and order n is induced (in the sense of 5.4) by some (n,t)-automorphism.*

The above results enables us to classify all the pairs (n, t) for which an inner transitive hering configuration of type t and order n exists (these are exactly the pairs (n, t) for which an (n, t)-automorphism exists). We conclude our brief survey with the following (see Theorem 5 in [14]).

Theorem 5.6 *Let $t \ge 3$ be a natural number and let $n = p_1^{e_1} p_2^{e_2} \cdots p_r^{e_r}$, where the p_is are all the distinct prime divisors of n. Then there exists an inner transitive Hering configuration of type t and order n if and only if t divides $p_i^{e_i} - 1$ for every $1 \le i \le r$.*

References

[1] B. Alspach, *The Oberwolfach problem*, in: *CRC handbook of Combinatorial Designs* (C. J. Colbourn and J. H. Dinitz, Eds.), pp. 394-395, CRC Press, Boca Raton, FL 1996.

[2] B. A. Anderson and E. C. Ihrig, *Every finite solvable group with a unique element of order two, except the quaternion group, has a symmetric sequencing*, J. Combin. Des. **1** (1993), 3-14.

[3] O. D. Artemovich, *Isolated elements of prime order in finite groups*, Ukra. Math. J. **40** (1988), 343-345.

[4] D. Bryant, *Large sets of hamiltonian cycles and path decompositions*, Congresus Numerantium **135** (1998), 147-151.

[5] J. A. Gallian, *A dynamic survey of graph labeling*, The Electronic Journal of Combinatorics, Dynamic Surveys (September 2000).

[6] G. Glaubermann, *Central elements in core free group*, J. Algebra **4** (1966), 402-420.
[7] G. Glaubermann, *Weakly closed elements of Sylow subgroups*, Math. Z. **107** (1968), 1-20.
[8] G. Glaubermann, Problem 4.21, in: *The Kourovka Notebook: Unsolved Problems in Group Theory*, Seventh Augmented Edition. American Math. Soc. Translations, Ser. 2, **121** (1983).
[9] F. Hering, *Balanced pairs*, Ann. Discrete Math. **20** (1984), 177-182.
[10] G. Kaplan, A. Lev and Y. Roditty, *On graph labeling problems and regular factorizations of complete graphs*, submitted.
[11] G. Kaplan, A. Lev and Y. Roditty, *Regular Oberwolfach problems and group aequencings*, J. Combin. Theory Ser. A, to appear.
[12] G. Kaplan, A. Lev and Y. Roditty, *Transitive large sets of disjoint decompositions and group sequencings*, submitted.
[13] D. Keedwell, *Complete mappings and sequencings of finite groups*, in: *CRC handbook of Combinatorial Designs* (C. J. Colbourn and J. H. Dinitz, Eds.), pp. 246-253, CRC Press, Boca Raton, FL, 1996.
[14] A. Lev and Y. Roditty, *On Hering decomposition of DK_n induced by group actions on conjugacy classes*, Europ. J. Combinatorics **21** (2000), 379-393.
[15] G. Ringel, Problem 25 in: *Theory of Graphs and its Applications*, Proc. Symposium Smolenice 1963, Prague **162** (1964).
[16] A. Rosa, *On certain valuations of the vertices of a graph*, Theory of Graphs (Internat. Symposium, Rome, July 1966), Gordon and Breach, N. Y. and Dunod Paris (1967), 349-355.
[17] T. W. Tilson, *A Hamiltonian decomposition of K^*_{2m}, $2m \geq 8$*, J. Comb. Theory, Ser. B **29** (1980), 68-74.

ORBITS IN FINITE GROUP ACTIONS

THOMAS MICHAEL KELLER

Department of Mathematics, Southwest Texas State University, 601 University Drive, San Marcos, TX 78666, USA

Abstract

We give an overview of results on the orbit structure of finite group actions with an emphasis on abstract linear groups. The main questions considered are the number of orbits, the number of orbit sizes and the existence of large orbits, particularly regular orbits. Applications of such results to some important problems in group and representation theory are also discussed, such as the $k(GV)$–problem and the Taketa problem.

1 Introduction

Ever since the beginning of abstract group theory the study of group actions has played a fundamental role in its development. In finite group theory this is all too obvious in that the fundamental theorems of Sylow – without which finite group theory would not get beyond its beginnings – are based on the study of various group actions. Group actions capture the fact that every group can be represented as a permutation group, and permutation groups were the first groups to be considered when the abstract term of a group had not yet been coined.

Naturally information on the orbits induced by a group action is vital to an understanding of the action which is why a wealth of such results is scattered throughout the literature. Rarely however are the orbits the main focus of the investigation, and most results on orbits are proved with applications to other questions in mind. This shows that – as in the case of Sylow's theorems – a result on the orbits of a group action often is at the core of a seemingly unrelated problem. Unfortunately, however, general questions on orbits tend to be very hard to answer and often can be solved under specialized restrictions only.

In this survey article we discuss some recent results on orbits in finite group actions, concentrating on abstract groups and focussing on one of the most important kinds of group actions, namely that of linear groups on finite modules. We also describe some of the applications of these results to problems in group theory as well as representation theory, such as the $k(GV)$–problem, the Taketa problem and bounding the Fitting height in coprime group actions, and others. All this is spiced up with a number of open problems for further research. Needless to say, in the choice of the topics treated here we have also been guided by our own research interests.

All groups throughout this paper are finite.

2 The number of orbits

Let G be a group acting on a G–module V. One of the most fundamental questions on the orbit structure is the one for the number of orbits of G on V. Here we denote this by $r(G,V)$. Of course, there is the beautiful, general and elementary Cauchy–Frobenius orbit counting formula (also known as Burnside's lemma) stating that

$$r(G,V) \;=\; \frac{1}{|G|}\sum_{g\in G}|C_V(g)|$$

which is extremely useful in many instances. To calculate (or at least bound) $r(G,V)$ with it, however, the group action must be known very well which in general situations is not always the case.

In the literature $r(G,V)$ is often studied in the setting of permutation groups. Let G be a permutation group with a regular elementary abelian normal subgroup V. (It is well–known that solvable primitive permutation groups always have this structure.) Then $G = HV$ is the semidirect product of a point stabilizer H and V, and $r(H,V)$ happens to be the *rank* of the permutation group.

One line of investigation on $r(G,V)$ is to describe (or, if possible, characterize) group actions with given $r(G,V)$. For instance, if $r(G,V) = 2$, then GV is a doubly transitive permutation group, and so the structure of GV is known; solvable doubly transitive permutation groups were classified by Huppert [47], and the general case was a consequence of the classification of finite simple groups (see e.g. [12, Theorem 4.11 and, in particular, Table 7.3]); the affine case, which is of interest to us here, was done by Hering [46]. The cases $r(G,V) = 3$ and 4 also have been studied intensely (see [22, 23, 29, 75, 79]).

Another – for our emphasis on abstract groups more relevant – line of investigation is the search for good lower bounds on $r(G,V)$ in terms of parameters of G and V. A very general and elementary result useful for induction arguments is the following

Lemma 2.1 [29] *Let V be imprimitive, so that $V = V_1 \oplus \ldots \oplus V_n$, where the V_i are permuted by G. Let $H = N_G(V_1)$. Then*

$$r(G,V) \;\geq\; \binom{r(H,V)+n-1}{n}.$$

This lemma is best possible in the sense that if $N = \bigcap_{i=1}^{n} N_G(V_i)$ is isomorphic to the direct product of all the $N_G(V_i)/C_G(V_i)$ $(i = 1,\ldots,n)$ and $G/N \cong S_n$ (= the symmetric group on n letters), then equality holds.

For solvable groups, however, a lot more can be said. Dornhoff [21] was the first to give an upper bound for $|V|$ in terms of $r(G,V)$ in the case that G acts primitively, but not semilinearly on V. Note that in case of semilinear action, i.e., $GV \lesssim A\Gamma(GF(|V|))$, where $A\Gamma(g^n) := \{x \to ax^\sigma + b \mid a,b \in GF(q^n),\ a \neq 0,\ \sigma \in \mathrm{Aut}(GF(q^n))\}$ for any prime q, no such bound exists. Later Seager [104] sharpened and quantified Dornhoff's result as follows:

Theorem 2.2 *Let p be a prime and let G be a solvable primitive subgroup of the linear group $GL(n,p)$, and let V be the natural underlying vector space. Suppose that GV is not isomorphic to a subgroup of $A\Gamma(p^n)$. Then*

$$r(G,V) \ > \ \frac{p^{\frac{n}{2}}}{12n} + 1$$

except possibly when $p^n \in \{17^4, 19^4, 7^6, 5^8, 7^8, 13^8, 7^9, 3^{16}, 5^{16}\}$.

In [105] Seager generalized this to include also imprimitive actions as follows:

Theorem 2.3 *Let G be a solvable group and V be a finite faithful irreducible G–module. Then one of the following holds:*
(i) $|V| \ \leq \ \left(\frac{r(G,V)+1.43}{24^{\frac{1}{3}}}\right)^c$ for $c = 36.435663$, or
(ii) for some integers m,k with $k \leq 0.157 \log_3 \left(\frac{r(G,V)+1.43}{24^{\frac{1}{3}}}\right)$ and a prime p we have $|V| = p^{mk}$ and $G \lesssim S_{p^m} \wr S_k$ (as permutation group on V) in its primitive action on $GF(p^m)^k$.

For solvable groups the derived length $\mathrm{dl}(G)$ is one of the most important parameters to measure the group's complexity and how far it is away from being abelian. One naturally would expect $r(G,V)$ to grow rapidly with $\mathrm{dl}(G)$. This is indeed the case, and Seager's results make it easy to obtain the qualitatively best possible bound.

Theorem 2.4 *Let G be solvable and let V be a finite faithful irreducible G–module. Then there exist universal constants C_1, C_2 such that*

$$\mathrm{dl}(G) \ \leq \ C_1 \log\log r(G,V) + C_2.$$

Proof By [81, Theorem 3.12(b)] we have $\mathrm{dl}(G) \leq 2\log_2 \dim V$, and as $\dim V \leq \log_2 |V|$, we see that there exist universal constants D_1, D_2 such that $\mathrm{dl}(G) \leq D_1 \log\log |V| + D_2$ $(*)$. Now the Corollary in [105, Section 6] yields that either $\mathrm{dl}(G) \leq 7.22 + \frac{5}{2}\log_2\log_3\left(\frac{r(G,V)+1.43}{24^{\frac{1}{3}}}\right)$ – in which case we are obviously done – or $|V| \leq \frac{r(G,V)+1.43}{24^{\frac{1}{3}}}$. But in the latter case using $(*)$ also implies the assertion, and so the theorem is proved. □

As an application of these results we mention that Theorem 2.2 plays an integral role in the answer to a question of R. Brauer in the modular representation theory of finite groups, namely, whether for $k \in \mathbb{N}$ there are only finitely many isomorphism types of defect groups of blocks with exactly k ordinary characters. This question (known as Brauer's Problem 21) is a generalization of an old theorem of Landau stating that for $k \in \mathbb{N}$ there are only finitely many isomorphism classes of finite groups with exactly k conjugacy classes. Using Theorem 2.2, Külshammer [70] could answer this question in the affirmative by proving

Theorem 2.5 *There exists a function $h : \mathbb{N} \to \mathbb{N}$ such that if B is a block of a solvable group with defect group D containing exactly k irreducible complex characters, then $|D| \leq h(k)$.*

Later Külshammer [72] extended his result to p–solvable groups by proving the following result [71]

Theorem 2.6 *Let V be a finite vector space of characteristic p, and let G be a p–solvable semilinear group on V. If G has exactly k orbits on the corresponding projective space, then G contains a solvable normal subgroup N such that $|G/N| \leq f(k)$ for some universal function $f : \mathbb{N} \to \mathbb{N}$.*

Problem: The function h and f in Theorems 2.5 and 2.6 are far from best possible. What would more "realistic" functions look like (asymptotically)?

3 The number of orbit sizes

In the previous section we saw that the number of orbits of a group on a finite vector space tends to be large as the size and the complexity of the group increases; for solvable groups this was particularly striking. Obviously one would expect the sizes of the orbits to vary quite a bit, that is, if $n(G, V) = |\{|v^G| \mid v \in V\}|$ denotes the number of orbit sizes of G on V, one would expect $n(G, V)$ to be large in some sense.

Also, as was the case with $r(G, V)$, group actions with $n(G, V)$ very small are of a limited structure, e.g. solvable groups acting faithfully and irreducibly on the finite G–module V with $n(G, V) = 2$ were characterized by Passman. Once again the problem is disguised as one in permutation groups: A permutation group on Ω is $\frac{3}{2}$–transitive if it is transitive and all orbits of a point stabilizer on Ω without that point are of the same length > 1. So studying affine $\frac{3}{2}$–transitive permutation groups amounts to looking at linear group group actions with $n(G, V) = 2$. Passman [93, 94] even characterized all solvable $\frac{3}{2}$–transitive permutation groups.

Next we ask – similarly as in Section 2 – for good lower bounds on $n(G, V)$ in terms of parameters of G and V. In general it is quite delicate to prove strong statements on $n(G, V)$. This is in part because if $v, w \in V$ and we know $|v^G|$

and $|w^G|$, basically nothing can be said about $|(v+w)|^G$. The only result in this direction in due to Yuster [127, Theorem 1.5]:

Theorem 3.1 *Let H, G be groups such that H acts on G by automorphisms. Suppose there exist $x, y \in G$ such that $(|x^H|, |y^H|) = 1$. Then*

$$|(xy)^H| \;=\; |x^H| \cdot |y^H|.$$

Unfortunately, the hypothesis of the orbit sizes being coprime is quite special.

Still intuitively it is clear that in group actions with many orbits one would expect many different orbit sizes, that is, $n(G, V)$ should grow rapidly (not as rapidly as $r(G, V)$, though) with the complexity of G. If G has an extraspecial normal subgroup, this is indeed the case, as the following general result [61, Theorem 1.5] shows:

Theorem 3.2 *Let G be a group and V be a finite G–module over $GF(p)$ for a prime p. Suppose that $E \trianglelefteq G$ is a q–group for a prime q such that $Z(E)$ acts fixed point freely on V. Suppose that $|E| = q^{2m+1}$ for some $m \geq 11$, and suppose that E is extraspecial of exponent q or 4. Then*

$$n(G, V) \;\geq\; m - 10.$$

The proof of this result explicitly exhibits representatives of the orbits whose centralizers form a chain of subgroups of G, so the proof is fairly "constructive". An even stronger statement should be true:

Conjecture: The conclusion of Theorem 3.2 essentially remains true if E is an arbitrary q–group of class at most 2 and of exponent q or 4.

The conjecture is unknown even for solvable groups; for groups of odd order and $q \geq 5$ it was proved in [61, Theorem 2.1], and this was generalized further in [62, Lemma 1.13] to a more technical statement.

We remark that once we assume that the field over which we work is sufficiently large relative to G, it is not hard to prove very strong results of this kind, because over large fields the set $\{C_G(v) \mid v \in V\}$ is closed under building intersections which makes it a lot easier to show that $\{|C_G(v)| \mid v \in V\} = n(G, V)$ is large. In [59, Theorem 2.5] Moretó and Jaikin–Zapirain show the following concerning the rank $rk(G) = \max\{d(H) \mid H \leq G\}$ (where $d(H)$ denotes the minimum number of generators of H).

Theorem 3.3 *Let G be a group and V a faithful FG–module, where F is a field of order greater than the number of subgroups of G and of characteristic zero or coprime to $|G|$. Then*

$$rk(G) \;\leq\; \frac{3}{2}n(G, V)^2 \;+\; \frac{1}{2}n(G, V) \;+\; 1.$$

Conjecture (Moretó, Jaikin–Zapirain): A linear bound holds in the above theorem.

For solvable groups we now present an analogue to Theorem 2.4. Considering the action of the group of, say, lower triangular matrices over $GF(p)$ on a faithful module of coprime characteristic readily shows that in general at best we can expect a logarithmic bound for $\mathrm{dl}(G)$ in $n(G,V)$ (see [64, Example 1.6]).

For A–groups, that is, solvable groups all of whose Sylow subgroups are abelian, Marshall [82] proved that $\mathrm{dl}(G)$ is bounded above by some function of $n(G,V)$ in case of G acting faithfully and completely reducibly on V. The function is not explicitly given, but is exponential. However, in [60] a much stronger bound is obtained, namely, essentially $\mathrm{dl}(G)$ is bounded above by $\frac{\log|n(G,V)|}{\log\log|n(G,V)|}$ which is even stronger that logarithmic. The best bound from below obtained (by constructing examples) is of the size $\left(\frac{\log|n(G,V)|}{\log\log|n(G,V)|}\right)^{\frac{1}{2}}$, so the upper bound is nearly best possible. It is not known, however, what the real bound is. (Note that these results are not stated explicitly in [60], but only in their character theoretic version, from which they easily follow).

The general case has finally been settled in [61, 62, 64] with the following result.

Theorem 3.4 *There exist universal constants C_1, C_2 such that for any solvable group G acting faithfully and irreducibly on a finite G–module V we have*

$$\mathrm{dl}(G) \ \leq \ C_1 \log n(G,V) + C_2.$$

The actual proof brings about $\log = \log_2$, $C_1 = 24$, $C_2 = 364$, but these constants certainly are not best possible. It falls into two major parts: first, a logarithmic bound for $\mathrm{dl}(G/F(G))$ in $n(G,V)$, second, a logarithmic bound for $\mathrm{dl}(F(G))$ in terms of $n(G,V)$. The latter result has an immediate predecessor due to Isaacs [55, Theorem 2.1] which only obtains a linear bound, but works for arbitrary (not only solvable) groups. Its most general version reads as follows:

Theorem 3.5 *Let the group G act by automorphisms on the abelian p'–group V (where p is a prime) and write $P = O_p(G)$. Then*

$$\mathrm{dl}(P/C_P(V)) \ \leq \ n(G,V) + 1.$$

Problem: Does even a logarithmic bound hold in this general situation?

One way of improving the constants in Theorem 3.4 would be to improve the inductive process underlying the key lemmas 1.7 and 3.21 in [62]. Both require the existence of nine nontrivial orbits of some group on a module, but analyzing their proofs shows that six would be sufficient if one could solve the following curious and purely combinatorial problem for $k = 6$.

Problem: Suppose that for $i, j, k \in \mathbb{N}$ we are given $M_{ijk} \subseteq \mathbb{N}$ with $|M_{ijk}| = k$. Find a $k \in \mathbb{N}$ and $a_{ij} = a_{ij}(k) \in \mathbb{N}$ $(i = 1, \dots, k,\ j \in \mathbb{N})$ such that the following hold:

(1) $a_{ij} \in M_{ijk}$ for $i = 1, \dots, k$, and $j \in \mathbb{N}$

(2) For any $i \in \{1, \dots, k\}$ we have

 (*a*) $a_{i1}, \dots, a_{i5}$ are mutually distinct

 (*b*) $a_{ij} \notin \{a_{i1}, a_{i2}, a_{i3}, a_{i4}\}$ for $j \geq 5$

(3) For any $1 \leq l < m \leq k$ and any $j \in \mathbb{N}$ we have

$$\{a_{l1}, \dots, a_{lj}\} \neq \{a_{m1}, \dots, a_{mj}\}$$

It is easy to see that there can be no general solutions for $k = 5$: If $M_{ij5} = \{1, \dots, 5\}$ for all i, j, then (2) and (3) contradict each other. Also, there is a general solution for $k = 9$ (which is the core of [62, Lemmas 1.7 and 3.21]):

Without loss let $a_{i1} = i$ for $i = 1, \dots, 9$, and then construct a_{ij} $(j \geq 2,\ i = 1, \dots, 9)$ observing (2) and such that $a_{ij} \notin \{i+5, i+6, i+7, i+8\}$ (to be read modulo 9) for all i. This is possible as $|M_{ij9}| = 9$ for all i, j.

But it is open whether there is a solution for $k = 6$. In the seemingly most restricted case that $M_{ij6} = \{1, 2, 3, 4, 5, 6\}$ for all i, j one can solve the problem by taking $a_{ij} = i + j - 1 \pmod 6$ for $i = 1, \dots, 6$ and $j = 1, \dots, 5$, and $a_{ij} = a_{i5}$ for all i and $j \geq 5$. This is why a general solution for $k = 6$ should exist – but none has been found so far.

We turn to the main application of theorems on orbit sizes for solvable groups which is a fairly old question in the character theory of such groups.

A well–known result by K. Taketa [115] states that M–groups, that is, groups all of whose irreducible complex characters are monomial (i.e., induced from some linear character of a subgroup of G), are solvable and more precisely, if G is an M–group, then $\mathrm{dl}(G) \leq |\mathrm{cd}(G)|$, where $\mathrm{cd}(G) = \{\chi(1) \mid \chi \in \mathrm{Irr}(G)\}$ is the set of the degrees of the complex irreducible characters of G. From this old result in the 1970s there originated the Isaacs–Seitz–conjecture that $\mathrm{dl}(G) \leq |\mathrm{cd}(G)|$ should hold for arbitrary solvable groups, and over the time work of Isaacs [53], Berger [2], and Gluck [33] proved the conjecture for groups of odd order and also that in general $\mathrm{dl}(G) \leq 2|\mathrm{cd}(G)|$.

While these results (proved using mostly character theoretic arguments) are quite satisfying, early on there has been a feeling that a linear bound asymptotically may be far too large. Studies of certain "prominent" families of p–groups (see [49], [56], [95], and [97] rep. [41]) all gave logarithmic bounds, so that at least for p–groups a logarithmic bound is expected. Moreover, groups with $\mathrm{dl}(G) = |\mathrm{cd}(G)|$ seem to be rare as $\mathrm{dl}(G)$ grows: Examples are known only up to 5 (see [50, 18.8 and §27] for constructions of such examples), and for $|\mathrm{cd}(G)| \leq 5$ it has also been shown that $\mathrm{dl}(G) \leq |\mathrm{cd}(G)|$ (see [54, Theorem 12.15], [89], [30], [73], [74]). All these results indicate that $\mathrm{dl}(G)$ might grow slower than linearly with $|\mathrm{cd}(G)|$.

As it happens, the results on orbit sizes presented above shed some more light on the question. Namely if V is a G–module, then so is $V_0 = \mathrm{Irr}(V)$, and we consider the semidirect product GV_0. By [54, Problem (6.18)] every $\lambda \in V_0$ can be extended to its inertia group in GV, from which it can be induced to an irreducible character of GV. This shows that every orbit size of G on V_0 is a complex irreducible character degree of GV. Using this allows a translation of results on orbit sizes to results on character degrees.

For A–groups this means that $\mathrm{dl}(G)$ is bounded above by a bound whose order of magnitude is $\frac{\ln|\mathrm{cd}G|}{\ln\ln|\mathrm{cd}G|}$ (see [60, Theorem 2.4]). More importantly, the character theoretic version of Theorem 3.4 reads as follows (see [64, Corollary 2.2]):

Theorem 3.6 *Let G be a solvable group. Then*

$$\mathrm{dl}(G/F(G)) \leq 24\log_2|\mathrm{cd}(G)| + 364.$$

(Observe that combining this with [58, Corollary 2.3] yields a linear bound for $\mathrm{dl}(G)$ in $|\mathrm{cd}(G)|$ which is smaller than Gluck's bound $2|\mathrm{cd}(G)|$ for large values of $|\mathrm{cd}(G)|$.)

From Theorem 3.6 it is obvious that in general one should expect $\mathrm{dl}(G)$ to be bounded logarithmically in $|\mathrm{cd}(G)|$, but amazingly the case of p–groups, for which the linear bound was proven first (as p–groups are M–groups) and from which the doubts about the linear bound originated, is as of yet still unsolved. To get a general logarithmic bound, it remains to prove the following

Conjecture: There exist universal constants C_1, C_2 such that if G is a solvable group, then

$$\mathrm{dl}(F(G)) \leq C_1\log|\mathrm{cd}(G)| + C_2.$$

(A linear bound can be found in [58, Corollary 2.3].)

So essentially this means that the problem has been reduced to the case of p–groups for which it is as open as it could be. A result of Noritzsch [90] demonstrated the bad inductive behaviour of $|\mathrm{cd}(G)|$ in p–groups; work of Slattery [114] shows how hard it is to even obtain tiny improvements on the bound $\mathrm{dl}(G) \leq |\mathrm{cd}(G)|$. As such the above conjecture presents one of the toughest challenges for group theorists working on p–groups for the years to come.

We conclude this section by mentioning some offspring of the Taketa problem. One is the "Taketa problem for modular character degrees" which asks for bounds for $\mathrm{dl}(G/O_p(G))$ in terms of $|\mathrm{cd}_p(G)|$, the number of irreducible p–Brauer character degrees. The best bound known to date is of the order of magnitude $|\mathrm{cd}_p(G)| \cdot \log |\mathrm{cd}_p(G)|$ (see [64, Theorem 2.3]; cf. also [55] and [16] for related results). However, also here a logarithmic bound is expected.

A generalization of the Taketa problem is to study the set of A–invariant characters if A acts on G coprimely. This setting usually involves character correspondences; results along these lines have been obtained by Wolf [124], Navarro [88], and Turull [118]. All bounds obtained on derived lengths in these papers are linear. But also here one would expect them to be logarithmic. Other interesting variations of the Taketa problem are presented by Isaacs and Knutson in [58], where $\mathrm{dl}(N)$ is bounded (for $N \trianglelefteq G$) in terms of the numbers of characters of G not having N in their kernel.

4 Large orbits

One of the most important and natural goals of many studies on orbits is to establish the existence of an orbit of a certain size, and of particular interest are extreme cases. However, there are hardly any results on the existence of very small orbits; the only result that comes to mind is due to Berger ([2], see also [81, Section 8]) and states if V is a symplectic $GF(p)G$–module (with respect to some nonsingular symplectic G–invariant form) for some odd prime p and an odd order group G, then G has an orbit of size at most $\frac{p^n+1}{2}$, a result that is not true without the oddness hypothesis. This result is used to prove the Isaacs–Seitz conjecture for odd–order groups. Sometimes orbits are considered that in some sense are small. In [81, Sections 9 and 10] linear group actions are studied where all orbits have small (normally trivial) p–part for a prime p, and in [45] a situation is studied where $v^G = v^H$ for all v, where H is a complement to a certain normal subgroup of G (see also [44] for an application).

On the other side, there has been a deep interest and need to examine the size of the largest possible orbits in linear group actions for a long time, and from this a large body of literature evolved. Obviously the size of an orbit can never exceed $|G|$. An orbit of size $|G|$ is called a *regular* orbit. If V is a G–module with a regular orbit, this means that there is a $v \in V$ with $C_G(v) = 1$. We remark that if V is a faithful finite–dimensional G–module over an infinite field, then $\bigcup_{1 \neq g \in G} C_V(g) \subset V$ and so always a regular orbit exists.

If G is abelian and V a completely reducible faithful G–module, it is easy to see that there always exists a regular orbit. If G is a p–group for some prime p and V is a finite faithful and irreducible G–module over $GF(q)$ for a prime q with $p \neq q$, then G always has a regular orbit on V unless one of the following holds:

(i) $p = 2$ and q is a Mersenne prime or a Fermat prime
(ii) p is a Mersenne prime and $q = 2$

The cases (i), and (ii) are true exceptions, as can be shown by suitable examples. It can also be shown that if G does not involve a section isomorphic to $C_p \wr C_p$ (where C_p is the cyclic group of order p), then G has a regular orbit on V. All these results and more are collected in [81, Section 4]; cf. also [51]. Even in the exceptional cases there always exist large orbits. Improving former results of Passman [92], Isaacs [57] proved the following result (and also some refinements of it):

Theorem 4.1 *Let G be a nontrivial p–group that acts faithfully on a group H, where p does not divide $|H|$. Then there exists an element $x \in H$ such that $|C_G(x)| \leq \left(\frac{|G|}{p}\right)^{\frac{1}{p}}$. In particular, $|C_G(x)| \leq |G|^{\frac{1}{2}}$.*

The theorem is quite general in that H need not be a vector space, and not even abelian. This is due to the fact, proved by Hartley and Turull [43, Lemma 2.6.2], that if A acts by automorphisms on G with $(|A|, |G|) = 1$, then there exists an abelian group G_0 of squarefree exponent such that the actions of G and G_0 are permutation isomorphic. So without loss in coprime situations it always suffices to consider linear group actions on G–modules.

Large orbits still exist if G is allowed to be supersolvable. Wolf [125] proved the existence of orbits of size at least $|G|^{\frac{1}{2}}$ for supersolvable G acting faithfully on a completely reducible G–module V. Isaacs asked the following:

Question: If V is a faithful G–module with $(|G|, |V|) = 1$ and G solvable, does there always exist an orbit of size at least $|G|^{\frac{1}{2}}$?

Related results are concerned with regular orbits of nilpotent Hall subgroups of groups. Espuelas ([25] or [81, Theorem 7.3]) proved that if G is solvable and V is a finite faithful G–module and $P \in \mathrm{Syl}_p G$ such that $O_p(G) = 1$ and p is odd, then P has a regular orbit on V. (For $p = 2$ this is also true if P is $C_2 \wr C_2$–free.) Under some more oddness hypotheses Carlip [15] extends this result to nilpotent Hall subgroups (see also [27] for an earlier version). A very detailed study has been carried out by Hargraves ([42], see also [28]): If A is nilpotent and acts on the solvable group G by automorphisms, and if V is a faithful irreducible AG–module over a field of characteristic not dividing $|A|$, then it is proved that in most cases A has a regular orbit on V, and the exceptional cases are characterized. In later work [5, 6, 7] this is extended to the question whether $V|_A$ contains copies of the regular module. Other elaborations on Hargraves' result can be found in [117], [39, Lemma 2.6], [24] and [91, Proposition 3].

Proving the existence of regular orbits usually is done in two steps: First, prove it in the (quasi–) primitive case. Then use an inductive argument to deal with the imprimitive case. We therefore look at some results tailored for this strategy.

It turns out that the inductive behaviour of the existence of regular orbits is poor, but not entirely bad, at least for solvable groups. If it were good, then if G acts faithfully and irreducibly on the finite G–module V such that V is induced from the primitive (or quasiprimitive) module W, and $\overline{H} = N_G(W)/C_G(W)$ has a regular orbit on W, we would expect to be able to conclude that also G has a regular orbit on V. In general unfortunately this is not the case. However, if $\overline{H}$ has s regular orbits on W for a suitable $s > 1$, then we often can conclude that G has s regular orbits on V as well. For p–groups, this is demonstrated in [81, Lemma 4.6] for $s = 3$, and it is also explained why $s = 1$ or 2 do not work. For odd order groups the argument is shown for $s = 2$ in [26, Lemma 1.2]. Furthermore, for arbitrary solvable groups the induction goes through for $s = 5$. This is a consequence of the fact that if G is a primitive solvable permutation group on the set Ω, then there is a partition of Ω into four subsets such that only the identity stabilizes all of them (cf. Section 6). For details we refer the reader to [126], where also a formula for the number of regular orbits in certain wreath products is given.

It also would be interesting to have an inductive argument for the existence of many regular orbits for nonsolvable groups. Since the solvable case is based on the existence of small bases (resp. partitions stabilized only by the identity) for solvable primitive permutation groups and since similar results are available in the nonsolvable case with some exceptions, one could proceed similarly as in the solvable case and in addition would have to handle the exceptional cases. We will explore the concept of a base and results on it in more detail in Section 6.

In any case, induction works really well only if there are several regular orbits which often is not the case, as there need not exist even a single one.

We now turn to results on regular orbits that can be applied if V is primitive, thus addressing the inductive base of an inductive argument.

If G is solvable, it usually suffices to study quasiprimitive instead of primitive modules. An irreducible G–module V is called *quasiprimitive* if V_N is homogeneous for all $N \trianglelefteq G$. For G this implies that all of its normal subgroups are cyclic, and such groups have been described by P. Hall (see e.g. [48, III, 13.10]). See [81, Section 1] for a detailed analysis in the solvable case. It turns out that the Fitting subgroup $F(G)$ of such groups is a central product of extraspecial groups (for various primes) and a group T which is almost cyclic (that is, it has a normal subgroup of index at most 2). Let $e \in \mathbb{N}$ such that $e^2 = |F(G)/T|$ is the order of the extraspecial part modulo its center. In [81, Proposition 4.10] and [25] we find the following result:

Theorem 4.2 *Let G be solvable, and let V be a finite faithful quasiprimitive G–module. If $e > 118$, or if $|G|$ is odd and $e > 1$, then G has at least two regular orbits on V.*

In general groups with not–too–small normal extraspecial subgroups tend to have regular orbits, as [61, Lemma 1.3] shows:

Theorem 4.3 *Let G be a group and V be a faithful $GF(q)$–module for a prime q. Suppose that $E \trianglelefteq G$ is extraspecial of order p^{2m+1} such that $C_G(E)$ acts frobeniusly on V (i.e., $C_G(E)V$ is a Frobenius group with kernel V). If $m \geq 10$, or if $|G|$ is odd and $m \geq 5$, then G has a regular orbit on V.*

The underlying principle of the proof of regular orbit theorems is to assume that there is no regular orbit, so $V \subseteq \bigcup_{1 \neq g \in G} C_V(g)$ and $|V| \leq \sum_{1 \neq g \in G} |C_V(g)|$ and then use upper bounds on $|C_V(g)|$ to obtain a contradiction. Theorem 4.3 is no exception. Similarly, if G does not have s regular orbits on V, this leads to $|V| - (s-1)|G| \leq \sum_{1 \neq g \in G} |C_V(g)|$, and so from the proof of Theorem 4.3 it becomes clear that as we increase m, we can get as many regular orbits as we want.

In many instances, one can go beyond assuming V is primitive. A typical chain of reductions (of a purported minimal counterexample to a theorem) is to successively show that V can be assumed to be irreducible, primitive, scalar–primitive, tensor–indecomposable and tensor–primitive (cf. [4, Section III]). After these reductions the generalized Fitting subgroup $F^*(G)$ is either extraspecial or quasisimple, and $F^*(G)$ acts absolutely irreducibly on V. In both cases at least for large fields regular orbit results exist. In the extraspecial case, Robinson and Thompson [103, Theorem 8] proved:

Theorem 4.4 *Let F be a finite field of characteristic p with $|F| \geq 2^{27}$, let G be a group of order not divisible by p, and let V be a faithful FG–module. Suppose further that, for some prime q, G has a normal q–subgroup Q which acts absolutely irreducibly on V, and all of whose characteristic abelian subgroups are central. Then G has a regular orbit on V.*

In the quasisimple case, Liebeck [76] showed:

Theorem 4.5 *Let p be a prime with $p > 5^{30}$. Suppose that G is a group of order not divisible by p and that V is a faithful $GF(p)G$–module such that G has a quasisimple normal subgroup H which acts irreducibly on V. Then one of the following holds:*
(i) G has a regular orbit on V.
(ii) $H = A_c$ (the alternating group on c letters), $c < p$, and $V = \{(v_1, \ldots, v_c) \mid v_i \in GF(p)$ for all i, and $\sum_{i=1}^{c} v_i = 0\}$ with coordinates permuted naturally by H.

(ii) is a real exception, as is demonstrated by examples in Liebeck's paper. Theorem 4.5 has been elaborated by Goodwin [37, 38] who weakened the restriction on p considerably (see also [77] for a short survey on this).

We mention a few miscellaneous results on large orbits. As outlined at the beginning of this section, odd–order symplectic groups in odd characteristic have small orbits. Espuelas [26] showed that they also have regular orbits:

Theorem 4.6 *Let G be a (solvable) group of odd order, p be a prime and V be a faithful and completely reducible $GF(p)G$–module with respect to a nonsingular symplectic G–invariant form. If $p = 2$, suppose in addition that 3 does not divide $|G|$. Then G has at least two regular orbits on V.*

If critical sections isomorphic to certain semilinear groups do not occur in the odd order group G and p is a divisor of $|G|$ with $O_p(G) = 1$, then it is not too difficult to show that G has a p–regular orbit on $O_{p'}(G)$, that is, there is an $x \in O_{p'}(G)$ with $|C_G(x)|$ not divisible by p see [128]. In a similar vein, Manz and Wolf (see [81, Theorem 11.4] prove the existence of arithmetically large orbits; more precisely, they show that if G is solvable and $H \leq G$ is such that the finite faithful irreducible G–module V is induced from a primitive irreducible H–module W with $H/C_H(W)$ not isomorphic to a semilinear group, then there exists a $v \in V$ such that $|v^G|$ is divisible by each prime divisor p of $|G|$ with $p \geq 5$.

An interesting and new point of view on the problem of regular orbits is introduced by Gluck and Magaard in [34]. For coprime group actions they consider the problem of how many copies of a G–module V are needed so that G has a regular orbit on their direct sum. For $m \in \mathbb{N}$ write mV for the direct sum of m copies of V.

Theorem 4.7 *Let G be a group and V be a faithful $GF(q)G$–module for a prime power q with $(|G|, q) = 1$. If $m \geq 94$, then G has at least 64 regular orbits on mV.*

Most of the results presented in this section so far have not been proved for their own sake, but to be applied to some problem in group and representation theory as already browsing the titles of the cited literature indicates. Particularly regular orbit theorems have found numerous applications in obvious and less obvious ways. We take a look at some of the applications that results in this section have found. One of the main fields of application of regular orbit theorems is a fairly old problem having its origin in results of Dade [17] and Thompson [116]. The problem is to bound the Fitting height $h(G)$ of a finite solvable group by the composition length $l(A)$ (i.e., the number of primes of dividing $|A|$, counting multiplicities) of a finite group A acting by automorphisms on G such that $(|A|, |G|) = 1$ and $C_G(A) = 1$. The problem has a long history which has been outlined in detail by A. Turull in an expository paper [121] which is why we do not repeat it here and instead refer to that paper for further reference. The conjecture $h(G) \leq l(G)$ is still open in general, but strong sufficient conditions have been developed which depend on the existence of large orbits of subgroups of A on sections of G. In [120] Turull proved:

Theorem 4.8 *Let A be a finite group and suppose it acts on the finite group G with $(|A|, |G|) = 1$ and $C_G(A) = 1$. (From the classification of finite simple groups it follows that G must be solvable in this setting.) Assume furthermore that for every proper subgroup B of A and every B–invariant section S of G such that B acts irreducibly on S, there is a $v \in S$ with $C_B(v) = C_B(S)$ (i.e., $B/C_B(S)$ has a regular orbit on S). Then $h(G) \leq l(A)$.*

In [122] Turull refined this result; see also M. Pettet's review of this work (MR98j: 20017) for an interesting discussion of it, and [123] for a new application. For solvable A Turull proved that $h(G) \leq 2l(A)$, as follows immediately from the following more general result [119]:

Theorem 4.9 *Let A, G be finite solvable groups such that A acts on G and* $(|A|, |G|) = 1$. *Then*

$$h(G) \ \leq \ 2l(A) + h(C_G(A)).$$

Results on regular orbit theorems have also been applied to obtain new theorems of Hall–Higman type (see [3] for a typical example), but going into this wide field would lead us too far astray.

Most other applications of results on large orbits are in the character theory of finite groups. As indicated before, orbit sizes often can be interpreted as character degrees, and so results on large orbits often are used to tackle problems in character theory. Typical examples are [25] or [81, Sections 14 and 17] and concern large character degrees, p–parts of character degrees or Huppert's ρ–σ–conjecture. Theorem 4.3 is essential for the proof of Theorem 3.2, and Theorem 4.7 can be translated into a result on base sizes (cf. Section 6). Theorems 4.4 and 4.5 played a significant role in work on the $k(GV)$–problem, a prominent and outstanding problem which therefore we will discuss separately in the following section.

5 The k(GV)–problem

This problem originates from a question of Brauer in the modular representation theory of finite groups. In 1956 Brauer [8] conjectured that if G is a finite group, p a prime and B any p–block of G, then the number $k(B)$ of irreducible characters in B is bounded above by the order $|D| = p^{d(B)}$ of the defect group D of B, i.e., $k(B) \leq p^{d(B)}$.

This became known as Brauer's $k(B)$–conjecture and so far has been proved in full generality for $d(B) \in \{0, 1, 2\}$. The best general bound to date is due to Brauer and Feit [9] and states that $k(B) \leq p^{2d(B)-2}$. For other results on Brauer's $k(B)$–conjecture that are not based on some solvability hypothesis see e.g. Michler's survey [83]. Now if one assumes the group G to be p–solvable, more can be done. In [87] Nagao proved that for p–solvable groups Brauer's $k(B)$–problem is equivalent to what henceforth has been called the

k(GV)–problem: If the group G acts faithfully and irreducibly on the finite G–module V and if $(|G|, |V|) = 1$, then the number of conjugacy classes of the semidirect product GV is bounded above by the order of V, i.e.,

$$k(GV) \ \leq \ |V|.$$

This infamous conjecture, as innocent as it looks, has stubbornly resisted attempts to prove it for a very long time and thus over the years has attracted the attention of many mathematicians in the field.

In 1984, Knörr [66] made the first significant breakthrough and proved the $k(GV)$–conjecture for supersolvable groups. Even more important than this actual result are the character–theoretic techniques that he developed in his influential paper to tackle the problem. He showed that if the generalized character $\delta(G,V) = \frac{|V|}{|C_V(g)|}$ "contains a square", i.e., if there is a generalized character Θ of G such that $\delta \geq |\Theta|^2$ on G and $\delta \equiv \Theta \pmod{p}$ on G, then this already implies that $k(GV) \leq |V|$. In the same paper he could prove the following:

Theorem 5.1 *Suppose that the group G acts faithfully and irreducibly on the finite G–module V of coprime characteristic. If there is a $v \in V$ such that $C_G(v)$ is abelian, then $k(GV) \leq |V|$.*
In particular, if G has a regular orbit on V, then $k(GV) \leq |V|$.

The strength of this amazing result is that it yields the wanted conclusions from a tiny piece of information, the mere fact that the centralizer of a single $v \in V$ is abelian. Also this is the place where the importance of regular orbit theorems to a solution of the $k(GV)$–problem becomes obvious, although it remains a little mysterious as to why things work this nicely as Knörr's proof is not too intuitive to the untrained eye. This raises the following

Question: Is there a purely group theoretical proof of Theorem 5.1?

See also Lemma 5.3 below for a weaker "elementary version" of Theorem 5.1.

Most remarkably, all other significant contributions to the solution of the $k(GV)$–problem are based on Knörr's results and generalize and extend his approach. Soon after the above result Knörr (unpublished) and Gluck [32] independently proved the $k(GV)$–conjecture in the case of $|G|$ being odd (and thus automatically solvable). Gow [40] proved the $k(GV)$–conjecture for self–dual V and Knörr [67] in case of the existence of a v with $V|_{C_G(v)}$ being a permutation module, and Robinson [101] generalized both results by solving the problem if $V|_{C_G(v)}$ is self–dual for some v. All these efforts ultimately led to the celebrated 1996 paper [103] by Robinson and Thompson in which they solved the problem if the characteristic p of the underlying field of V is large enough, more precisely, if $p > 5^{30}$. They extend Knörr's result by showing the following:

Theorem 5.2 *Let G be a group acting faithfully and irreducibly on the finite G–module V such that $(|G|,|V|) = 1$. If there is a $v \in V$ such that $V|_{C_G(v)}$ contains a faithful and self–dual summand, then $k(GV) \leq |V|$.*

The proof of this theorem is based on very technical character theoretic arguments. Then the $k(GV)$–problem is solved by verifying that the hypothesis of the theorem is satisfied if $p > 5^{30}$. For this, the inductive procedure mentioned before

Theorem 4.4 is used, and since the existence of a regular orbit implies the hypothesis of Theorem 5.2, Theorems 4.4 and 4.5 are used to settle the base cases.

Vectors $v \in V$ such that $V|_{C_V(g)}$ contains a self–dual faithful submodule are sometimes called *weakly real*, and if $V|_{C_G(v)}$ is self–dual, v is called *real*. So Theorem 5.2 says that the existence of weakly real vectors implies that $k(GV) \leq |V|$. The point of this is that the existence of weakly real vectors is strong enough to yield the wanted conclusion $k(GV) \leq |V|$, but it is still weak enough to be true for all GV at least if p is sufficiently large. And its key property is that unlike the property $k(GV) \leq |V|$ it has a good inductive behaviour which makes the inductive proof possible. So it is just about the "right" property to look at. Other properties like the existence of regular orbits, which also imply $k(GV) \leq |V|$, are too strong and even for large p simply are not always satisfied, and correspondingly their inductive behaviour is not good enough.

Not surprisingly, almost all further work on the $k(GV)$–problem focussed on proving the existence of (weakly) real vectors for smaller primes, and in detailed studies by Robinson [102], Goodwin [37, 38], Gluck and Magaard [35], Koehler and Pahlings [68], Riese und Schmid [99] and Riese [98] established the existence of real vectors for all p except for $p \in \{3, 5, 7, 11, 13, 19, 31\}$, and if G is solvable, real vectors exist except for $p \in \{3, 5, 7, 13\}$. So by Theorem 5.2 this solved the $k(GV)$–problem for all primes different from the exceptional primes. On the other hand, in general as well as in the solvable case examples exist for each exceptional prime such that even no weakly real vectors exist. Hence Theorem 5.2 cannot be used to solve the $k(GV)$–problem completely in the exceptional cases, and so to complete the solution of the $k(GV)$–problem these few primes have to be studied in a case–by–case analysis. Very recently, Riese and Schmid [100] did this and were able to give an affirmative answer to the $k(GV)$–problem for $p = 3, 7, 11, 13, 19$ and 31. (The case $p = 31$ had also been solved previously by Gluck and Magaard [36].) So currently the $k(GV)$–problem remains open only for $p = 5$, where Riese and Schmid obtained $k(GV) \leq \left(\frac{7}{5}\right)^n \cdot |V|$ with $n = \frac{1}{2}\dim(V)$, and there is reason for optimism that this last case will be solved fairly soon as well.

While the results obtained so far are very satisfying, there are two striking aspects in the development of the $k(GV)$–problem.

The first one is that there are some beautiful and elementary formulas for $k(GV)$ in case of coprime action of G on V which are well known and fairly easy to establish, and each of them can easily be transformed into the other using the Cauchy–Frobenius orbit counting formula. They are as follows:

If $k(U)$ denotes the number of conjugacy classes for any group U and v_i, $i = 1, \ldots, r(G, V)$ (where $r(G, V)$ is as in Section 2), are representatives of the orbits of G on V, then

$$(1) \qquad k(GV) = \sum_{i=1}^{r(G,V)} k(C_G(v_i)) = \frac{1}{|G|}\sum_{v\in V} |C_G(v)|\, k(C_G(v)).$$

Alternatively, if g_i, $i = 1, \dots, k(G)$ are representatives of the conjugacy classes of G, then

$$(2) \qquad k(GV) = \sum_{i=1}^{k(G)} r(C_G(g_i), C_V(g_i)) = \frac{1}{|G|}\sum_{g\in G} |C_G(g)|\, r(C_G(g), C_V(g)).$$

These elegant formulas look meaningful and seem to be a key to the solution of the $k(GV)$–problem, but surprisingly in the development up to this point they have hardly been used at all, except for settling easy cases such as G abelian (see [109]) or more recently in [36] to solve the case $p = 31$. A closer look at the formulas reveals the reason for this. To calculate $k(GV)$ with one of the formulas, very detailed information on the group is required, which usually is not available. And if one estimates the terms in the sums in a trivial way, e.g. $k(C_G(v)) \leq |G|$, then one clearly exceeds $|V|$.

We remark, however, that by formula (2) in many general situations, if one can prove the existence of a regular orbit, there is no need to apply the highly nontrivial Theorem 5.1 to obtain the conclusion $k(GV) \leq |V|$. Namely, as seen in Section 4, instead of proving the existence of a regular orbit explicitly, most proofs of regular orbit theorems actually prove the much stronger statement that $\sum_{1\neq g\in G} |C_V(g)| < |V|$. By slightly modifying the proof (and possibly some parameter in the hypothesis) the same proof will also yield the slightly stronger result that $\sum_{1\neq g\in G} |C_V(g)| < \frac{1}{3}|V|$, and then the following elementary lemma will yield the wanted conclusion immediately:

Lemma 5.3 *Let $G > 1$ be a group acting faithfully on the finite G–module V with $(|G|, |V|) = 1$ and $C_V(G) = 1$. If $\sum_{1\neq g\in G} |C_V(g)| < \frac{|G|-1}{|G|+1}\,|V| = \left(1 - \frac{2}{|G|+1}\right)|V|$, then $k(GV) \leq |V|$. In particular, if $\sum_{1\neq g\in G} |C_V(g)| < \frac{1}{3}|V|$, then $k(GV) \leq |V|$.*

Proof By the Cauchy–Frobenius orbit counting formula we know that

$$r(G, V) = \frac{1}{|G|}\sum_{g\in G} |C_V(g)|,$$

and so with formula (2) we obtain

$$\begin{aligned} k(GV) &= \sum_{g\in G} \frac{|C_G(g)|}{|G|}\, r(C_G(g), C(V(g)) \\ &= \frac{1}{|G|}\left(\sum_{g\in G}|C_V(g)| + \sum_{1\neq g\in G} |C_G(g)|\, r(C_G(g), C_V(g))\right) \\ &= \frac{1}{|G|}\left(|V| + \sum_{1\neq g\in G}\Big(|C_V(g)| + |C_G(g)|\, r(C_G(g), C_V(g))\Big)\right). \end{aligned}$$

Now for all $g \in G$ we see that

$$|C_V(g)| + |C_G(g)|\, r(C_G(g), C_V(g)) \;\leq\; (|G|+1)\, |C_V(g)|$$

and thus with the above and our hypothesis we find that

$$k(GV) \leq \frac{|V|}{|G|} + \frac{|G|+1}{|G|} \sum_{1\neq g\in G} |C_V(g)| \;\leq\; \frac{|V|}{|G|} + \frac{|G|-1}{|G|}\, |V| \;=\; |V|,$$

as wanted. □

So the proofs of the regular orbit theorems in Section 4 in case of coprime action also (almost) show that $k(GV) \leq |V|$ in these situations in an elementary way. In this sense Lemma 5.3 can be considered a weak version of Theorem 5.1.

The second curiosity about the $k(GV)$–problem are Knörr's intriguing methods to solve it themselves. In his review of Gow's paper Gluck called them "powerful, but rather mysterious" (MR94i: 20020), and they remain quite mysterious – or miraculous – even today. A more intuitive and "down–to–earth" approach to the problem is still desirable. For weaker bounds in more general settings see e.g. [78] and its predecessors [69].

A first attempt towards an easier solution to the $k(GV)$–problem is made in [65] where a completely new "Knörr–free" approach to the problem is developed which is far more elementary (albeit still quite technical) and mostly group theoretical. It is based on the (bad) inductive behaviour of $k(GV)$ itself and makes extensive use of formula (1) above. Thus it deals with both curiosities of the history of the problem mentioned above. In [65] the power of the new approach is demonstrated by solving the $k(GV)$–problem for solvable groups and large p. It is hoped that the new ideas will stimulate further research on the problem. Also, the new approach makes the fascinating $k(GV)$–problem accessible to a greater mathematical audience.

6 Orbits of permutation groups on the power set

In this last section we are concerned with permutation groups on a finite set Ω and their action on $\mathcal{P}(\Omega)$, the power set of Ω. Note that if for $A, B \in \mathcal{P}(\Omega)$ we define

$$A + B \;:=\; (A \cup B) - (A \cap B) \;=\; (A - B) \cup (B - A),$$

then $\mathcal{P}(\Omega)$ with this addition becomes a $GF(2)$–module, and thus the action of G on $\mathcal{P}(\Omega)$ fits within the framework of this article of studying linear group actions, and as we will see, it has raised some interest and found important applications in the past, some of which are related to the previously discussed topics. Clearly the action of G on the $GF(2)$–module $\mathcal{P}(\Omega)$ is faithful, but not irreducible, and so some of the results of the previous sections also can be applied to this special action, but often more can be said and proofs tend to be more combinatorial.

As to the number of orbits on $\mathcal{P}(\Omega)$, Babai and Pyber proved in [1] the following

Theorem 6.1 *Let G be a permutation group on Ω with $|\Omega| = n$, let $t \geq 4$ and assume that no composition factor of G is isomorphic to A_k (the alternating group on k letters) for any $k > t$. Then*

$$r(G, \mathcal{P}(\Omega)) \;\geq\; a^{\frac{n}{t}}$$

for some absolute constant $a > 1$.

So for solvable groups we have $r(G, \mathcal{P}(\Omega)) \geq a^{\frac{n}{4}}$ which is no surprise in view of Theorem 2.3. Also, combining this result with logarithmic bounds on the derived length of a solvable group in terms of its degree (see e.g. [81, Theorem 3.9(a)]) will lead – in analogy to Theorem 2.4 – to a double logarithmic bound for $\mathrm{dl}(G)$ in terms of $r(G, \mathcal{P}(\Omega))$. For results on the number of orbits on the k–subsets (i.e., subsets of cardinality k) of Ω see e.g. [111].

We next turn to the question of the number of orbit sizes. While for solvable groups it is not difficult to obtain a logarithmic bound for $\mathrm{dl}(G)$ in terms of $n(G, \mathcal{P}(\Omega))$ from Theorem 3.4, it should be possible to obtain such a result with much better constants and significantly less effort directly. A first step in this direction is done in [62, Theorems 2.7 and 2.8], where the following general analogue to Theorem 3.2 is proved:

Theorem 6.2 *Let G be a permutation group on the finite set Ω. Let $E \trianglelefteq G$ be elementary abelian of order p^m for a prime p and an integer $m \geq 9$ and suppose that E acts semi–regularly on Ω (i.e., no element of Ω is fixed by a nontrivial element of E). Then*

$$n(G, \mathcal{P}(\Omega)) \;\geq\; m - 8.$$

Again, of course, we conjecture that essentially it should be possible to replace E by an arbitrary p group of class 1 or 2 (and to drop the hypothesis of semi–regular action). The conjecture is open in general, but has been proved in case that $|G|$ is odd (and thus solvable) in [63, Theorem 3], using Theorem 6.4 below.

A completely different (mostly combinatorial) approach to the problem of orbit sizes is presented by Mnukhin in [86]:

Theorem 6.3 *Let G be a permutation group on the finite set Ω. If $\Sigma \subseteq \Omega$ with $|\Sigma| = k \geq 2$, then there is a $\Delta \subset \Sigma$ with $|\Delta| = k-1$ such that*

$$|\Delta^G| \geq \frac{2}{k^2}\,|\Sigma^G|^{\frac{k-1}{k}}.$$

This result, while very general, is not strong enough to allow conclusions on the number of distinct orbit sizes. For relations between the orbits on $(k-1)$–sets and orbits on k–sets see also Mnukhin's other work (such as [84] and [85]) or work of Siemons and Wagner [113].

As is the case for linear groups in general, also in the special case of interest in this section large orbits are of particular interest, and so this question has been studied intensely. For solvable groups the best possible answer is [81, Corollary 5.7]:

Theorem 6.4 *Let G be a solvable permutation group on the finite set Ω. Then there exists a $\Delta \subseteq \Omega$ such that $C_G(\Delta) := \{g \in G \mid \Delta^g = \Delta\}$ is a $\{2,3\}$–group. If $|G|$ is odd, then there exists a regular orbit of G on $\mathcal{P}(\Omega)$.*

This result follows from Gluck's characterization of primitive solvable permutation groups with no regular orbit on the power set (see [31]). Seress [107], using the classification of finite simple groups, took Gluck's work and other predecessors such as [10] or [14] further by removing the solvability hypothesis. In particular, with the exception of S_n, A_n and some other primitive permutation groups of degree ≤ 32, every primitive permutation group has a regular orbit on the power set. In [129] Zhang provides a classification of primitive solvable permutation groups that have fewer than eight orbits on the power set.

Based on Seress' work, Dolfi [19] investigated the action of a permutation group G on the set $\mathcal{P}_n(\Omega)$ of ordered partitions $(\Lambda_1, \ldots, \Lambda_{n+1})$ of Ω into $n+1$ parts (with possibly $\Lambda_i = \emptyset$ for some of the i). (Observe that $\mathcal{P}_1(\Omega)$ may be identified with $\mathcal{P}(\Omega)$.) For instance, he classifies the groups that have at most two regular orbits on $\mathcal{P}_2(\Omega)$ and also the groups that have at most three regular orbits on $\mathcal{P}_4(\Omega)$. This is used to show that for any permutation group not involving A_n for all $n \geq 5$ there exists a partition of Ω into five subsets such that only the identity stabilizes each of these sets. For solvable groups, this result had been obtained earlier by Seress (see [106, Theorem 1.2]), and for primitive solvable groups the result with "four" instead of "five" is a consequence of Gluck's result [31]. An application of Dolfi's work can be found in [20].

Related to the study of regular orbits is the study of bases. A *base* for a permutation group on Ω is a subset $B \subseteq \Omega$ such that only the identity fixes all the elements of B. It is often important to find a minimal base, i.e., a base B with $|B|$ as small as possible. Therefore the size $b(G)$ of a minimal base of the permutation group G on Ω is an important parameter. It is well–known (and easy to see) that

$2^{b(G)} \leq |G| \leq |\Omega|^{b(G)}$.

Note that if G if a permutation group on Ω with a base B of size m, this means that G has a regular orbit on Ω^m (= the cartesian product of m copies of Ω). Even more important for the focus of this section, it means that the setwise stabilizer of B is a subgroup of S_m, and if m is small, this means that B^G is a large (though usually not regular) orbit of G on $\mathcal{P}(\Omega)$.

Because of the many applications of minimal bases in various contexts, particularly computational group theory, they have attracted quite a bit of attention in recent years, and thus their basic properties as well as recent breakthroughs in this area have been presented in many places such as [12, Section 4.13], [18, Chapter 3], [13, Section 7], [11, Section 8], [96, Section 3] and [108, Section 5.3]. So to avoid repetition, we refer the reader to those references and mention only a few results here.

Seress [106] showed that for solvable primitive permutation groups G always $b(G) \leq 4$ and even $b(G) \leq 3$ if $|G|$ is odd. This result is crucial to some inductive arguments in the proof of Theorem 3.4. Liebeck and Shalev [80], confirming a conjecture of Cameron, prove that for some constant c, with known exceptions an almost simple permutation group always has a base of size at most c. In the same paper they prove:

Theorem 6.5 *There is a linear function $f : \mathbb{N} \to \mathbb{R}$ such that if G is a primitive permutation group not involving A_n, then $b(G) \leq f(n)$.*

We finish this section by mentioning some miscellaneous results. In [112] and [52] Siemons, Wagner and Inglis discuss how far a primitive permutation group is determined by its orbits on the power set, and a complete answer is obtained, not using the classification of finite simple groups. Furthermore in [110] it is shown that if two groups have the same orbits on the k–subsets of Ω, they also have the same orbits on the $(k-1)$–subsets of Ω.

This concludes our survey on orbits in finite group actions. While some classical problems such as the search for regular orbits have been treated almost exhaustively by now, it is clear that many other questions in this area are still at the beginning of their exploration and will provide exciting challenges for the future.

References

[1] L. Babai, L. Pyber, *Permutation groups without exponentially many orbits on the power set*,
J. Combin. Theory Ser. A **66** (1994), 160–168.

[2] T. R. Berger, *Character degrees and derived length in groups of odd order*, J. Algebra **39** (1976), 199–207.

[3] T. R. Berger, *Hall–Higman type theorems, III*, Trans. Amer. Math. Soc. **228** (1977), 47–83.

[4] T. R. Berger, *Representation theory and solvable groups: length type problems*, Proceedings of Symposia in Pure Mathematics **37**, American Mathematical Society, Providence, 1980, 431–441.

[5] T. R. Berger, B. B. Hargraves, C. Shelton, *The regular module problem I*, Trans. Amer. Math. Soc. **333** (1992), 251–274.

[6] T. R. Berger, B. B. Hargraves, C. Shelton, *The regular module problem II*, Comm. Algebra **18** (1990), 74–91.

[7] T. R. Berger, B. B. Hargraves, C. Shelton, *The regular module problem III*, J. Algebra **131** (1990), 74–91.

[8] R. Brauer, *Number theoretical investigations on groups of finite order*, Proceedings of the International Symposium on Algebraic Number Theory, Tokyo, (1956), 55–62.

[9] R. Brauer, W. Feit, *On the number of irreducible characters of finite groups in a given block*, Proc. Nut. Acad. Sci. USA **45** (1959), 361–365.

[10] P. J. Cameron, *Regular orbits of permutation groups on the power set*, Discrete Math. **62** (1986), 307–309.

[11] P. J. Cameron, *Permutation groups*, Handbook of Combinatorics, Vol. 1, Elsevier, Amsterdam, 1995, 611–645.

[12] P. J. Cameron, *Permutation groups*, London Mathematical Society Student Texts **45**, Cambridge University Press, Cambridge, 1999.

[13] P. J. Cameron, *Permutations*, Preprint, 1999 (for Erdös Memorial Conference, Budapest, 1999), available at www.maths.qmw.ac.uk/˜pjc/homepage.html

[14] P. J. Cameron, P. M. Neumann, J. Saxl, *On groups with no regular orbits on the set of subsets*, Arch. Math. **43** (1984), 295–296.

[15] W. Carlip, *Regular orbits of nilpotent subgroups of solvable groups*, Illinois J. Math. **38** (1994), 198–222.

[16] M. Cazzola, *Partial character degrees and length of solvable groups: the p–length 1 case*, Comm. Algebra **27** (1999), 3363–3375.

[17] E. C. Dade, *Carter subgroups and Fitting heigths of finite solvable groups*, Illinois J. Math. **13** (1969), 449–514.

[18] J. D. Dixon, B. Mortimer, *Permutation groups*, Graduate Texts in Mathematics **163**, Springer–Verlag, New York, 1996.

[19] S. Dolfi, *Orbits of permutation groups on the power set*, Arch. Math. **75** (2000), 321–327.

[20] S. Dolfi, *Finite groups in which p'–classes have q'–length*, Israel J. Math. **122** (2001), 93–115.

[21] L. Dornhoff, *The rank of primitive solvable permutation groups*, Math. Z. **109** (1969), 205–210.

[22] L. Dornhoff, *On imprimitive solvable rank 3 permutation groups*, Illinois J. Math. **14** (1970), 692–707.

[23] L. Dornhoff, *Some highly homogeneous groups*, Trans. Amer. Math. Soc. **182** (1973), 275-301.

[24] A. Espuelas, *The existence of regular orbits*, J. Algebra **127** (1989), 259–268.

[25] A. Espuelas, *Large character degrees of groups of odd order*, Illinois J. Math. **35** (1991), 499–505.

[26] A. Espuelas, *Regular orbits on symplectic modules*, J. Algebra **138** (1991), 1–12.

[27] A. Espuelas, G. Navarro, *Regular orbits of Hall π–subgroups*, Manuscripta Math. **70** (1991), 255–260.

[28] P. Fleischmann, *Finite groups with regular orbits on vector spaces*, J. Algebra **103** (1986), 211–215.

[29] D. A. Foulser, *Solvable primitive permutation groups of low rank*, Trans. Amer. Math. Soc. **143** (1969), 1–54.

[30] S. C. Garrison, *On groups with a small number of character degrees*, Ph. D. thesis, University of Wisconsin, Madison, 1973.
[31] D. Gluck, *Trivial set stabilizers in finite permutation groups*, Canad. J. Math. **35** (1983), 59–67.
[32] D. Gluck, *On the $k(GV)$–problem*, J. Algebra **89** (1984), 46–55.
[33] D. Gluck, *Bounding the number of character degrees of a solvable group*, J. London Math. Soc. (2) **31** (1985), 457–462.
[34] D. Gluck, K. Magaard, *Base sizes and regular orbits for coprime affine permutation groups*, J. London Math. Soc. **58** (1998), 603–618.
[35] D. Gluck, K. Magaard, *The extraspecial case of the $k(GV)$–problem*, Trans. Amer. Math. Soc., to appear.
[36] D. Gluck, K. Magaard, *The $k(GV)$–conjecture for modules in characteristic 31*, Preprint, 2001.
[37] D. P. M. Goodwin, *Regular orbits of linear groups with an application to the $k(GV)$–problem I*, J. Algebra **227** (2000), 395–432.
[38] D. P. M. Goodwin, *Regular orbits of linear groups with an application to the $k(GV)$–problem II*, J. Algebra **227** (2000), 433–473.
[39] R. Gow, *On the number of characters in a p–block of a p–solvable group*, J. Algebra **65** (1980), 421–426.
[40] R. Gow, *On the number of characters in a block and the $k(GV)$–problem for self–dual V*, J. London Math. Soc. **48** (1993), 441–451.
[41] A. Hanaki, T. Okuyama, *Groups with some combinatorial properties*, Osaka J. Math. **34** (1997), 337–356.
[42] B. B. Hargraves, *The existence of regular orbits for nilpotent groups*, J. Algebra **72** (1981), 54–100.
[43] B. Hartley, A. Turull, *On characters of coprime operator groups and the Glauberman character correspondence*, J. reine angew. Math. **451** (1994), 175–219.
[44] T. Hawkes, J. Humphreys, *A character–theoretic criterion for the existence of normal complements to subgroups of finite groups*, J. Algebra **94** (1985), 382–387.
[45] T. Hawkes, G. Jones, *On the structure of a group whose orbits on a finite module are the orbits of a proper subgroup*, J. Algebra **94** (1985), 364–381.
[46] C. Hering, *Transitive linear groups and linear groups which contain irreducible subgroups of prime order II*, J. Algebra **93** (1985), 151–164.
[47] B. Huppert, *Zweifach transitive, auflösbare Permutationsgruppen*, Math. Z. **68** (1957), 126–150.
[48] B. Huppert, Endliche Gruppen I, Springer–Verlag, Berlin–Heidelberg–New York, 1967.
[49] B. Huppert, *A remark on the character degrees of some p–groups*, Arch. Math. **59** (1992), 313–318.
[50] B. Huppert, Character theory of finite groups, de Gruyter, Berlin, 1998.
[51] B. Huppert, O. Manz, *Orbit sizes of p–groups*, Arch. Math. **54** (1990), 105–110.
[52] N. F. J. Inglis, *On orbit equivalent permutation groups*, Arch. Math. **43** (1984), 297–300.
[53] I. M. Isaacs, *Character degrees and derived length of a solvable group*, Canad. J. Math. **27** (1975), 146–151.
[54] I. M. Isaacs, Character theory of finite groups, Academic Press, New York, 1976.
[55] I. M. Isaacs, *Number of modular character degrees and lengths for solvable groups*, J. Algebra **148** (1992), 264–273.
[56] I. M. Isaacs, *Characters of groups associated with finite algebras*, J. Algebra **177** (1995), 708–730.
[57] I. M. Isaacs, *Large orbits in actions of nilpotent groups*, Proc. Amer. Math. Soc. **127** (1999), 45–50.

[58] I. M. Isaacs, G. Knutson, *Irreducible character degrees and normal subgroups*, J. Algebra **199** (1998), 302–326.

[59] A. Jaikin–Zapirain, A. Moretó, *Character degrees and nilpotence class of finite p–groups: an approach via pro–p groups*, Preprint, 2001.

[60] T. M. Keller, *On the asymptotic Taketa–bound for A–groups*, J. Algebra **191** (1997), 127–140.

[61] T. M. Keller, *Orbit sizes and character degrees*, Pacific J. Math. **187** (1999), 317–332.

[62] T. M. Keller, *Orbit sizes and character degrees, II*, J. reine angew. Math. **516** (1999), 27–114.

[63] T. M. Keller, *On the orbit sizes of permutation groups on the power set*, Algebra Colloq. **7** (2000), 27–32.

[64] T. M. Keller, *Orbit sizes and character degrees, III*, J. reine angew. Math., to appear.

[65] T. M. Keller, *A new approach to the $k(GV)$–problem*, Preprint, 2001, submitted.

[66] R. Knörr, *On the numbers of characters in a p–block of a p–solvable group*, Illinois J. Math. **28** (1984), 181–210.

[67] R. Knörr, *A remark on Brauer's $k(B)$–conjecture*, J. Algebra **131** (1990), 444–454.

[68] C. Koehler, H. Pahlings, *Regular orbits and the $k(GV)$–problem*, Preprint, 1999.

[69] L. G. Kovács, G. R. Robinson, *On the number of conjugacy classes of a finite group*, J. Algebra **160** (1993), 441–460.

[70] B. Külshammer, *Blocks, solvable permutation groups, and Landau's theorem*, J. reine angew. Math. **398** (1989), 180–186.

[71] B. Külshammer, *Solvable subgroups of p–solvable semilinear groups*, J. reine angew. Math. **404** (1990), 171–188.

[72] B. Külshammer, *Landau's theorem for p–blocks of p–solable groups*, J. reine angew. Math. **404** (1990), 189–191.

[73] M. L. Lewis, *Derived lengths of solvable groups having five irreducible character degrees I*, Algebras and Representation Theory, to appear.

[74] M. L. Lewis, *Derived lengths of solvable groups having five irreducible character degrees II*, Algebras and Representation Theory, to appear.

[75] M. W. Liebeck, *The affine permutation groups of rank three*, Proc. London Math. Soc. **54** (1987), 477–516.

[76] M. W. Liebeck, *Regular orbits of linear groups*, J. Algebra **184** (1996), 1136–1142.

[77] M. W. Liebeck, *Regular orbits and the $k(GV)$–problem*, Groups and Geometries (Siena, 1996), 145–148, Trends Math., Birkhäuser, Basel, 1998.

[78] M. W. Liebeck, L. Pyber, *Upper bounds for the number of conjugacy classes of a finite group*, J. Algebra **198** (1997), 538–562.

[79] M. W. Liebeck, J. Saxl, *The finite primitive permutation groups of rank three*, Bull. London Math. Soc. **18** (1986), 165–172.

[80] M. W. Liebeck, A. Shalev, *Simple groups, permutation groups, and probability*, J. Amer. Math. Soc. **12** (1999), 497–520.

[81] O. Manz, T. R. Wolf, Representations of solvable groups, London Math. Soc. Lect. Notes Ser. **185**, Cambridge University Press, 1993.

[82] M. K. Marshall, *Numbers of conjugacy class sizes and derived lengths for A–groups*, Can. Math. Bull. **39** (1996), 346–351.

[83] G. O. Michler, *Contributions to modular representation theory of finite groups*, Progress in Mathematics **95** (1991), 99–140.

[84] V. B. Mnukhin, *Reconstruction of the k–orbits of a permutation group (Russian)*, Mat. Zametki **42** (1987), 863–872, 911; translation in Math. Notes **42** (1987), 975–980.

[85] V. B. Mnukhin, *On the construction of $(k+1)$–orbits of a permutation group from its k–orbits (Russian)*, Mat. Zametki **51** (1992), 81–84, 142; translation in Math. Notes **51** (1992), 382–384.

[86] V. B. Mnukhin, *Some relations for the lengths of orbits on k–sets and $(k+1)$–sets*, Arch. Math. **69** (1997), 275–278.

[87] H. Nagao, *On a conjecture of Brauer for p–solvable groups*, J. Math. Osaka City Univ. **13** (1962), 35–38.

[88] G. Navarro, *Character degrees, derived length and Sylow normalizers*, Arch. Math. **68** (1997), 450–453.

[89] T. Noritzsch, *Groups having three complex irreducible character degrees*, J. Algebra **175** (1995), 767–798.

[90] T. Noritzsch, *A note on character degrees of p–groups and their normal subgroups*, Arch. Math. **59** (1992), 319–321.

[91] P. P. Pálfy, L. Pyber, *Small groups of automorphisms*, Bull. London Math. Soc. **30** (1998), 386–390.

[92] D. S. Passman, *Groups with normal solvable Hall p'–subgroups*, Trans. Amer. Math. Soc. **123** (1966), 99–111.

[93] D. S. Passman, *Solvable* 3/2*–transitive permutation groups*, J. Algebra **7** (1967), 192–207.

[94] D. S. Passman, *Exceptional* 3/2*–transitive permutation groups*, Pacific J. Math. **29** (1969), 669–713.

[95] A. Previtali, *Orbit lengths and character degrees of p–groups and their normal subgroups*, Arch. Math. **59** (1992), 319–321.

[96] L. Pyber, *Asymptotic results for permutation groups*, Groups and computation, DIMACS Ser. Discrete Math. Theoret. Comput. Sci. **11**, Providence, (1991), 197–219.

[97] J. M. Riedl, *Character degrees, class sizes, and normal subgroups of a certain class of p–groups*, J. Algebra **218** (1999), 190–215.

[98] U. Riese, *The quasisimple case of the $k(GV)$–conjecture*, J. Algebra **235** (2001), 45–65.

[99] U. Riese, P. Schmid, *Self–dual modules and real vectors for solvable groups*, J. Algebra **227** (2000), 159–171.

[100] U. Riese, P. Schmid, *Real vectors for linear groups and the $k(GV)$–problem*, Preprint, 2001.

[101] G. R. Robinson, *Some remarks on the $k(GV)$–problem*, J. Algebra **172** (1995), 159–166.

[102] G. R. Robinson, *Further reductions for the $k(GV)$–problem*, J. Algebra **195** (1997), 141–150.

[103] G. R. Robinson, J. G. Thompson, *On Brauer's $k(B)$–problem*, J. Algebra **184** (1996), 1143–1160.

[104] S. M. Seager, *The rank of a finite primitive solvable permutation group*, J. Algebra **105** (1987), 389–394.

[105] S. M. Seager, *A bound on the rank of primitive solvable permutation groups*, J. Algebra **116** (1988), 342–352.

[106] Á. Seress, *The minimal base size of primitive solvable permutation groups*, J. London Math. Soc. (2) **53** (1996), 243–255.

[107] Á. Seress, *Primitive groups with no regular orbits on the set of subsets*, Bull. London Math. Soc. **29** (1997), 697–704.

[108] A. Shalev, *Probabilistic group theory*, Groups St. Andrews 1997 in Bath, II, London Mathematical Society Lecture Note Series **261**, Cambridge University Press, 1999, 648–678.

[109] S. M. Shi, Y. Fan, L. Hui, *On the number of conjugacy classes in the semidirect product of certain finite groups*, Algebra Colloq. **1** (1994), 307–311.

[110] J. Siemons, *On partitions and permutation groups on unordered sets*, Arch. Math. **38** (1982), 391–403.

[111] J. Siemons, *Permutation groups on unordered sets.I.*, Arch. Math. **43** (1984), 483–487.

[112] J. Siemons, A. Wagner, *On finite permutation groups with the same orbits on unordered sets*, Arch. Math. **45** (1985), 492–500.

[113] J. Siemons, A. Wagner, *On the relationship between the lengths of orbits on k–sets and $(k+1)$–sets*, Abh. Math. Sem. Univ. Hamburg **58** (1988), 267–274.

[114] M. C. Slattery, *Character degrees and derived length in p–groups*, Glasgow Math. J. **30** (1988), 221–230.

[115] K. Taketa, *Über die Gruppen, deren Darstellungen sich sämtlich auf monomiale Gestalt transformieren lassen*, Proc. Acad. Tokyo **6** (1930), 31–33.

[116] J. G. Thompson, *Automorphisms of solvable groups*, J. Algebra **1** (1964), 259–267.

[117] A. Turull, *Supersolvable automorphism groups of solvable groups*, Math. Z. **183** (1983), 47–73.

[118] A. Turull, *Centralizers and character degrees*, Arch. Math. **74** (2000), 410–413.

[119] A. Turull, *Fitting height of groups and of fixed points*, J. Algebra **86** (1984), 555–566.

[120] A. Turull, *Fixed point free action with regular orbits*, J. reine angew. Math. **371** (1986), 67–91.

[121] A. Turull, *Character theory and length problems, Finite and locally finite groups*, NATO Adv. Sci. Inst. Ser. C Math. Phys. Sci. **471**, Kluwer, Dordrecht, 1995, 377–400.

[122] A. Turull, *Fixed point free action with some regular orbits*, J. Algebra **194** (1997), 362–377.

[123] A. Turull, *Cyclic by prime fixed point free action*, Proc. Amer. Math. Soc. **125** (1997), 3465–3470.

[124] T. R. Wolf, *Character correspondences, degrees and derived length in solvable groups*, J. Algebra **191** (1997), 653–667.

[125] T. R. Wolf, *Large orbits of supersolvable linear groups*, J. Algebra **215** (1999), 235–247.

[126] T. R. Wolf, *Regular orbits of induced modules*, in preparation.

[127] T. Yuster, *Orbit sizes under automorphism actions in finite groups*, J. Algebra **82** (1982), 342–352.

[128] J. Zhang, *p–regular orbits and p–blocks of defect zero*, Commun. Algebra **21** (1993), 299–307.

[129] J. Zhang, *Finite groups with few regular orbits on the power set*, Algebra Colloq. **4** (1997), 471–480.

GROUPS WITH FINITELY GENERATED INTEGRAL HOMOLOGIES

DESSISLAVA H. KOCHLOUKOVA[1]

IMECC, UNICAMP, Cx. P. 6065,
13083-970 Campinas, SP, Brasil
E-mail: desi@ime.unicamp.br

Abstract

Suppose A is an abelian normal subgroup of a finitely generated group G such that G/A is abelian and $H_i(G,\mathbb{Z})$ is finitely generated for all i . We show that A is of finite (Prüfer) rank. This generalises the main result of [5] that deals with the same problem for split extension metabelian groups.

1 Introduction

In [5] J. R. J. Groves shows that if G is a finitely generated group, a split extension of an abelian group A by an abelian group Q and the homology group $H_i(G,\mathbb{Z})$ is finitely generated for all i then A is of finite rank i.e. $A \otimes_{\mathbb{Z}} \mathbb{Q}$ is finite dimensional over $\mathbb{Q}$ and the torsion part of A is finite. We generalise this result to the non-split case.

Theorem A *Suppose that A is a normal abelian subgroup of a finitely generated group G such that G/A is abelian and $H_i(G,\mathbb{F}_p)$ is finite for all i and all primes p. Then A is of finite (Prüfer) rank.*

Corollary B *Suppose that A is a normal abelian subgroup of a finitely generated group G such that G/A is abelian and $H_i(G,\mathbb{Z})$ is finitely generated for all i. Then A is of finite (Prüfer) rank.*

Our proofs substantially use the method and the main tools from [5]: the geometric invariant for modules over finitely generated abelian groups defined in [2], Cartan's formula for $H_*(A,\mathbb{F}_p)$ for abelian groups A and the finite field $\mathbb{F}_p$ with p elements (in the case $p=2$ the formula holds only for groups A of exponent 2) and close examination of the LHS spectral sequence in homology. The new ingredients are a lemma that gives sufficient conditions when $H_0(Q,H_j(A,\mathbb{F}_p))$ finite implies that $H_i(Q,H_j(A,\mathbb{F}_p))$ is finite for all $i \geq 0$ and a generalisation of Cartan's formula for the homologies of abelian groups with trivial coefficients $\mathbb{F}_2$. We note that our proof does not require all the results and techniques from [5], for example we do not use the results from the sections 4.3 and 5.2 in [5].

[1]Supported by a post-doctoral grant no. 98/00482-3 from FAPESP, Brazil.

2 Preliminaries on homologies of abelian groups with trivial coefficients and on the geometric invariant for modules over finitely generated abelian groups

The geometric invariant $\Sigma_A(Q)$ for a finitely generated $\mathbb{Z}[Q]$-module A was first defined in [2]. By definition

$$\Sigma_A(Q) = \{[\chi] = \mathbb{R}_{>0}\chi \mid \chi \in Hom(Q,\mathbb{R}) \setminus \{0\}, A \text{ is finitely generated over } \mathbb{Z}[Q_\chi]\},$$

$$\Sigma_A^c(Q) = S(Q) \setminus \Sigma_A(Q)$$

where $Q_\chi = \{g \in Q \mid \chi(g) \geq 0\}$ and $S(Q) = \{[\chi] \mid \chi \in Hom(Q,\mathbb{R}) \setminus \{0\}\} \simeq S^{n-1}$ where n is the torsion-free rank of Q. A is said to be m-tame if every m-point subset of $\Sigma_A^c(Q)$ is contained in an open half subspace of $S(Q)$. One important property of tameness is that whenever A is m-tame the m-fold tensor power of A over $\mathbb{Z}$ is finitely generated over $\mathbb{Z}[Q]$ via the diagonal Q-action [1, Section 3.5]. Note that in general the converse does not hold but in the case when A is of prime exponent, m-tameness and the finite generation (over $\mathbb{Z}[Q]$) of the m-th tensor power of A are equivalent [1, Thm 3.4].

We discuss now a result of H. Cartan [4] that will play an important role in the proof of our main theorem. Suppose A is an abelian group. By definition $\widetilde{S}^j({}_pA)$ is the set of the elements in the j-th tensor power of ${}_pA = \{a \in A \mid pa = 0\}$ over $\mathbb{F}_p$ which are invariant under the action of the symmetric group on j elements via permutation of the factors of the j-th tensor power. Note $\widetilde{S}({}_pA) = \oplus_{i \geq 0}\widetilde{S}^i({}_pA)$ is a graded algebra with multiplication given by the shuffle product $*$ of the tensor algebra of ${}_pA$ i.e.

$$(a_1 \otimes \ldots \otimes a_s) * (a_{s+1} \otimes \ldots \otimes a_{s+k}) = \sum_\sigma a_{\sigma(1)} \otimes \ldots \otimes a_{\sigma(s+k)}$$

where $a_i \in {}_pA$ and the sum is over all permutation σ such that $\sigma(1) < \sigma(2) < \ldots < \sigma(s)$ and $\sigma(s+1) < \ldots < \sigma(s+k)$.

In general for a commutative ring with unity k and an arbitrary group G the homology group $H_*(G,k) = \oplus_{i \geq 0} H_i(G,k)$ does not have multiplicative structure though the cohomology group $H^*(G,k) = \oplus_{i \geq 0} H^i(G,k)$ has one via the cup product. Still for A abelian there is a Pontryagin product in $H_*(A,k)$ defined as the composite:

$$H_i(A,k) \times H_j(A,k) \xrightarrow{\alpha} H_{i+j}(A \times A, k) \xrightarrow{\beta} H_{i+j}(A,k)$$

where α is the homology cross product and β is induced by the multiplication map $A \times A \to A$. With the Pontryagin product $H_*(A,k)$ becomes a strictly anticommutative ring equipped with divided powers, for details see [3, Ch. 5, Section 5]. We remind the reader the axioms of divided powers. There is a family of functions ${}^{(i)} : H_{2j}(A,k) \to H_{2ij}(A,k)$ for all $i \geq 0, j \geq 1$ with the following properties:

1. $x^{(0)} = 1, x^{(1)} = x$
2. $x^{(i)}x^{(j)} = \binom{i+j}{i} x^{(i+j)}$
3. $(x^{(i)})^{(j)} = e_{i,j}x^{(ij)}$ for all $i,j > 0$ where $e_{i,j} = \prod_{2 \leq t \leq j} \binom{ti-1}{i-1}$.

4. $(x+y)^{(i)} = \sum_{j+k=i} x^{(j)} y^{(k)}$
5. For $i \geq 2$

$$(xy)^{(i)} = x^i y^{(i)} \text{ whenever } x, y \text{ are even elements and } deg(y) > 0,$$

$$(xy)^{(i)} = 0 \text{ if } x, y \text{ are odd elements.}$$

Proposition 2.1 *[4, Ch. 9 , Ch. 10] Suppose p is a prime and $\mathbb{F}_p$ is the field with p elements.*

1. If p is odd

$$H_*(A, \mathbb{F}_p) \simeq (\wedge_{\mathbb{F}_p}(A/pA)) \otimes_{\mathbb{F}_p} \widetilde{S}({}_pA),$$

where A/pA has weight 1 and ${}_pA$ has weight 2 i.e.

$$H_n(A, \mathbb{F}_p) \simeq \oplus_{0 \leq k \leq [\frac{n}{2}]} (\wedge_{\mathbb{F}_p}^{n-2k}(A/pA)) \otimes_{\mathbb{F}_p} \widetilde{S}^k({}_pA).$$

2. If p = 2 and A is of exponent 2

$$H_*(A, \mathbb{F}_2) \simeq \widetilde{S}(A),$$

where A has weight 1 i.e.

$$H_n(A, \mathbb{F}_2) \simeq \widetilde{S}^n(A).$$

In both cases the isomorphism is a natural isomorphism of graded divided powers algebras i.e. preserves grading, the multiplicative and the divided power structures.

Whenever possible we prefer using homologies with coefficients in $\mathbb{F}_p$ rather than $\mathbb{Z}$. The Pontryagin product gives a natural embedding of the exterior algebra of $H_1(A, \mathbb{Z}) \simeq A$ in $H_*(A, \mathbb{Z})$ and it is an isomorphism if A is $\mathbb{Z}$-torsion-free [3, Ch. 5, Thm 6.4]. The problem is that in general this embedding does not naturally split as in the case of coefficients $\mathbb{F}_p$ for p odd. In [6, Thm C] some results linking the integral homology groups of A and finite generation (over a commutative ring) of tensor powers of A are established but they are not directly applicable to the proof of Theorem A.

3 More on the homology with coefficients in $\mathbb{F}_2$

In [3, Thm 6.6] it is stated that there is a non-natural isomorphism

$$H_*(A, \mathbb{F}_2) \simeq \wedge(A/2A) \otimes \widetilde{S}({}_2A) \tag{1}$$

and that the above isomorphism could be proved as in the proof of [3, Thm 6.4]. The generating space ${}_2A = \{a \in A \mid 2a = 0\}$ of the symmetric algebra comes from a non-natural splitting of the exact sequence

$$0 \to \wedge^2(A/2A) \to H_2(A, \mathbb{F}_2) \to {}_2A \to 0.$$

Still it is not made clear in [3] how to combine the non-naturality of (1) with the ideas of the proof of [3, Thm 6.4] that deals with a natural description of homology groups.

In this section we show how such an isomorphism could be proved and in fact we give a natural description of $H_*(A, \mathbb{F}_2) = \oplus_{i\geq 0} H_i(A, \mathbb{F}_2)$ in terms of a filtration with quotients isomorphic to the direct summands of (1). Naturality is important for two purposes. First it is needed in the proof of the fact that the filtration is exhausting. Secondly in the applications considered in this paper A is endowed with the structure of a $\mathbb{Z}[Q]$-module and we are interested not only in the underlying additive structure of the homology $H_*(A, \mathbb{F}_2)$ but in its structure as $\mathbb{Z}[Q]$-module. As before $H_*(A, \mathbb{F}_2)$ is equipped with strictly anticommutative Pontryagin product and divided power structure.

Note that

$$H_1(A, \mathbb{F}_2) \simeq A/2A,$$

and by the exact universal coefficient sequence

$$0 \to H_2(A, \mathbb{Z}) \otimes \mathbb{F}_2 \to H_2(A, \mathbb{F}_2) \to Tor_1^{\mathbb{Z}}(H_1(A, \mathbb{Z}), \mathbb{F}_2) \to 0.$$

As $H_2(A, \mathbb{Z}) \simeq \wedge^2 A$ and $Tor_1^{\mathbb{Z}}(H_1(A, \mathbb{Z}), \mathbb{F}_2) \simeq {}_2A = \{a \in A \mid 2a = 0\}$ we have

$$\frac{H_2(A, \mathbb{F}_2)}{\wedge^2 H_1(A, \mathbb{F}_2)} \simeq {}_2A.$$

Theorem 3.1 *Let*

$$F^i(H_*(A, \mathbb{F}_2)) = \sum_{k\geq 0, j_1+\ldots+j_t\leq i} H_1(A, \mathbb{F}_2)^k H_2(A, \mathbb{F}_2)^{(j_1)} \ldots H_2(A, \mathbb{F}_2)^{(j_t)},$$

where $H_2(A, \mathbb{F}_2)^{(j)}$ is the subspace spanned by all elements $\lambda^{(j)}$ for $\lambda \in H_2(A, \mathbb{F}_2)$. Then $\cup_{i\geq 1} F^i(H_*(A, \mathbb{F}_2))$ is an exhausting filtration of the graded algebra $H_*(A, \mathbb{F}_2)$ with quotients

$$F^i(H_*(A, \mathbb{F}_2))/F^{i-1}(H_*(A, \mathbb{F}_2)) \simeq \wedge(A/2A) \otimes \widetilde{S}^i({}_2A). \tag{2}$$

Proof Note that to prove that the filtration is exhausting i.e.

$$\cup_{i\geq 1} F^i(H_*(A, \mathbb{F}_2)) = H_*(A, \mathbb{F}_2) \tag{3}$$

and that (2) holds it is sufficient to consider the following cases:

1. A is cyclic;

2. if (2) and (3) hold for the finitely generated abelian groups A_1 and A_2 then (2) and (3) hold for $A = A_1 \oplus A_2$;

Then (2) and (3) hold for all finitely generated abelian groups. Finally as every group A is the direct limit of its finitely generated subgroups and our filtration commutes with direct limits both (2) and (3) hold in general. In more details if $A = \lim_{\overrightarrow{i}} A_i$ we have

$$F^j(H_*(A, \mathbb{F}_2)) \simeq \lim_{\overrightarrow{i}} F^j(H_*(A_i, \mathbb{F}_2))$$

since

1. $\lim\limits_{\overrightarrow{i}} H_2(A_i, \mathbb{F}_2)^{(j)} \simeq H_2(A, \mathbb{F}_2)^{(j)}$ with isomorphism induced by the maps $H_2(A_i, \mathbb{F}_2)^{(j)} \to H_2(A, \mathbb{F}_2)^{(j)}$ that themselves are induced by the maps $A_i \to A$ given by $A = \lim\limits_{\overrightarrow{i}} A_i$ and

2. by the definition of Pontryagin product, the fact that homology commutes with direct limits (i.e. $H_*(A, \mathbb{F}_2) \simeq \lim\limits_{\overrightarrow{i}} H_*(A_i, \mathbb{F}_2)$) and 1 we get

$$\lim_{\overrightarrow{i}} H_1(A_i, \mathbb{F}_2)^k H_2(A_i, \mathbb{F}_2)^{(j_1)} \dots H_2(A_i, \mathbb{F}_2)^{(j_t)} \simeq$$

$$(\lim_{\overrightarrow{i}} H_1(A_i, \mathbb{F}_2)^k)(\lim_{\overrightarrow{i}} H_2(A_i, \mathbb{F}_2)^{(j_1)}) \dots (\lim_{\overrightarrow{i}} H_2(A_i, \mathbb{F}_2)^{(j_t)}) \simeq$$

$$H_1(A, \mathbb{F}_2)^k H_2(A, \mathbb{F}_2)^{(j_1)} \dots H_2(A, \mathbb{F}_2)^{(j_t)}.$$

Now we show that (2) and (3) hold for A cyclic. If A is infinite then $H_i(A, \mathbb{F}_2) = 0$ for all $i \geq 2$ and there is nothing to prove. Now let A be cyclic of order m. Consider the free resolution $\mathcal{F}$ of the trivial module $\mathbb{Z}$ over $\mathbb{Z}[A]$

$$\mathcal{F} : \dots \to \mathbb{Z}[A]e_n \xrightarrow{\partial_n} \mathbb{Z}[A]e_{n-1} \xrightarrow{\partial_{n-1}} \dots \xrightarrow{\partial_1} \mathbb{Z}[A]e_0 \xrightarrow{\epsilon} \mathbb{Z} \to 0$$

with ∂_n being multiplication with $1 + t + \dots + t^{m-1}$ for n odd and multiplication with $t - 1$ for n even. The Pontryagin product in $H_*(A, \mathbb{F}_2)$ is induced by an admissible product in $\mathcal{F}$ i.e. product satisfying the following

$$\epsilon(xy) = \epsilon(x)\epsilon(y) \text{ and } \partial(xy) = \partial(x)y + (-1)^{degx} x\partial(y).$$

By [3, p.119] we have

$$e_{2i} = e_2^i/i! \ , e_{2i}e_{2j} = (i,j)e_{2i+2j} \ , e_{2i}e_{2j+1} = (i,j)e_{2i+2j+1} \ , e_{2i+1}e_{2j+1} = 0$$

For m odd, $i \geq 1$ the homology group $H_i(A, \mathbb{F}_2)$ has exponent dividing 2 and m, hence is zero. If m is even $\mathbb{F}_2 \otimes_{\mathbb{Z}[A]} \mathcal{F}$ has trivial differentials and $H_*(A, \mathbb{F}_2) = H_*(\mathbb{F}_2 \otimes_{\mathbb{Z}[A]} \mathcal{F}) = (\wedge \mathbb{F}_2 e_1) \otimes_{\mathbb{F}_2} \Gamma_{\mathbb{F}_2}(e_2)$, where $\Gamma_{\mathbb{F}_2}(e_2)$ is the divided polynomial algebra over $\mathbb{F}_2$ as defined in [3, p.119]. Then $H_1(A, \mathbb{F}_2) = \mathbb{F}_2 e_1$, $H_2(A, \mathbb{F}_2) = \mathbb{F}_2 e_2$, $H_2(A, \mathbb{F}_2)^{(j)} = \mathbb{F}_2 e_2^{(j)} = \mathbb{F}_2 e_{2j}$ and

$$F^i(H_*(A, \mathbb{F}_2)) = \sum_{k \geq 0, j_1 + \dots + j_t \leq i} H_1(A, \mathbb{F}_2)^k H_2(A, \mathbb{F}_2)^{(j_1)} \dots H_2(A, \mathbb{F}_2)^{(j_t)} =$$

$$\sum_{\epsilon = 0,1; j_1 + \dots + j_t \leq i} \mathbb{F}_2 e_1^\epsilon e_{2j_1} \dots e_{2j_t} = \sum_{j \leq 2i+1} \mathbb{F}_2 e_j,$$

the latter equality holds as $e_1 e_{2i} = e_{2i+1}$ and $e_1^\epsilon e_{2j_1} \dots e_{2j_t} = \lambda e_{\epsilon + 2j_1 + \dots + 2j_t}$ for $\lambda \in \mathbb{F}_2$. Thus (3) follows immediately.

Finally for A cyclic of even order m we have $\widetilde{S}^i({}_2A) = \widetilde{S}^i(\mathbb{F}_2) =^{\otimes^i} \mathbb{F}_2 \simeq \mathbb{F}_2 \simeq \mathbb{F}_2 e_{2i}$ and $F^{(i)}(H_*(A, \mathbb{F}_2))/F^{(i-1)}(H_*(A, \mathbb{F}_2)) \simeq \sum_{2i \leq j \leq 2i+1} \mathbb{F}_2 e_j = \mathbb{F}_2 e_{2i} \oplus \mathbb{F}_2 e_1 e_{2i} = (\mathbb{F}_2 \oplus \mathbb{F}_2 e_1).\mathbb{F}_2 e_{2i} \simeq (\wedge A/2A) \otimes \widetilde{S}^i({}_2A)$. In particular (2) holds.

Now we prove that the filtration is exhausting for $A = A_1 \oplus A_2$ provided the same holds for A_1 and A_2. Consider the commutative diagram

$$\begin{array}{ccc} (\cup_{i\geq 1}F^i(H_*(A_1,\mathbb{F}_2)))\otimes(\cup_{i\geq 1}F^i(H_*(A_2,\mathbb{F}_2))) & \stackrel{\beta_1\otimes\beta_2}{\longrightarrow} & (\cup_{i\geq 1}F^i(H_*(A,\mathbb{F}_2))) \\ \downarrow \alpha_1\otimes\alpha_2 & & \downarrow\alpha \\ H_*(A_1,\mathbb{F}_2)\otimes H_*(A_2,\mathbb{F}_2) & \stackrel{\varphi}{\longrightarrow} & H_*(A,\mathbb{F}_2) \end{array}$$

The maps α_1, α_2 and α are the obvious inclusions. By assumptions α_1 and α_2 are isomorphisms. The map

$$\beta_i : \cup_{s\geq 1}F^s(H_*(A_i,\mathbb{F}_2)) \to \cup_{s\geq 1}F^s(H_*(A,\mathbb{F}_2))$$

is induced by the maps

$$H_j(A_i,\mathbb{F}_2) \to H_j(A,\mathbb{F}_2)$$

for $j = 1, 2$ both given by the inclusion of A_i in A. The map φ is the inclusion given by the Künneth formula and as $\mathbb{F}_2$ is a field φ is an isomorphism. Then $\varphi(\alpha_1\otimes\alpha_2)$ is an isomorphism and hence α is surjective. By construction α is injective and so α is an isomorphism, as required. Then $\beta_1\otimes\beta_2$ is an isomorphism too.

Claim $(\beta_1\otimes\beta_2)(\sum_{i+k=m} F^i(H_*(A_1,\mathbb{F}_2))\otimes F^k(H_*(A_2,\mathbb{F}_2))) = F^m(H_*(A,\mathbb{F}_2))$

Proof Assume x_1, y_1 are elements of degree one and x_2, y_2 are elements of degree two. Then by the axioms of divided powers for $j \geq 2$ the element $(x_2+y_2+x_1y_1)^{(j)}$ equals

$$\sum_{\alpha_1+\alpha_2+\alpha_3=j} x_2^{(\alpha_1)}y_2^{(\alpha_2)}(x_1y_1)^{(\alpha_3)} = \sum_{\alpha_1+\alpha_2=j} x_2^{(\alpha_1)}y_2^{(\alpha_2)} + \sum_{\alpha_1+\alpha_2=j-1} x_2^{(\alpha_1)}y_2^{(\alpha_2)}x_1y_1.$$

This together with the Künneth formula

$$H_2(A,\mathbb{F}_2) \simeq \oplus_{0\leq i\leq 2}H_i(A_1,\mathbb{F}_2)\otimes H_{2-i}(A_2,\mathbb{F}_2)$$

implies that $H_2(A,\mathbb{F}_2)^{(j)}$ is a subset of

$$\sum_{j_1=0,1;j_0+j_1+j_2=j} H_2(A_1,\mathbb{F}_2)^{(j_0)}H_1(A_1,\mathbb{F}_2)^{j_1}H_1(A_2,\mathbb{F}_2)^{j_1}H_2(A_2,\mathbb{F}_2)^{(j_2)}$$

$$\subseteq \wedge H_1(A,\mathbb{F}_2)\sum_{j_0+j_2\leq j} H_2(A_1,\mathbb{F}_2)^{(j_0)}H_2(A_2,\mathbb{F}_2)^{(j_2)} \subseteq H_*(A,\mathbb{F}_2).$$

This together with the definition of the filtration implies

$$F^m(H_*(A,\mathbb{F}_2)) \subseteq (\beta_1\otimes\beta_2)(\sum_{i+k=m} F^i(H_*(A_1,\mathbb{F}_2))\otimes F^k(H_*(A_2,\mathbb{F}_2))).$$

The inverse inclusion is obvious.

□

By the claim and the fact that $\beta_1 \otimes \beta_2$ is an isomorphism we have

$$F^m(H_*(A, \mathbb{F}_2))/F^{m-1}(H_*(A, \mathbb{F}_2)) \simeq$$
$$\oplus_{i+k=m} F^i(H_*(A_1, \mathbb{F}_2))/F^{i-1}(H_*(A_1, \mathbb{F}_2)) \otimes F^k(H_*(A_1, \mathbb{F}_2))/F^{k-1}(H_*(A_1, \mathbb{F}_2)).$$

By assumption

$$F^i(H_*(A_1, \mathbb{F}_2))/F^{i-1}(H_*(A_1, \mathbb{F}_2)) \simeq \wedge(A_1/2A_1) \otimes \widetilde{S}^i({}_2(A_1)),$$
$$F^{m-i}(H_*(A_1, \mathbb{F}_2))/F^{m-i-1}(H_*(A_1, \mathbb{F}_2)) \simeq \wedge(A_2/2A_2) \otimes \widetilde{S}^{m-i}({}_2(A_2)).$$

Furthermore

$$\wedge(A_1/2A_1) \otimes \wedge(A_2/2A_2) \simeq \wedge((A_1/2A_1) \oplus (A_2/2A_2)) = \wedge(A/2A).$$

Finally it remains to note that for abelian groups M and N of exponent 2

$$\oplus_{0 \leq i \leq m}(\widetilde{S}^i(M) \otimes \widetilde{S}^{m-i}(N)) \simeq \widetilde{S}^m(M \oplus N)$$

and apply this for $M = {}_2(A_1), N = {}_2(A_2)$, $M \oplus N \simeq {}_2A = \{a \in A \mid 2a = 0\}$ to obtain the required isomorphism

$$F^m(H_*(A, \mathbb{F}_2))/F^{m-1}(H_*(A, \mathbb{F}_2)) \simeq (\wedge(A/2A)) \otimes \widetilde{S}^m({}_2A).$$

The formula about M and N could be verified either by hand or by Proposition 2.1 (remember both M and N have exponent 2) it is transformed to a special case of the Künneth formula

$$H_*(M, \mathbb{F}_2) \otimes H_*(N, \mathbb{F}_2) \simeq H_*(M \oplus N, \mathbb{F}_2).$$

This completes the proof of the theorem. □

To illustrate the above theorem we consider the case when A has exponent 2. By Proposition 2.1

$$H_*(A, \mathbb{F}_2) \simeq \widetilde{S}(A).$$

Any symmetric element of $\otimes^i A$ is a linear combination of elements of the form

$$(a_1^{\otimes^{k_1}}) * (a_2^{\otimes^{k_2}}) * \ldots * (a_s^{\otimes^{k_s}})$$

for some $s \leq i$, some pairwise different elements $a_1, \ldots, a_s$ of A where $\sum k_j = i$ and $*$ is the shuffle product. The Pontryagin product is the shuffle product $*$ and the divided power structure is given by

$$(a \otimes a)^{(i)} = a^{\otimes^{2i}}.$$

Then

$$a^{\otimes^k} = a^\epsilon * (a \otimes a)^{([k/2])} \in H_1(A, \mathbb{F}_2)^\epsilon H_2(A, \mathbb{F}_2)^{([k/2])},$$

where $\epsilon = k - 2[k/2]$ and the element $(a_1^{\otimes^{k_1}}) * (a_2^{\otimes^{k_2}}) * \ldots * (a_s^{\otimes^{k_s}})$ belongs to

$$(\wedge H_1(A, \mathbb{F}_2)) H_2(A, \mathbb{F}_2)^{([\frac{k_1}{2}])} H_2(A, \mathbb{F}_2)^{([\frac{k_2}{2}])} \ldots H_2(A, \mathbb{F}_2)^{([\frac{k_s}{2}])}.$$

Thus

$$F^j(H_*(A, \mathbb{F}_2)) \cap \widetilde{S}^i(A) = \sum (a_1^{\otimes^{k_1}}) * (a_2^{\otimes^{k_2}}) * \ldots * (a_s^{\otimes^{k_s}}),$$

where the sum is over all s-tuples $(k_1, \ldots, k_s)$ of non-negative integers such that $[k_1/2] + \ldots + [k_s/2] \leq j$ and $k_1 + \ldots + k_s = i$.

4 Some results about homology

Lemma 4.1 *Suppose Q is a finitely generated abelian group, B is a left $\mathbb{Z}[Q]$-module equipped with a finite filtration of $\mathbb{Z}[Q]$-submodules*

$$B = B_1 \supset B_2 \supset \ldots \supset B_k \supset B_{k+1} = 0$$

such that B_j/B_{j+1} is a cyclic R_j-module for some commutative Noetherian ring R_j, Q embeds in R_j and the action of Q on B_j/B_{j+1} via the embedding of Q in R_j is the original action of Q. Suppose further that $H_0(Q, B)$ is finite. Then $H_i(Q, B)$ is finite for all i.

Proof To prove the lemma we induct on the length k of the filtration. Assume first that $k = 1$ and so B is a cyclic R_1-module and a commutative Noetherian ring. Let $\mathcal{F}$ be a resolution of the trivial right module $\mathbb{Z}$ over $\mathbb{Z}[Q]$ with all modules finitely generated. Consider the complex $\mathcal{F} \otimes_{\mathbb{Z}[Q]} B$. Its modules could be viewed as (right) B-modules via the multiplication in B. Note this B-action is compatible with the differentials because B is commutative. As B is a Noetherian ring and all modules in $\mathcal{F} \otimes_{\mathbb{Z}[Q]} B$ are finitely generated over B we deduce that all homology groups

$$H_i(\mathcal{F} \otimes_{\mathbb{Z}[Q]} B) \simeq H_i(Q, B)$$

are finitely generated B-modules. Furthermore the action of Q on $H_i(Q, B)$ is trivial and so $H_i(Q, B)$ is a finitely generated module over $\mathbb{Z} \otimes_{\mathbb{Z}[Q]} B \simeq H_0(Q, B)$. As by assumption $H_0(Q, B)$ is finite we are done.

If $k \geq 2$ consider the short exact sequence of modules

$$0 \to B_2 \to B \to B/B_2 \to 0.$$

It induces a long exact sequence in homology

$$\ldots \to H_i(Q, B_2) \to H_i(Q, B) \to H_i(Q, B/B_2) \to \ldots$$

$$\to H_1(Q, B/B_2) \to H_0(Q, B_2) \to H_0(Q, B) \to H_0(Q, B/B_2) \to 0$$

As $H_0(Q, B)$ is finite $H_0(Q, B/B_2)$ is finite and by induction $H_i(Q, B/B_2)$ is finite for all i. In particular $H_1(Q, B/B_2)$ is finite and hence $H_0(Q, B_2)$ is finite. Again by induction $H_i(Q, B_2)$ is finite for all i and using the long exact sequence we see that $H_i(Q, B)$ is finite for all i, as required.

□

Lemma 4.2 *Suppose A is a finitely generated $\mathbb{Z}[Q]$-module and for some prime p and some j the homology group $H_0(Q, H_j(A, \mathbb{F}_p))$ is finite. Then $H_i(Q, H_j(A, \mathbb{F}_p))$ is finite for all i.*

Proof By Proposition 2.1 (for p odd) and Theorem 3.1 (for $p = 2$) there is a filtration of $H_j(A, \mathbb{F}_p)$ with quotients isomorphic to some $B_{\alpha,\beta} = \wedge^\alpha(A/pA) \otimes \widetilde{S}^\beta({}_pA)$ for $\alpha + 2\beta = j$, $\alpha \geq 0, \beta \geq 0$ and the action of Q on $H_j(A, \mathbb{F}_p)$ corresponds to the diagonal action of Q on $B_{\alpha,\beta}$. Note that $B_{\alpha,\beta}$ is a module over $\Pi_\alpha \otimes \Pi_\beta$, where Π_k is the invariant subring of $\mathbb{F}_p[Q^k]$ under the action of the symmetric group S_k that permutes the factors of Q^k. The $\mathbb{F}_p$-algebra Π_k is finitely generated and contains the diagonal subgroup of Q^k. Then $\Pi_\alpha \otimes \Pi_\beta$ is a Noetherian commutative ring containing the diagonal subgroup of $Q^{\alpha+\beta}$. Finally it remains to prove that $B_{\alpha,\beta}$ is finitely generated over $\Pi_\alpha \otimes \Pi_\beta$ and then apply the previous lemma.

Note that $\otimes^k C$ for $C = A/pA$ or ${}_pA$ is finitely generated module over $\mathbb{F}_p[Q^k]$. As every element f of Q^k is a root of the polynomial $\prod_{\sigma \in S_k}(x - \sigma(f)) \in \Pi_k[x]$ we see that f is integral over Π_k. Furthermore $\mathbb{F}_p[Q^k]$ is a finitely generated algebra and so $\mathbb{F}_p[Q^k]$ is finitely generated as Π_k-module. Thus $\otimes^k C$ is finitely generated over Π_k and hence $B_{\alpha,\beta}$ is finitely generated over $\Pi_\alpha \otimes \Pi_\beta$. □

Proposition 4.3 *Suppose some extension G of A by Q is finitely generated, p is a prime and $H_t(G, \mathbb{F}_p)$ is finite for all t. Then $H_i(Q, H_t(A, \mathbb{F}_p))$ is finite for all t and all i.*

Proof We prove the proposition by induction on t. The case $t = 1$ is very easy, as $H_1(A, \mathbb{F}_p) \simeq A/pA$ and A is finitely generated over $\mathbb{Z}[Q]$. Thus $H_0(Q, H_1(A, \mathbb{F}_p))$ is finite. By Lemma 4.2 $H_i(Q, H_1(A, \mathbb{F}_p))$ is finite for all i.

For the inductive step assume $H_i(Q, H_j(A, \mathbb{F}_p))$ is finite for all $0 \leq j \leq t-1$ and all $i \geq 0$ and consider the first quadrant Lyndon-Hochshild-Serre spectral sequence for the trivial module $\mathbb{F}_p$

$$E^2_{i,j} = H_i(Q, H_j(A, \mathbb{F}_p))$$

with differentials

$$d^r : E^r_{i,j} \to E^r_{i-r,j+r-1}$$

By induction $E^2_{i,k}$ is finite for $0 \leq k \leq t-1$ and all $i \geq 0$. This together with the fact that d^r has bidegree $(-r, r-1)$ and our spectral sequence is a first quadrant spectral sequence, in particular $E^\infty_{0,t} = E^{t+2}_{0,t}$, implies that

1. $E^s_{i,k}$ is finite for all $s \geq 2, 0 \leq k \leq t-1$ and $i \geq 0$;
2. $E^\infty_{0,t}$ is finite if and only if $E^2_{0,t}$ is finite.

At the same time since $H_{i+j}(G, \mathbb{F}_p)$ is finite we have that $E^\infty_{i,j}$ is finite for every i, j. Thus $E^2_{0,t} = H_0(Q, H_t(A, \mathbb{F}_p))$ is finite. Finally by Lemma 4.2 $H_i(Q, H_t(A, \mathbb{F}_p))$ is finite for all i.

□

5 Proof of Theorem A

Assume that A is not of finite rank. By [5, Section 3.1] there exists a prime number p such that A/pA is infinite and A has an epimorphic image B which is a just-infinite $\mathbb{F}_p[Q]$-module of exponent p. Let M be a non-trivial cyclic $\mathbb{F}_p[Q]$-submodule of B and A_1 be the preimage of M in A. Then M is a just-infinite cyclic $\mathbb{F}_p[Q]$-module of exponent p and A_1 has finite index in A.

Lemma 5.1 *There exists a normal subgroup G_1 of finite index in G such that $G_1 \cap A = A_1$. In particular $H_i(G_1, \mathbb{F}_p)$ is finite for all i and all primes p.*

Proof 1. Let $\pi : Q \to G$ be a section. Then $G_1 = A_1 \pi(Q^m)$ is a normal subgroup of finite index in G if $[\pi(Q^m), \pi(Q)] \subseteq \{q^{m-1} + q^{m-2} + \ldots + 1 \mid q \in Q\} A \subseteq A_1$. Note that as A/A_1 is finite there is m as above.

2. As G/G_1 is metabelian, finite we can use induction on the order of G/G_1 to show that $H_i(G_1, \mathbb{F}_p)$ is finite and restrict to the case when G/G_1 is cyclic of prime order q. If $q = p$ by [5, Prop.2.3] $H_j(G, \mathbb{F}_p)$ is finite for all j. Suppose $q \neq p$. Consider the LHS spectral sequence $E^2_{i,j} = H_i(G/G_1, H_j(G_1, \mathbb{F}_p))$ converging to $H_{i+j}(G, \mathbb{F}_p)$. For $i \geq 1$ the exponent of $E^2_{i,j}$ divides p and q, hence $E^2_{i,j} = 0$. Then the spectral sequence collapses and induces an isomorphism $H_j(G_1, \mathbb{F}_p) \simeq H_j(G, \mathbb{F}_p)$. □

Lemma 5.2 *For p odd $H_0(Q, \wedge^i M)$ is finite for all i and for $p = 2$ the image of $H_0(Q, \wedge^i M)$ in $H_0(Q, H_i(M, \mathbb{F}_2)) \simeq H_0(Q, \widetilde{S}^i(M))$ is finite for all i.*

Proof Let G_1 be the group given by the previous lemma and $Q_1 = G_1/A_1$. By Lemma 5.1 $H_i(G_1, \mathbb{F}_p)$ is finite for all i and all prime p. Hence by Proposition 4.3 $H_0(Q_1, H_i(A_1, \mathbb{F}_p))$ is finite for all p.

We consider first the case p odd. Indeed by Proposition 2.1 $\wedge^i_{\mathbb{F}_p}(A_1/pA_1)$ is a direct summand of $H_i(A_1, \mathbb{F}_p)$. Thus the embedding of the exterior algebra of $H_1(A_1, \mathbb{F}_p) \simeq A_1/pA_1$ in $H_*(A_1, \mathbb{F}_p)$ is natural and split and compatible with the multiplicative structure on the strictly anticommutative algebra $H_*(A_1, \mathbb{F}_p)$. Hence the action of Q_1 on the homology group induces on the exterior algebra of A_1/pA_1 the diagonal Q_1-action. Then $\mathbb{F}_p \otimes_{\mathbb{F}_p[Q_1]} (\wedge^i(A_1/pA_1))$ embeds in $\mathbb{F}_p \otimes_{\mathbb{F}_p[Q_1]} H_i(A_1, \mathbb{F}_p)$. As $\mathbb{F}_p \otimes_{\mathbb{F}_p[Q_1]} H_i(A_1, \mathbb{F}_p)$ is finite $\mathbb{F}_p \otimes_{\mathbb{F}_p[Q_1]} (\wedge^i(A_1/pA_1))$ is finite. As M is a surjective image of A_1/pA_1 we deduce that $H_0(Q, \wedge^i M)$ is finite.

If $p = 2$ consider the commutative diagram

$$\begin{array}{ccccc}
H_0(Q_1, H_i(A_1, \mathbb{F}_2)) & \to & H_0(Q_1, H_i(M, \mathbb{F}_2)) & \to & H_0(Q, H_i(M, \mathbb{F}_2)) \\
\uparrow \alpha_1 & & \uparrow \alpha_2 & & \uparrow \alpha_3 \\
H_0(Q_1, (\wedge^i_{\mathbb{F}_2}(A_1/2A_1))) & \xrightarrow{\varphi_1} & H_0(Q_1, \wedge^i M) & \xrightarrow{\varphi_2} & H_0(Q, \wedge^i M)
\end{array}$$

where the rows are induced by the projection $A_1 \to M$ and by the inclusion $Q_1 \to Q$ and the columns are induced by the inclusions $\wedge^i H_1(C, \mathbb{F}_2) \to H_i(C, \mathbb{F}_2)$ for $C = A_1$ and $C = M$, where wedging corresponds to the Pontryagin product. As $H_0(Q, H_i(A, \mathbb{F}_2))$ is finite α_1 has finite image. The commutativity of the above diagram together with the surjectivity of φ_1 and φ_2 imply that the image of α_3 is finite, as required.

□

From now on we will forget the existence of the group G and will deal only with just -infinite cyclic $\mathbb{Z}[Q]$-modules M of additive exponent p with the property that for p odd $\mathbb{F}_p \otimes_{\mathbb{F}_p[Q]} (\wedge^i_{\mathbb{Z}} M)$ is finite for all i and for $p = 2$ the image of $\mathbb{F}_2 \otimes_{\mathbb{F}_2[Q]} (\wedge^i_{\mathbb{Z}} M)$ in $\mathbb{F}_2 \otimes_{\mathbb{F}_2[Q]} \widetilde{S}^i(M)$ is finite for all i. By [5, Section 5.2] there exists a series

$$Q = Q_0 \subseteq Q_1 \subseteq \ldots \subseteq Q_t$$

of multiplicative subgroups of the field of fractions K of M and a series

$$M = M_0 \subseteq M_1 \subseteq \ldots \subseteq M_t$$

of additive subgroups of K defined by $M_i = Q_i M$ with the following properties:

1. $Q_{i+1} = Q_i \times < \alpha_i >$;
2. there exists $r_i \in \mathbb{Z}[Q_i]$ with $(r_i - \alpha_i)M_{i+1} = 0$;
3. M_i is fully tame for $i \geq 1$ i.e. M_i is n_i-tame as a module over $\mathbb{F}_p[Q_i]$ where n_i is the torsion free rank of Q_i.
4. The centraliser $C_{Q_i}(M_i)$ of M_i in Q_i is the same as $C_{Q_0}(M_0)$.

Lemma 5.3 *If p is odd and $\mathbb{F}_p \otimes_{\mathbb{Z}[Q_i]} (\wedge^j M_i)$ is finite then $\mathbb{F}_p \otimes_{\mathbb{Z}[Q_{i+1}]} (\wedge^j M_{i+1})$ is finite.*

If $p = 2$ and the image of $\mathbb{F}_2 \otimes_{\mathbb{F}_2[Q]} (\wedge^j_{\mathbb{Z}} M_i)$ in $\mathbb{F}_2 \otimes_{\mathbb{F}_2[Q]} \widetilde{S}^j(M_i)$ is finite then the image of $\mathbb{F}_2 \otimes_{\mathbb{F}_2[Q]} (\wedge^j_{\mathbb{Z}} M_{i+1})$ in $\mathbb{F}_2 \otimes_{\mathbb{F}_2[Q]} \widetilde{S}^j(M_{i+1})$ is finite.

Proof First let p be odd. Suppose $f_1, \ldots, f_s$ are elements of $\wedge^j_{\mathbb{F}_p} M_i$ such that

$$\wedge^j_{\mathbb{F}_p} M_i = \mathbb{F}_p f_1 + \ldots + \mathbb{F}_p f_s + Aug(\mathbb{F}_p[Q_i])(\wedge^j_{\mathbb{F}_p} M_i),$$

where Aug denotes the augmentation ideal. We claim that

$$\wedge^j_{\mathbb{F}_p} M_{i+1} = \mathbb{F}_p f_1 + \ldots + \mathbb{F}_p f_s + Aug(\mathbb{F}_p[Q_{i+1}])(\wedge^j_{\mathbb{F}_p} M_{i+1}).$$

Let f be an element from $\wedge^j_{\mathbb{F}_p} M_{i+1}$. By the construction of M_{i+1} for some large positive integer β we have $\alpha_i^\beta f \in \wedge^j_{\mathbb{F}_p} M_i$. Then

$$\alpha_i^\beta f = z_1 f_1 + \ldots + z_s f_s + w$$

where $z_i \in \mathbb{F}_p$ and $w \in Aug(\mathbb{F}_p[Q_i])(\wedge^j M_i)$ and hence

$$f = \alpha_i^{-\beta}(z_1 f_1 + \ldots + z_s f_s + w) \in z_1 f_1 + \ldots + z_s f_s + Aug(\mathbb{F}_p[Q_{i+1}])(\wedge^j_{\mathbb{F}_p} M_{i+1}),$$

as required.

If $p = 2$ consider the commutative diagram

$$\begin{array}{ccc} \varphi_i : \wedge^j_{\mathbb{F}_2} M_i & \to & \widetilde{S}^j(M_i) \\ \downarrow & & \downarrow \\ \varphi_{i+1} : \wedge^j_{\mathbb{F}_2} M_{i+1} & \to & \widetilde{S}^j(M_{i+1}) \end{array}$$

where φ_i is the natural inclusion of $\wedge^j H_1(M_i, \mathbb{F}_2)$ in $H_j(M_i, \mathbb{F}_2)$, the column maps are induced by the inclusion of M_i in M_{i+1}. Since the image of $\mathbb{F}_2 \otimes_{\mathbb{F}_2[Q_i]} \wedge^j_{\mathbb{F}_2} M_i$ in $\mathbb{F}_2 \otimes_{\mathbb{F}_2[Q_i]} \widetilde{S}^j(M_i)$ is finite there exist elements $f_1, \ldots, f_s \in \wedge^j_{\mathbb{F}_2} M_i$ such that

$$\varphi_i(\wedge^j_{\mathbb{F}_2} M_i) \subseteq \varphi_i(\mathbb{F}_2 f_1 + \ldots + \mathbb{F}_2 f_s) + Aug(\mathbb{Z}[Q_i])(\widetilde{S}^j(M_i)).$$

An obvious modification of the first part of the proof gives

$$\varphi_{i+1}(\wedge^j_{\mathbb{F}_2} M_{i+1}) \subseteq \varphi_{i+1}(\mathbb{F}_2 f_1 + \ldots + \mathbb{F}_2 f_s) + Aug(\mathbb{F}_2[Q_{i+1}])(\widetilde{S}^j(M_{i+1}))$$

and hence the image of $\mathbb{F}_2 \otimes_{\mathbb{F}_2[Q_{i+1}]} \wedge^j_{\mathbb{F}_2} M_{i+1}$ in $\mathbb{F}_2 \otimes_{\mathbb{F}_2[Q_{i+1}]} \widetilde{S}^j(M_{i+1})$ is finite. □

Thus to prove the main theorem it is sufficient to work with $M = M_t$ and $Q = Q_t/C_{Q_t}(M_t)$, so we can assume that

1. M is a cyclic $\mathbb{F}_p[Q]$-module and Q acts faithfully on M. Note we have dropped the assertion that M is just-infinite;
2. M is n-tame where n is the torsion free rank of Q;
3. As t could be chosen arbitrary large we can assume that $n+1$ is a multiple of the order of the torsion part of Q.
4. If p is odd $\mathbb{F}_p \otimes_{\mathbb{F}_p[Q]} \wedge^j M$ is finite for all j. If $p = 2$ the image of $\mathbb{F}_2 \otimes_{\mathbb{F}_2[Q]} (\wedge^j_{\mathbb{Z}} M)$ in $\mathbb{F}_2 \otimes_{\mathbb{F}_2[Q]} \widetilde{S}^j(M)$ is finite for all j.

Then by [5, Proposition 4.3] conditions 1,2 and 3 imply $\mathbb{F}_p \otimes_{\mathbb{F}_p[Q]} (\otimes^{n+1}_{\mathbb{F}_p} M)$ is infinite. At the same time as shown in [5, Section 5.4] the fourth property of M together with the n-tameness of M implies that $\mathbb{F}_p \otimes_{\mathbb{F}_p[Q]} (\otimes^{n+1}_{\mathbb{F}_p} M)$ is finite, a contradiction.

6 Proof of Corollary B

For p prime consider the short exact sequence

$$0 \to H_i(G, \mathbb{Z}) \otimes_{\mathbb{Z}} \mathbb{F}_p \to H_i(G, \mathbb{F}_p) \to Tor_1^{\mathbb{Z}}(H_{i-1}(G, \mathbb{Z}), \mathbb{F}_p) \to 0$$

given by the universal coefficients theorem. Note that $Tor_1^{\mathbb{Z}}(H_{i-1}(G, \mathbb{Z}), \mathbb{F}_p) = \{\lambda \in H_{i-1}(G, \mathbb{Z}) \mid p\lambda = 0\}$. Then whenever $H_i(G, \mathbb{Z})$ and $H_{i-1}(G, \mathbb{Z})$ are finitely generated the homology group $H_i(G, \mathbb{F}_p)$ is finite. Now Corollary B follows immediately from Theorem A.

Acknowledgements The author thanks the referee for the careful reading of the paper.

References

[1] R. Bieri, J. R. J. Groves, *Metabelian groups of type FP_∞ are virtually of type FP*, Proc. London Math. Soc. (3)45, (1982), 365–384.

[2] R. Bieri, R. Strebel, *Valuations and finitely presented metabelian groups*, Proc. London Math. Soc. (3)41 (1980), 439–464.

[3] K. S. Brown, *Cohomology of Groups*, Springer-Verlag, 1982.

[4] H. Cartan, *Algebres d'Eilenberg-MacLane*, Seminaire H. Cartan, Ecole Normale Superieure, included in Collected Works, Volume 3, Springer–Verlag, 1979.

[5] J. R. J. Groves, *Metabelian groups with finitely generated integral homologies*, Quart. J. Math. Oxford (2), 33 (1982), 405–420.

[6] D. H. Kochloukova, *Geometric invariants and modules of type FP_∞ over constructible nilpotent-by-abelian groups*, J. Pure Appl. Algebra, (2-3) 159 (2001), 187-202.

INVARIANTS OF DISCRETE GROUPS, LIE ALGEBRAS AND PRO-p GROUPS

DESSISLAVA H. KOCHLOUKOVA[1]

IMECC, UNICAMP, Cx. P. 6065,
13083-970 Campinas, SP, Brasil
E-mail: desi@ime.unicamp.br

1 Introduction

The first geometric invariant of groups was introduced by R. Bieri and R. Strebel for the special class of finitely generated metabelian groups [20]. Their work was motivated by an earlier result that every finitely presented soluble group with infinite cyclic quotient is an ascending HNN-extension with stable letter corresponding to a generator of the cyclic quotient and a finitely generated base [22]. It turned out that the new geometric invariant classifies the finitely presented groups in the class of all finitely generated metabelian groups [20]. Later on the definition of Bieri-Strebel was generalised for any finitely generated discrete group [18] and higher dimensional homological and homotopical analogues of this invariant were introduced by R. Bieri and B. Renz [19], [64]. We will discuss the precise definitions of these invariants in the following section. They are important as they determine the homological and homotopical types FP_m and F_m of subgroups containing the derived subgroup.

In general the geometric invariants are very hard to compute and there are very few cases when they are calculated. One of them is the class of right angled Artin groups [48] which gave the first known examples of groups of type FP_∞ which are not finitely presented [7]. Even in the class of metabelian groups the structure of the geometric invariants is not completely understood. There are two open conjectures in the metabelian case: the FP_m-Conjecture and the Σ^m-Conjecture. Both are known to hold in small dimensions $m = 2$ [20], [16] and $m = 3$ when the metabelian group in consideration is a split extension of abelian groups [42], [43]. It is somewhat surprising that even the case of metabelian groups presents such difficulties.

It is interesting to note that the few successful calculations of the geometric invariants completed by now are achieved by using mixtures of geometric with algebraic techniques. It is quite often in this subject that the existence of a complex with higher connectivity on which a group G acts with "small" stabilizers and cocompactly that gives methods for calculating the homotopy type of the groups and its geometric invariants [25], [44], [4]. The only known construction of such a complex in the metabelian case that works in every dimension (at the expense of considering a special class of metabelian groups) depends on valuation trees and valuation theoretic properties of commutative rings [4], [56].

[1]Supported by a post-doctoral grant no. 98/00482-3 from FAPESP, Brazil.

Recently questions about the homological types of pro-p groups and Lie algebras have been addressed in [30], [31],[51]. A characterisation of these types in the metabelian case (in the spirit of the FP_m-Conjecture) has been obtained in [57],[58] (with the restriction in the Lie case that the considered Lie algebra is a split extension of abelian Lie algebras), though the similar question for discrete groups is still open. It is interesting to note that with few exceptions [32],[65] nothing is known outside the metabelian case. The basic problem is that in the case of metabelian Lie algebras and metabelian pro-p groups the invariants are defined in pure algebraic terms and their properties are consequences of valuation theoretic arguments. The absence of geometric methods makes it difficult to obtain results beyond the metabelian case. Furthermore in the pro-p and Lie cases [65] the HNN-construction does not play a vital role in the classification of finite presentability, contrary to the situation in the discrete case.

This essay is written with very few proofs, its aim is to give an overview of the recent developments in the subject, especially the links via valuation theory of the theory of discrete groups with the pro-p and Lie theory. In the last section we discuss briefly some aspects of the geometric invariants and the homological finiteness properties in the discrete case that could not get a place in the main body of the paper due to lenght restrains.

2 Definitions of the geometric invariants and basic properties

Everywhere in sections 2-5 G will denote a finitely generated discrete group. For any such group the set $Hom(G, \mathbb{R})$ of the real characters of G is an abelian group isomorphic to $\mathbb{R}^n$, where n is the torsion-free rank of the abelianization of G. Two non-trivial characters χ_1 and χ_2 are equivalent if they are obtained one from the other by multiplication with a positive real number. The class $\mathbb{R}_{>0}\chi$ of a non-zero character χ is denoted by $[\chi]$ and the set of all $[\chi]$ equivalence classes

$$S(G) = \{[\chi] \mid \chi \in Hom(G, \mathbb{R}) \setminus \{0\}\}$$

is identified with the unit sphere S^{n-1} in $\mathbb{R}^n \simeq Hom(G, \mathbb{R})$.

By definition a group G is of type FP_m if there is a projective resolution (over $\mathbb{Z}[G]$) of the trivial module $\mathbb{Z}$ with all terms in dimension $\leq m$ finitely generated. For a $\mathbb{Z}[G]$-module A we say that it has the homological type FP_m if there is a projective resolution of A with finitely generated modules in all dimensions $\leq m$. Finally we are ready to define the homological geometric invariants $\Sigma^m(G, A)$ for a finitely generated $\mathbb{Z}[G]$-module A. The case $A = \mathbb{Z}$ plays a special role in the theory.

$$\Sigma^m(G, A) = \{[\chi] \mid A \text{ is of homological type } FP_m \text{ over } \mathbb{Z}[G_\chi]\},$$

$$\Sigma^m(G, A)^c = S(G) \setminus \Sigma^m(G, A)$$

where $G_{\chi \geq d}$ is the set $\{g \mid \chi(g) \geq d\}$ and G_χ is the monoid $G_{\chi \geq 0}$.

The homological geometric invariants $\{\Sigma^m(G, \mathbb{Z})\}_{m \geq 1}$ have homotopical counterparts $\{\Sigma^m(G)\}_{m \geq 1}$. The homotopical invariant $\Sigma^m(G)$ is defined only for groups

G of type F_m i.e. groups for which there is a $K(G,1)$-complex Y with finite m-skeleton. Note that the universal cover $\widetilde{Y}$ of Y is $(m-1)$-connected and G acts cocompactly and freely on its m-skeleton. A continuous function $h : \widetilde{Y} \to \mathbb{R}$ is called a χ-equivariant height function if $h(vg) = h(v) + \chi(g)$ for every point in $\widetilde{Y}$. In case the restriction of h on every cell of $\widetilde{Y}$ attains its extremes on the boundary of the cell the height function is called regular. The maximal subcomplex of $h^{-1}[d, +\infty)$ is denoted by $\widetilde{Y}_{h \geq d}$. By definition $\Sigma^m(G)$ contains all classes $[\chi] \in S(G)$ for which there is a $K(G,1)$-complex Y (depending on χ) and a χ-equivariant height function $h : \widetilde{Y} \to \mathbb{R}$ such that $\widetilde{Y}_{h \geq 0}$ is $(m-1)$-connected. By definition $\Sigma^m(G)^c = S(G) \setminus \Sigma^m(G)$.

There is an equivalent definition for the geometric invariant $\Sigma^m(G)$. Consider a fixed $K(G,1)$-complex Y with finite m-skeleton and its universal cover $\widetilde{Y}$ together with a χ-equivariant height function $h : \widetilde{Y} \to \mathbb{R}$. Then $[\chi] \in \Sigma^m(G)$ if and only if $\widetilde{Y}_{h \geq 0}$ is essentially $(m-1)$-connected, i.e. exists $d < 0$ such that the maps $\pi_i(\widetilde{Y}_{h \geq 0}) \to \pi_i(\widetilde{Y}_{h \geq d})$ given by inclusion are trivial for all $i \leq m-1$. The equivalence between both definitions is discussed in [44, Sec.2]

At first sight the definitions of the geometric homological invariants for $A = \mathbb{Z}$ look quite different from the definition of the geometric homotopical invariant but this impression is false. There is a strong link between $\Sigma^m(G)$ and $\Sigma^m(G,\mathbb{Z})$ in the way the homology and homotopy groups are related. Though the homological invariant $\Sigma^m(G,\mathbb{Z})$ is defined for any finitely generated group G (this is group of type F_1) if $\Sigma^m(G,\mathbb{Z})$ is non-empty then the group G is of homological type FP_m (see Lemma 5.2).

Suppose there is a $(m-1)$–acyclic complex X on which G acts freely and cocompactly. Then $[\chi] \in \Sigma^m(G,\mathbb{Z})$ if and only if for some χ-equivariant height function $h : X \to \mathbb{R}$ the complex $X_{h \geq 0}$ is essentially $(m-1)$-acyclic, i.e. exists $d < 0$ such that the maps $H_i(\widetilde{Y}_{h \geq 0}) \to H_i(\widetilde{Y}_{h \geq d})$ given by inclusion are trivial for all $i \leq m-1$. This gives the inclusion $\Sigma^m(G) \subseteq \Sigma^m(G,\mathbb{Z})$ for groups G of type F_m. A Hurewicz type argument gives immediately

$$\Sigma^m(G) = \Sigma^m(G,\mathbb{Z}) \cap \Sigma^2(G)$$

for groups of type F_m and $m \geq 2$. Thus every problem about the geometric invariant in dimension $m \geq 3$ can always be reduced to its homological version provided we know the homotopical counterpart of the problem in dimension 2.

In general the structure of the invariants can be very complicated. We state some of the few results that hold for arbitrary group (provided that the invariant in consideration is well-defined).

Theorem 2.1 *[19, Thm A and Sec.6.5] For any finitely generated group G and any $\mathbb{Z}[G]$-module A the geometric homological invariant $\Sigma^m(G, A)$ is open in $S(G)$. Furthermore if G is of type F_m then $\Sigma^m(G)$ is open in $S(G)$.*

Theorem 2.2 *[19, Thm B and Sec. 6.5] Suppose G is a group of type FP_m (type F_m) and H is a subgroup of G containing the derived subgroup. Then H is of type*

FP_m (type F_m) if and only if

$$\{[\chi] \mid \chi(H) = 0\} \subseteq \Sigma^m(G, \mathbb{Z}) \text{ (resp. } \Sigma^m(G)).$$

3 Groups acting on spaces

One of the main criteria for groups of type F_m and FP_m is due to K. Brown. It characterises the homological and homotopical types of a group via its action on a space. In fact in [25, Thm 4] only the case of groups of type F_2 (finitely presented) is considered, the general theorem is obvious corollary of the homological part of the theorem plus the fact that a group is of type F_m if and only if it is of type FP_m and is finitely presented.

Theorem 3.1 *[25, Thm 4] [27, Prop. 1.1] Suppose G is a finitely generated group acting on a complex X via permutation of the cells and such that X is $(m-1)$-connected (acyclic), the stabilizer of any i-dimensional cell is of type F_{m-i} (resp. FP_{m-i}) and G acts cocompactly on X. Then G is of homotopical type F_m (homological type FP_m).*

Thus Brown's criterion reduces the problem of finding the homotopical and homological type of a group to the construction of an appropriate complex. The problem is that it is really difficult to find such complexes. For finitely generated metabelian groups G there are two outstanding conjectures that suggest when the group is of type FP_m and how to calculate $\Sigma^m(G, \mathbb{Z})$.

The FP_m-Conjecture *A finitely generated metabelian group is of type FP_m if and only if every subset of $\Sigma^1(G)^c$ with at most m points lies in an open hemisphere of $S(G)$. The latter geometric condition is called m-tameness.*

The characterisation of the finitely presented metabelian groups, given in [20], implies that metabelian groups of type FP_2 are always finitely presented, i.e. F_2. This, together with the fact that a group is F_m if and only if it is FP_m and finitely presented, implies that for metabelian groups type F_m is equivalent to type FP_m. In general, this is false [7] though it is not known whether there are examples of soluble groups of type FP_2 that are not finitely presented.

The Σ^m-Conjecture *For a finitely generated metabelian group G of type FP_m both invariants $\Sigma^m(G)^c$ and $\Sigma^m(G, \mathbb{Z})^c$ coincide and are equal to*

$$[conv_{\leq m}(\mathbb{R}_{>0}\Sigma^1(G)^c)] = \{[\chi] \mid \chi = \chi_1 + \ldots + \chi_m \text{ where all } [\chi_j] \in \Sigma^1(G)^c\}.$$

In [4] a special insight into the case of metabelian groups G of finite Prüfer rank (i.e. groups where there is an upper bound on the number of the generators for the finitely generated subgroups) is achieved by constructing a special subcomplex X of a finite product of trees such that G acts cocompactly on X with polycyclic stabilizers. It turns out that if any m points in $\Sigma^1(G)^c$ lie in an open hemisphere of $S(G)$ then the space X is $(m-1)$-connected. This combined with the above criterion shows one of the directions of the FP_m-Conjecture for metabelian groups of finite Prüfer rank. Later in [56] the same idea was used to prove the torsion version of Åberg's result. More precisely, suppose G is an extension $1 \to A \to G \to Q \to 1$ with G finitely generated, A and Q abelian such that A is torsion as abelian group and of Krull dimension 1 as a module over $\mathbb{Z}[Q]$ (via conjugation). Then there exists an Åberg's type complex with all properties described in the original Åberg's construction. The starting point in adapting Åberg's approach for this new class of groups is the consideration of a (very general) construction of a tree associated to a character χ whose class $[\chi]$ is in $\Sigma^1(G)^c$ [21, Sect.II.2.5, Sect.II.4.3] (unfortunately [21] is not accessible by a general reader, but good places to learn about the link between $\Sigma^1(G)$ and the actions of G on a tree are [38] and [28]). In the class of groups G described above, $\Sigma^1(G)^c$ has only finitely many points and this gives finitely many trees on which G acts. Then a subcomplex X of the product of these trees can be defined following Åberg's recipe. Although the Åberg's original construction and its version in the torsion case look quite geometrical the fact that the cell stabilizers are polycyclic is equivalent to solving a system of independent valuation equations in $\mathbb{Z}[Q]/ann(A)$ and involves non-trivial commutative algebra arguments.

There is a Σ-version of Brown's criterion due to H. Meinert. We discuss here only the Σ^2-case, as the higher dimensional version is needed only in its homological form [45, Sec.3.2] (remember the general philosophy that homotopy is needed only in dimension 2, the higher dimensional cases can be substituted with their homological counterparts).

Theorem 3.2 *[44, Thm A] Suppose a group G acts on a 1-connected G-finite 2-complex X such that all vertex stabilizers are finitely presented and all edge stabilizers are finitely generated. Let $\chi : G \to \mathbb{R}$ be a homomorphism and assume that $h : X \to \mathbb{R}$ is a χ-equivariant height function. Then $[\chi] \in \Sigma^2(G)$ if and only if there is a constant $d \geq 0$ such that the maps*

$$\pi_i(X_{h\geq 0}) \to \pi_i(X_{h\geq -d}),$$

induced by inclusion, are trivial for $i = 0, 1$.

It is not difficult to see that χ-equivariant height function h as above always exists if all cell stabilizers G_σ of a cell σ are contained in the kernel of χ and if G_σ fixes the cell σ pointwise. The following result deals with the other extreme case when none of the cell stabilizers are in the kernel of χ.

Theorem 3.3 *[44, Thm B] Let a group G act on a 1-connected G-finite 2-complex X. If $\chi : G \to \mathbb{R}$ is a homomorphism such that $0 \neq \chi \mid_{G_\sigma}, [\chi \mid_{G_\sigma}] \in \Sigma^{2-dim(\sigma)}(G_\sigma)$ for all cells σ of X, then $[\chi] \in \Sigma^2(G)$.*

Theorems 3.1, 3.2 and 3.3 (together with their homological form for arbitrary dimension [45]) imply the FP_m-Conjecture and the Σ^m-Conjecture in the case of metabelian groups of finite Prüfer rank and their torsion analogue.

In general both conjectures are still open though the small dimensional cases are well understood. For example both conjectures hold for $m = 2$ [20],[42] and for $m = 3$ in case G is a split extension of abelian groups [16], [43]. Some higher dimensional cases for specific metabelian groups can be found in [4], [34], [63], [54], [56]. Furthermore, the above conjectures are likely to hold only in the case of metabelian groups. Already for constructible (nilpotent of class 2)-by-abelian groups there is an example of a group G such that $\Sigma^2(G,\mathbb{Z})^c = [conv_{\leq 3}(\mathbb{R}_{>0}\Sigma^1(G)^c)]$ [62]. This example turns out to be a counter example to a conjecture of Meinert that for a constructible nilpotent-by-abelian group $\Sigma^m(G,\mathbb{Z})^c = \Sigma^\infty(G,\mathbb{Z})^c$ implies that the dimension of the span of $\Sigma^1(G)^c$ in $Hom(G,\mathbb{R})$ is at most m [46].

4 Finite generation of modules and m-tameness

As we will see later in this paper there is a strong link between metabelian groups of type FP_m and tensor and exterior powers of finitely generated modules A over $\mathbb{Z}[Q]$ for Q a finitely generated abelian group. We view the tensor power $\otimes^i A$ as a $\mathbb{Z}[Q]$-module via the diagonal Q-action, i.e. $(a_1 \otimes \ldots \otimes a_i)q = (a_1 q) \otimes \ldots \otimes (a_i q)$. This action induces a Q-action on the exterior power $\wedge^i A$. The classification of the modules A for which the tensor power $\otimes^i A$ is finitely generated over Q (via the diagonal action) is given in terms of an invariant $\Delta_v(Q,A)$ [15].

Let v be a fixed valuation (in the sense of Bourbaki) of the image of $\mathbb{Z}$ in $\mathbb{Z}[Q]/ann(A)$; this valuation is either trivial in the sense that its image is the set $\{\infty, 0\}$ or up to multiplication is the p-adic valuation. By definition $\Delta_v(Q,A)$ contains all real characters χ of Q including the trivial character, such that χ can be lifted to a real valuation of $\mathbb{Z}[Q]/ann(A)$, whose restriction to the image of $\mathbb{Z}$ inside $\mathbb{Z}[Q]/ann(A)$ is the fixed valuation v. Using methods from valuation theory R. Bieri and J. Groves prove the following important characterization.

Theorem 4.1 *[15] Suppose Q is a finitely generated abelian group and A is a finitely generated $\mathbb{Z}[Q]$-module. Then $\otimes^i A$ is not finitely generated over $\mathbb{Z}[Q]$ via the diagonal action if and only if there exist characters $\chi_1, \ldots, \chi_i$ in $\Delta_v(Q,A)$ for some non-negative valuation v of the image of $\mathbb{Z}$ in $\mathbb{Z}[Q]/ann(A)$ such that $\chi_1 + \ldots + \chi_i = 0$ and at least one χ_j is non-zero.*

The invariant $\Delta_v(Q,A)$ turns out to be closely related to the invariant $\Sigma^1(G)^c$ for G an extension of A by Q, the type of the extension being irrelevant. This can be seen via the bijection $\mu : \Sigma^0(Q,A)^c \to \Sigma^1(G)^c$ sending $[\chi]$ to $[\chi\pi]$, where $\pi : G \to Q$ is the canonical projection, together with the following result.

Theorem 4.2 *[14] For G an extension of A by Q we have*

$$\Sigma^0(Q,A)^c = \cup_v[\Delta_v(Q,A) \setminus \{0\}]$$

where the union is over all valuations v of the image of $\mathbb{Z}$ in $\mathbb{Z}[Q]/ann(A)$.

Furthermore, the above description of $\Sigma^1(G)^c$ together with deep valuation theory arguments show that $\Sigma^1(G)^c$ has a special structure [14, Thm.E] : is a rationally defined spherical polyhedron, i.e. finite union of sets, each of which is finite intersection of finitely many closed hemispheres in $S(G)$ defined by rational points in $S(G)$. The rationality in this description is very important as it implies that the rational points in $\Sigma^1(G)^c$ form a dense subset of $\Sigma^1(G)^c$ and in many problems allows one to restrict to the case of rational characters i.e. characters with cyclic image. Note that such a description does not hold outside the class of metabelian groups. There are examples in [18, Sec.8] of finitely generated subgroups G of the group of PL homeomorphism of the unit interval [0,1] for which $\Sigma^1(G)^c$ has an isolated non-discrete point.

Now we discuss how the finite generation over $\mathbb{Z}[Q]$ of exterior powers of a $\mathbb{Z}[Q]$-module A is linked with the integral homology of A. First we note that $H_*(A,\mathbb{Z}) = \oplus_i H_i(A,\mathbb{Z})$ is equipped with divided powers and with a anticommutative product structure via the Pontryagin product. More precisely for a commutative ring k the Pontryagin product on $H_*(A,k)$ is the composite

$$H_*(A,k) \otimes H_*(A,k) \to H_*(A \times A, k \otimes k) \xrightarrow{\mu_*} H_*(A,k)$$

where $\mu : (A \times A, k \otimes_{\mathbb{Z}} k) \to (A,k)$ is the multiplication map and the first map is induced by tensoring resolutions of k over $\mathbb{Z}[A]$. More details about the divided power structure of $H_*(A,k)$ can be found in [26, p. 124]. It is well known that the exterior algebra of A naturally embeds in $H_*(A,\mathbb{Z})$, the wedge product in the exterior algebra corresponding to the Pontryagin product in the homology group and the action of Q on the homology group restricted to the the exterior power of A being the diagonal one. In case A is $\mathbb{Z}$-torsion-free this embedding is an isomorphism [26, Ch.4, Thm 6.4]. In general it is really difficult to find a natural description of the integral homology of A. In case of coefficients $\mathbb{F}_p$ the integral homology is described in [35] except for the case $p = 2$ and A not of exponent 2.

By definition for an abelian group B of prime exponent $\widetilde{S}(B)$ is the subalgebra of all symmetric elements in the tensor algebra $\oplus_{i \geq 0} \otimes^i B$ i.e. $\widetilde{S}(B) = \oplus_{i \geq 0} \widetilde{S}^i(B)$ and $\widetilde{S}^i(B)$ is the subspace of the elements of $\otimes^i B$ invariant under the action of the symmetric group on i elements, the symmetric group acts via permutation of the factors of $\otimes^i B$.

Theorem 4.3 *[35, Ch.9, Ch.10] Suppose p is a prime and $\mathbb{F}_p$ is the field with p elements.*

1. If p is odd

$$H_*(A,\mathbb{F}_p) \simeq (\wedge_{\mathbb{F}_p}(A/pA)) \otimes_{\mathbb{F}_p} \widetilde{S}(A_p)$$

where A/pA has weight 1 and A_p has weight 2. Here $A_p = \{a \in A \mid pa = 0\}$.

2. If $p = 2$ and A is of exponent 2

$$H_*(A,\mathbb{F}_2) \simeq \widetilde{S}(A)$$

where A has weight 1.

In both cases the isomorphism is a natural isomorphism of graded divided powers algebras i.e. preserves grading, the multiplicative and the divided power structures.

The case $p = 2$ is completed in [59] where the following result is obtained

Theorem 4.4 *Let A be an arbitrary abelian group. Then there is a natural filtration $\{F^i(H_*(A,\mathbb{F}_2))\}_{i\geq 0}$ of the graded algebra $H_*(A,\mathbb{F}_2)$ with quotients*

$$F^i(H_*(A,\mathbb{F}_2))/F^{i-1}(H_*(A,\mathbb{F}_2)) \simeq \wedge(A/2A) \otimes \widetilde{S}^i(A_2).$$

In [59] Theorem 4.4 is used to answer a question raised by J. Groves in [37] : for a finitely generated metabelian group G if the homology groups $H_i(G,\mathbb{Z})$ are finitely generated for all $i \geq 0$ then G is of finite rank, i.e. there is natural number d such that every finitely generated subgroup of G can be generated by d elements. In [60] the link between tensor powers and homology of abelian groups is further explored.

Theorem 4.5 *[60] Suppose A is a finitely generated $\mathbb{Z}[G]$–module, where G is a finitely generated group with a normal subgroup H that acts nilpotently on A and G/H is abelian. Assume furthermore that all tensor powers of A are finitely generated over $\mathbb{Z}[G]$ via the diagonal G–action Then*

1. $H_i(A,\mathbb{Z})$ has a finite filtration with factors isomorphic to $\mathbb{Z}[G]$–subquotients of some (possibly different) tensor powers of A, where the action of G on the tensor powers is the diagonal one;

2. for $i \geq 1$ all tensor powers of $H_i(A,\mathbb{Z})$ are finitely generated over $\mathbb{Z}[G]$ via the diagonal G–action, $\Sigma^0(G,H_i(A,\mathbb{Z}))^c \subseteq conv(\mathbb{R}_{>0}\Sigma^0(G,A)^c)$ and H acts nilpotently on $H_i(A,\mathbb{Z})$.

As a corollary of this result we obtain an interesting fact about sections of constructible nilpotent-by-abelian groups.

Theorem 4.6 *[60] If $[N_1,N_1] \subseteq N_2 \subseteq N_1$ are normal nilpotent subgroups of a constructible nilpotent–by–abelian group G then N_1/N_2 is a module of type FP_∞ over $\mathbb{Z}[G]$, where G acts via conjugation, and $\Sigma^\infty(G,N_1/N_2)^c$ lies in an open hemisphere of $S(G)$. In particular abelian normal subgroups of G are of type FP_∞ as $\mathbb{Z}[G]$–modules via conjugation.*

5 Homological methods

There are powerful homological methods that can help in determining whether a group is of type FP_m. Note that in general homology commutes with direct limits and cohomology with direct products i.e. $H_*(G,\varinjlim V_j) = \varinjlim H_*(G,V_j)$ and $H^i(G,\prod V_j) = \prod_j H^i(G,V_j)$. As the following criterion, due to R. Bieri, shows when G is of type FP_m the homology $H_i(G,\ )$ commutes with direct products for $i \leq m-1$ and the converse is true, too. Note that as $\mathbb{Z}[G]$ is a free G-module we have $H_i(G,\mathbb{Z}[G]) = Tor_i^{\mathbb{Z}[G]}(\mathbb{Z},\mathbb{Z}[G]) = 0$ for $i \geq 1$.

Lemma 5.1 *[12, Thm 1.3 + Remarks] Let G be a group and A a finitely presented (right) module over $\mathbb{Z}[G]$. Then A is of type FP_m over $\mathbb{Z}[G]$ if and only if*

$$Tor_i^{\mathbb{Z}[G]}(A,\prod \mathbb{Z}[G]) = 0$$

for all $1 \leq i \leq m-1$ *and any direct product* $\prod \mathbb{Z}[G]$ *of arbitrary many copies of* $\mathbb{Z}[G]$. *In particular for* $A = \mathbb{Z}$ *and* G *finitely generated we obtain that* G *is of type* FP_m *if and only if* $H_i(G, \prod \mathbb{Z}[G]) = 0$ *for all* $1 \leq i \leq m-1$ *and all direct products.*

Lemma 5.2 *Suppose* G *is a finitely generated group and* A *is a finitely generated (right)* $\mathbb{Z}[G]$*-module. If* $\Sigma^m(G, A) \neq \emptyset$ *then* A *is of type* FP_m.

Proof Suppose $[\chi] \in \Sigma^m(G, A)$ i.e. A is of type FP_m over $\mathbb{Z}[G_\chi]$. Let

$$\mathcal{F} : \ldots \to F_m \xrightarrow{\partial_m} F_{m-1} \xrightarrow{\partial_{m-1}} \ldots \to F_0 \xrightarrow{\partial_0} A \to 0$$

be a projective resolution of A over $\mathbb{Z}[G_\chi]$ with all F_i finitely generated for $i \leq m$ and g an element of G with $\chi(g) < 0$. The complex $\mathcal{F}g^k$ with modules $F_i g^k$ is a resolution of $A = Ag^k$ over $\mathbb{Z}$ with differentials $d_i(f_i g^k) = \partial_i(f_i) g^k$ for $f_i \in F_i$, $i \geq 0$. Then the direct limit of the complexes $\mathcal{F}g^k$ when k goes to infinity is a projective resolution of A over $\mathbb{Z}[G]$ with finitely generated modules in dimension $\leq m$. □

As a consequence of Lemma 5.1 it is easy to deduce a fact that is often overlooked, a remark to it together with a proof can be found well hidden in [4, p.280].

Lemma 5.3 *Let* G *be a group of type* FP_m *and* H *be a quotient of* G *that splits, i.e.* $G = N \rtimes H$ *for some normal subgroup* N *of* G. *Then* H *is of type* FP_m.

Proof It is easy to check that $H_i(G, \prod \mathbb{Z}[G])$ is functorial with respect to G. As H is a split-quotient of G we get that $H_i(H, \prod \mathbb{Z}[H])$ is a split-quotient of $H_i(G, \prod \mathbb{Z}[G]) = 0$ for $1 \leq i \leq m-1$. In particular $H_i(H, \prod \mathbb{Z}[H]) = 0$ for $1 \leq i \leq m-1$ and by Lemma 5.1 H is of type FP_m. □

There is a homotopic version of the above result due to J. Alonso stating that G finitely presented implies H finitely presented provided H is a split quotient of G [5]. It is worth noting that the condition that the projection $G \to H$ splits is necessary in Lemma 5.3. By [20, Thm B] metabelian quotients of finitely presented groups without non-cyclic free subgroups are finitely presented too. But in general quotients of finitely presented groups (resp. type FP_2) are not finitely presented (of type FP_2) [2].

Now we discuss some homological corollaries of the FP_m property. Though simple the following lemma has important applications for the structure of the groups of type FP_m.

Lemma 5.4 *[30] Let* G *be a group of type* FP_m *and* N *a normal subgroup such that* G/N *is polycyclic. Then* $H_i(N, k)$ *is finitely generated as a (right) module over* $k[G/N]$ *(with action induced by conjugation) for* $k = \mathbb{Z}$ *or* $k = \mathbb{F}_p$.

Proof Let $\mathcal{F}$ be a free resolution of $\mathbb{Z}$ over $\mathbb{Z}[G]$ with all modules in dimensions $\leq m$ finitely generated. Then $H_i(N, k) \simeq H_i(\mathcal{F} \otimes_{\mathbb{Z}[N]} k)$. Note that $\mathcal{F} \otimes_{\mathbb{Z}[N]} k$ is a complex of $k[G/N]$-modules and in dimension $i \leq m$ these modules are finitely generated. Furthermore since G/N is polycyclic the group algebra $k[G/N]$ is right Noetherian and hence any subquotient of a finitely generated right $k[G/N]$-module is finitely generated. In particular the homology groups $H_i(\mathcal{F} \otimes_{\mathbb{Z}[N]} k)$ are finitely generated over $k[G/N]$ for $i \leq m$. □

Lemma 5.5 *[30] Suppose $1 \to A \to G \to Q \to 1$ is a short exact sequence of groups with A and Q abelian, A of prime exponent p and G of type FP_m. Then any m points in $\Sigma^1(G)^c$ lie in an open hemisphere of $S(G)$, i.e. G is m-tame.*

Proof By the previous lemma $H_i(A, \mathbb{F}_p)$ is finitely generated over $\mathbb{F}_p[Q]$ for $i \leq m$. By the results in the previous section $\wedge^i A$ naturally embeds in $H_i(A, \mathbb{F}_p)$ and the action of Q on the exterior power is diagonal i.e. $(a_1 \wedge \ldots \wedge a_i)q = (a_1 q) \wedge \ldots \wedge (a_i q)$. By [13] in the case of modules of prime exponent the finite generation of $\wedge^i A$ over $\mathbb{F}_p[Q]$ via the diagonal action is equivalent to the finite generation of $\otimes^i A$ over $\mathbb{F}_p[Q]$ via the diagonal action. This together with Theorem 4.1 and Theorem 4.2 completes the proof. □

The above result has been generalised. Using homological methods that extend the ideas introduced in [4], [63] it has been shown in [56] that whenever G is FP_m, extension of A by Q with A, Q abelian and either the extension splits or A is $\mathbb{Z}$-torsion i.e. of finite exponent as abelian group, then G is m-tame. The main idea is to prove that whenever G is not m-tame, m minimal with this property, there is a special element in $H^0(Q', H_0(A, \prod \mathbb{Z}[G]))$ whose image in $H_0(Q'', H_0(A, \prod \mathbb{Z}[G]))$ is non-trivial. Here Q is decomposed as $Q' \times Q''$, Q' has torsion-free rank $m-1$ and Q has torsion free rank n. As finitely generated abelian groups are duality groups and by the above remark we have a non-trivial composition map $H^0(Q', E) \simeq H_{m-1}(Q', E) \to H_{m-1}(Q, E) \simeq H^{n-m+1}(Q, E) \to H^{n-m+1}(Q'', E) \simeq H_0(Q'', E)$ for $E = H_0(A, \prod \mathbb{Z}[G])$. In particular the isomorphism $H_{m-1}(Q, E) \simeq H_{m-1}(G, \prod \mathbb{Z}[G])$ given by the Lyndon-Horschild-Serre spectral sequence is non-zero and hence G is not of type FP_m, a contradiction.

6 Homological properties of pro-p groups

6.1 Basic properties of pro-p modules

In this section we consider only pro-p groups G, i.e. inverse limits of finite p-groups. In the category of pro-p groups there are free pro-p groups with the usual universal property. And finite presentability can be defined as in the discrete case i.e. G is finitely presented as a pro-p group if there is a free pro-p group F with basis a finite set X and a finite subset R of $F(X)$ such that G is the quotient of F by the smallest closed normal subgroup of F that contains R. As X is finite F can be identified with the inverse limit of all quotients $\widetilde{F}/N$, where $\widetilde{F}$ is the free discrete group with basis X and N is any normal subgroup of index a power of p.

In the case of pro-p groups G the analogue of the group algebra is the so called completed group algebra $\mathbb{Z}_p[[G]]$. By definition it is the inverse limit of $(\mathbb{Z}/p^k\mathbb{Z})[G/U]$ of ordinary group algebras over all positive integers k and open subgroups U of G. Note $\mathbb{Z}_p[[G]]$ contains the ordinary group algebra of G over $\mathbb{Z}_p$ as a dense subset and $\mathbb{Z}_p[[G]]$ is a local compact ring; its unique (right) ideal is the kernel of the canonical map $\mathbb{Z}_p[[G]] \to \mathbb{F}_p$.

Definition 6.1 A (right) pro-p $\mathbb{Z}_p[[G]]$–module M is an abelian pro-p group together with a continuous map $M \times \mathbb{Z}_p[[G]] \to M$, that makes M a $\mathbb{Z}_p[[G]]$-module in the usual sense and whose restriction to $M \times \mathbb{Z}_p \to M$ is given by the natural action of $\mathbb{Z}_p$ on M.

We are interested in the homological finiteness properties of pro-p modules over $\mathbb{Z}_p[[G]]$. Before defining these properties we discuss some basic properties of the category of pro-p $\mathbb{Z}_p[[G]]$-modules. As shown in [66] every pro-p $\mathbb{Z}_p[[G]]$–module has a system of open neighbourhoods of 0 consisting of additive subgroups that are invariant under the action of G. In the category of pro-p $\mathbb{Z}_p[[G]]$–modules all morphisms $f : A \to B$ are continuous module homomorphisms.

Theorem 6.2 *[33, 1.5] Suppose M is a (right) pro-p $\mathbb{Z}_p[[G]]$–module and X is a subset of M. Then X generates M as a topological $\mathbb{Z}_p[[G]]$–module if and only if the image of X generates $M/M\Delta$ as a topological $\mathbb{Z}_p[[G]]/\Delta \simeq \mathbb{F}_p$–module, where Δ is the maximal right ideal of $\mathbb{Z}_p[[G]]$. Furthermore if $\bar{x}_1, \ldots, \bar{x}_n$ is a set of topological generators of $M/M\Delta$ over $\mathbb{F}_p$ then M is abstractly generated by $x_1, \ldots, x_n$ over $\mathbb{Z}_p[[G]]$.*

In the category of the pro-p modules there are enough projectives i.e. every pro-p $\mathbb{Z}_p[[G]]$–module is a quotient of a free pro-p $\mathbb{Z}_p[[G]]$–module (= module with free universal property) [33]. The free pro-p $\mathbb{Z}_p[[G]]$–module $F(X)$ with basis X can be constructed as the inverse limit of $L(X)/U$ over all abstract $\mathbb{Z}_p[[G]]$-submodules U of finite index in $L(X)$ such that U contains all but finitely many elements of X. Here $L(X)$ is the abstract free $\mathbb{Z}_p[[G]]$–module with basis X. A pro-p group G is said to be of type FP_m if the trivial $\mathbb{Z}_p[[G]]$–module $\mathbb{Z}_p$ has a free resolution over $\mathbb{Z}_p[[G]]$ (note all differentials are morphisms in the category of pro-p modules over $\mathbb{Z}_p[[G]]$) with all modules in dimensions $\leq m$ finitely generated (topologically or abstractly is the same by Theorem 6.2). Note that a pro-p group G is of type FP_1 if and only if it is (topologically) finitely generated.

The completed tensor product plays basic role in the investigation of metabelian pro-p groups of type FP_m. By definition if A is a right and B a left pro-p $\mathbb{Z}_p[[G]]$–module (possibly $G = 1$) the completed tensor product $A \hat{\otimes}_{\mathbb{Z}_p[[G]]} B$ is the inverse limit $A/U_i \otimes_{\mathbb{Z}_p[[G]]} B/V_j$ of ordinary tensor products over all open pro-p $\mathbb{Z}_p[[G]]$–submodules U_i and V_j of A and B respectively. The completed tensor product satisfies a standard universal property : there is a map $\alpha : A \times B \to A \hat{\otimes}_{\mathbb{Z}_p[[G]]} B$ such that every continuous $\mathbb{Z}_p[[G]]$–bihomomorphic map β from $A \times B$ to a pro-p $\mathbb{Z}_p[[G]]$–module factors through α. The completed n-th exterior power $\hat{E}^n_{\mathbb{Z}_p[[G]]}(A)$ or the completed n-th symmetric power $\hat{S}^n_{\mathbb{Z}_p[[G]]}(A)$ can be defined in a similar way either as inverse limits of the exterior or symmetric powers or by the corresponding universal properties.

By definition if A is a right and B is a left pro-p $\mathbb{Z}_p[[G]]$–module $\widehat{Tor}_n^{\mathbb{Z}_p[[G]]}(A, B)$ is the n–th left derived functor of $T(A, B) = A \hat{\otimes}_{\mathbb{Z}_p[[G]]} B$. We write $\hat{H}_n(G, A)$

for the pro-p homology group $\widehat{Tor}_n^{\mathbb{Z}_p[[G]]}(A, \mathbb{Z}_p)$. If M is a pro-p $\mathbb{Z}_p[[G]]$–module of exponent p we view it as a module over $\mathbb{F}_p[[G]] = \mathbb{Z}_p[[G]]/p\mathbb{Z}_p[[G]]$. For such modules the functor $-\hat{\otimes}_{\mathbb{F}_p[[G]]}-$ defined in a similar way to the completed tensor product over $\mathbb{Z}_p[[G]]$ coincides with $-\hat{\otimes}_{\mathbb{Z}_p[[G]]}-$. The following theorem is crucial for calculating homology groups in the category of pro-p groups and modules.

Theorem 6.3 *[33, 4.3] Suppose G is an inverse limit of pro-p groups G_i, A is an inverse limit of (right) pro-p $\mathbb{Z}_p[[G_i]]$–modules A_i. Then $\hat{H}_n(G, A)$ is an inverse limit of $\hat{H}_n(G_i, A_i)$ over i.*

Finally we note that there is a convenient criterion when a pro-p group G is of homological type FP_m. It happens precisely when the homology groups $\hat{H}_i(G, \mathbb{F}_p)$ are finite for all $i \leq m$. Furthermore a pro-p group is finitely presented if and only if it is of type FP_2. This shows how differently the pro-p groups behave compared to the discrete case where finite presentability is not equivalent to the property FP_2. Furthermore by Lemma 5.4 for a discrete group H of type FP_m the homology $H_i(H, \mathbb{F}_p)$ is finite for all $i \leq m$ but the converse does not hold.

6.2 Homological finiteness properties of metabelian pro-p groups

Let Q be a finitely generated abelian pro-p group and $T(Q)$ denotes the set of continuous homomorphisms from Q to the group of units of the power series algebra $K[[T]]$, where K is the algebraic closure of $\mathbb{F}_p$. Every finitely generated closed multiplicative subgroup of $1 + TK[[T]]$ is a free abelian pro-p group and the continuous homomorphisms from Q to $K[[T]]^\times$ have images inside $1 + TK[[T]]$. Now every element $v \in T(Q)$ extends to a unique (continuous) algebra homomorphism $\overline{v} : \mathbb{Z}_p[[Q]] \to K[[T]]$. Suppose A is a non-zero finitely generated pro-p $\mathbb{Z}_p[[Q]]$–module. Then the invariant of A as defined in [51], [52] is

$$\Delta_A(Q) = \{v \in T(Q) \mid A^o \subset ker\,\overline{v}\} \cup \{1\}$$

where $A^o = Ann_{\mathbb{Z}[[Q]]}A$ is the annihilator ideal of A in $\mathbb{Z}_p[[Q]]$. We say that A is m-tame over $\mathbb{Z}_p[[Q]]$ if whenever $v_1, \ldots, v_m \in \Delta_A(Q)$ have the property $v_1 \ldots v_m = 1$ then $v_1 = \ldots = v_m = 1$. In [52] the above invariant is called geometric, and is associated to any pro-p extension G of A by Q. But there is little justification for classifying this invariant as geometric. In fact the absence of geometric idea makes it difficult to generalise this invariant to the case of an arbitrary finitely generated pro-p group G. Still this invariant is strikingly important in the theory of metabelian pro-p groups and as the following theorem shows plays the same role as the Bieri-Strebel invariant $\Sigma^1(G)^c$ in the theory of discrete groups.

Theorem 6.4 *[51, sec.15],[52] If $1 \to A \to G \to Q \to 1$ is a short exact sequence of pro-p groups such that G is (topologically) finitely generated, A and Q are abelian then G is finitely presented as a pro-p group if and only if A is 2–tame over $\mathbb{Z}_p[[Q]]$.*

As shown in [51, Thm 9.2], [53] if G is a pro–p group, N a normal closed subgroup such that the completed group algebra $\mathbb{F}_p[[G/N]]$ is topologically (right)

Noetherian (i.e. every ascending sequence of closed right ideals stabilizes), then G is of type FP_m over $\mathbb{Z}_p$ if and only if $\hat{H}_i(N, \mathbb{F}_p)$ is finitely generated as a pro-p $\mathbb{F}_p[[G/N]]$–module for all $i \leq m$.

Note that by [66, Prop.8.6.7] and [51, Cor.7.6] for a pro-p group H of finite rank the completed group algebra $\mathbb{F}_p[[H]]$ is topologically and abstractly (right) Noetherian. Then to classify the abelian-by-finite rank pro–p groups of type FP_m we have to understand the module structure of $\hat{H}_m(A, \mathbb{F}_p)$ for (right) pro-p $\mathbb{Z}_p[[H]]$–modules A, H pro-p group of finite rank. The following theorem gives a description of the homology groups of A in terms of completed tensor products.

Corollary 6.5 *[58, Cor.C] Let A be a (right) finitely generated pro-p $\mathbb{Z}_p[[H]]$–module, where H is a pro-p group of finite rank. Then $\hat{H}_n(A, \mathbb{F}_p)$ has a filtration of (right) $\mathbb{Z}_p[[H]]$–submodules with quotients that are pro-p $\mathbb{Z}_p[[H]]$–subquotients of*

$$\hat{H}_{\alpha_1}(A_1, \mathbb{F}_p) \hat{\otimes}_{\mathbb{Z}_p} \hat{H}_{\alpha_2}(A_2, \mathbb{F}_p) \hat{\otimes}_{\mathbb{Z}_p} \ldots \hat{\otimes}_{\mathbb{Z}_p} \hat{H}_{\alpha_s}(A_s, \mathbb{F}_p),$$

where $\alpha_1 + \ldots + \alpha_s = n$,

A has a filtration $B_1 = A \supset B_2 = tor A \supset B_3 \supseteq \ldots \supseteq B_s \supset B_{s+1} = 0$, $A_j = B_j / B_{j+1}$, A_j has exponent p for $j \geq 2$ and the action of H on the above completed tensor product is the diagonal one.

In particular if the m–th tensor product $\hat{\otimes}^m_{\mathbb{Z}_p} A$ is finitely generated as a pro-p $\mathbb{Z}_p[[H]]$–module via the diagonal H-action for all $m \leq n$ then $\hat{H}_n(A, \mathbb{F}_p)$ is finitely generated as a pro-p $\mathbb{Z}_p[[H]]$–module (abstractly or topologically is the same).

It is interesting to note that the proof of the above result uses substantially the Cartan's Theorem 4.3 discussed in section 4. Corollary 6.5 together with Theorem A from [58] that states that finite generation over $\mathbb{Z}_p[[Q]]$ (via the diagonal Q-action, Q finitely generated abelian pro-p group) of completed tensor, symmetric and exterior powers of finitely generated pro-p $\mathbb{Z}_p[[Q]]$-module A are equivalent imply the conjectured by J. King classification of metabelian groups of type FP_m [51]. Note that an extra equivalent condition is included to the original hypothesis, namely the finite generation of the symmetric powers and that the original conjecture contained a more restrictive form of condition 2. It required that all i–th completed exterior powers for $i \leq m$, not only $i = m$, are finitely generated.

Theorem 6.6 *[58, Thm.D] Suppose $1 \to A \to G \to Q \to 1$ is a short exact sequence of pro-p groups, G is (topologically) finitely generated, A and Q are abelian. We view A as a (right) pro-p $\mathbb{Z}_p[[Q]]$–module via the action of Q given by (right) conjugation. Then the following are equivalent:*

1. G is of type FP_m over $\mathbb{Z}_p$;

2. the completed m–th exterior tensor power of A is a finitely generated pro-p $\mathbb{Z}_p[[Q]]$–module via the diagonal Q–action;

3. the completed m-th tensor power of A is a finitely generated pro-p $\mathbb{Z}_p[[Q]]$–module via the diagonal Q–action;

4. the completed m–th symmetric tensor power of A is a finitely generated pro-p $\mathbb{Z}_p[[Q]]$–module via the diagonal Q–action;

5. A is m–tame over $\mathbb{Z}_p[[Q]]$.

In the case $m = 2$ the equivalence between the conditions 1., 2., 3., and 5. is established in [51, section 15].

7 Homological properties of Lie algebras

In this section we discuss homological finiteness properties of Lie algebras L. By definition a Lie algebra L over a field K is of homological type FP_m if the trivial module K is of homological type FP_m over the universal enveloping algebra $U(L)$ of L. There is a deep analogy between the group theoretic case where we study the trivial module $\mathbb{Z}$ over the group algebra $\mathbb{Z}[G]$ and the Lie algebra case where $\mathbb{Z}$ is substituted with the field K and $U(L)$ plays the role of the group algebra. Still the Lie algebra case differs from the group theoretic case by its lack of geometric methods. All the results in the Lie case obtained by now depend on valuation theoretic arguments. Furthermore all known results for groups that have a homological proof hold in the Lie case and there are results as the generalised FP_m-Conjecture for metabelian group [55, Conj.1] whose version for Lie algebras holds for split extensions [57] but the same question for groups is still open.

Now we continue with a short description of the results of Bryant and Groves that lead to the classification of the finitely presented metabelian groups in [30], [31]. This classification depends on the invariant $\Delta(Q, A)$, where A is an abelian ideal of L with abelian quotient $Q = L/A$. Let $K[Q]$ be the polynomial algebra on n commuting variables where n is the dimension of Q. For every abelian Lie algebra L_0 over a field K the universal enveloping algebra $U(L_0)$ of L_0 is isomorphic to the symmetric algebra $S(L_0)$ of L_0 i.e. $S(L_0)$ is the quotient of the tensor algebra of L_0 over K through the ideal generated by $x \otimes y - y \otimes x$ for $x, y \in L_0$. In particular, $K[Q]$ is isomorphic to the universal enveloping algebra $U(Q)$ of Q. By definition for a finitely generated $K[Q]$-module A

$$\Delta(Q, A) = \{[\chi] \mid \chi \in Hom_K(Q, \overline{K}((t))), \chi \text{ is extendable to a ring homomorphism}$$

$$\chi' : K[Q]/Ann(A) \to \overline{K}((t))\} \subset Hom_K(Q, \overline{K}((t)))/Hom_K(Q, \overline{K}[[t]]),$$

where $[\chi] = \chi + Hom(Q, \overline{K}[[t]])$, $\overline{K}$ is the algebraic closure of K, $\overline{K}((t))$ is the field of fractions of $\overline{K}[[t]]$ and $Ann(A)$ is the annihilator of A in $K[Q]$. The proof of the following lemma was pointed out to me by J. Groves.

Lemma 7.1 *Suppose* $0 \to V_1 \to V \to V_2 \to 0$ *is a short exact sequence of finitely generated* $K[Q]$*–modules. Then*

$$\Delta(Q, V) = \Delta(Q, V_1) \cup \Delta(Q, V_2).$$

Proof Observe firstly, that $\Delta(Q, V)$ is defined in terms of the annihilator of V. More precisely, it is defined in terms of the prime ideals containing the annihilator, since the kernel of the extension χ' of χ for $[\chi] \in \Delta(Q, V)$ is a prime ideal. Thus, if $Ann(V) \subseteq Ann(W)$, then $\Delta(Q, W) \subseteq \Delta(Q, V)$. More precisely, if $\chi' : K[Q]/ann(W) \to \overline{K}((t))$ is the map showing that $[\chi] \in \Delta(Q, W)$ then we can combine this with the natural surjection $\overline{K}[Q]/ann(V) \to \overline{K}[Q]/ann(W)$ to show that $[\chi] \in \Delta(Q, V)$. Since the annihilator of any proper submodule or quotient contains the annihilator of V, this shows that $\Delta(Q, V_1) \cup \Delta(Q, V_2) \subseteq \Delta(Q, V)$.

For the converse, suppose that I_1, I_2 are the annihilators of V_1 and V_2. Then $I_1 I_2$ annihilates V. Suppose that $[\chi] \in \Delta(Q, V)$ and that the corresponding map induced on $K[Q]$ is χ' with kernel P. Then $I_1 I_2 \subseteq Ann(V) \subseteq P$ and so, as P is a prime ideal, $I_1 \subseteq P$ or $I_2 \subseteq P$. It follows, as in the previous paragraph, that $[\chi] \in \Delta(Q, V_1)$ or $[\chi] \in \Delta(Q, V_2)$. □

The main result of [30], [31] is the following characterization of the finitely presented metabelian Lie algebras.

Theorem 7.2 *Suppose L is a finitely generated Lie algebra over a field K, A is an abelian ideal in L with $Q = L/A$ abelian. Then the following conditions are equivalent:*

1. *L is finitely presented as a Lie algebra;*
2. *L is of type FP_2 over K;*
3. *the exterior square of A is finitely generated over $K[Q]$ via the diagonal adjoint action;*
4. *whenever $[\chi_1], [\chi_2] \in \Delta(Q, A) \setminus \{[0]\}$ we have $[\chi_1] + [\chi_2] \neq [0]$ i.e. $\Delta(Q, A)$ has no non-zero antipodal elements.*

We remind the reader that a module over a Lie algebra L (over a field K) is a module over its universal enveloping algebra $U(L)$. We are primarily interested in the homological finiteness properties of modules B over L such that some abelian ideal A of L with L/A abelian has the property that A acts trivially on B. This includes the case of the trivial module K.

Theorem 7.3 *[57]Suppose L is a finitely generated Lie algebra over a field K, A is an abelian ideal in L with $Q = L/A$ abelian and B is a finitely generated (right) module over the universal algebra $U(Q)$ of Q. Then the following are equivalent:*

1. *B is finitely presented as a (right) module over L (i.e. as a module over the universal algebra $U(L)$ of L) where the action of L is via the canonical projection $\pi : L \to Q$.*
2. *$A \otimes_K B$ is finitely generated over the universal algebra $U(Q)$, where $U(Q)$ acts via the diagonal homomorphism $\partial : U(Q) \to U(Q) \otimes U(Q)$ sending $q \in Q$ to $q \otimes 1 + 1 \otimes q$.*
3. *$\Delta(Q, A) \cap -\Delta(Q, B) = [0]$.*

Theorem 7.3 together with the Bryant-Groves characterization of the finitely presented metabelian Lie algebras implies the following surprising result.

Corollary 7.4 *[57] Suppose L is a finitely generated Lie algebra over a field with an abelian ideal A such that $Q = L/A$ is abelian. Then L is finitely presented as a Lie algebra if and only if A is finitely presented as a module over the universal algebra $U(L)$ of L.*

We finish this section with a Lie algebra version of the generalised FP_m-Conjecture suggested in [61, Conj.7] and [55, Conj.1]. In the Lie algebra case the Bryant-Groves invariant Δ plays the role of the Bieri-Strebel invariant $\Sigma^1(G)^c$. Note the result is known only for split extensions metabelian Lie algebras. And the group theoretic analogue of the equivalence of 1 and 3 in the following theorem is still an open problem.

Theorem 7.5 *[57] Suppose L is a finitely generated Lie algebra over a field K, A is an abelian ideal in L with $Q = L/A$ abelian and B is a finitely generated (right) module over the universal algebra $U(Q)$ of Q. We further assume that L is a split extension of A by Q. Then the following are equivalent:*

1. B is of type FP_m over L;

2. $B \otimes (\otimes^m A)$ is finitely generated over $U(Q)$ via the diagonal map $U(Q) \to \otimes^{m+1} U(Q)$ sending $q \in Q$ to $\sum_{0 \leq k \leq m} (\otimes^k 1) \otimes q \otimes (\otimes^{m-k} 1)$;

3. whenever $[v_2], \ldots, [v_{m+1}] \in \Delta(Q, A), [v_1] \in \Delta(Q, B)$ and $[0] = [v_1] + \ldots + [v_{m+1}]$ then all $[v_i] = [0]$.

The proof of the above theorem is very different from the proof of the pro-p version for B being the trivial $\mathbb{Z}_p[[G]]$-module $\mathbb{Z}_p$ i.e. Thm 6.6. In the Lie case it is not sufficient to study homology groups. The most difficult part of the proof of Theorem 7.5 is the building of resolution of B with the required homological finiteness properties.

8 Valuation theory : link between the Lie, pro-p and discrete cases

We have already emphasized the point that all three cases considered in this survey paper : discrete groups, pro-p groups and Lie algebras can be united (in the metabelian case) by valuation-theoretic arguments. In this short section we will draw the similarities and differences between these cases. Most of the arguments have been already mentioned in the different parts of the paper but putting them together can give a better insight into the theory.

1. First we outline several points about the discrete case. Here G is a finitely generated group, an extension of A by Q, A and Q both abelian. The finite presentability of G is equivalent to $\Sigma^1(G)^c = \Sigma^0(G, A)^c$ not containing antipodal elements. In general the finite presentability of G implies that $H_2(A, \mathbb{Z})$ is finitely generated over $\mathbb{Z}[Q]$ via the diagonal action (see Lemma 5.4), but the converse holds only in restrictive cases. Still finite generation (over $\mathbb{Z}[Q]$) of the exterior power of A is equivalent to the finite generation of the tensor square of A (over $\mathbb{Z}[Q]$) which is equivalent to the fact that for every non-negative valuation v the invariant $\Delta_v(Q, A)$ does not have antipodal non-zero characters. The proof of the last statement is based on two facts. The first is the direct product formula [15, Thm 4.2]

$$\Delta_v(Q_1 \times Q_2, A_1 \otimes_{\mathbb{Z}} A_2) = \Delta_v(Q_1, A_1) \times \Delta_v(Q_2, A_2)$$

The second is the following criterion [15, Lemma 5.1] : for a finitely generated $\mathbb{Z}[H]$-module M, $T \subseteq H$ finitely generated abelian groups, M is finitely generated over

$\mathbb{Z}[T]$ if and only if for every discrete non-negative valuation $v : \mathbb{Z}/(\mathbb{Z} \cap ann(M)) \to \mathbb{R}$ we have

$$\Delta_v(H, M) \cap res_T^{-1}(0) \subseteq \{0\} \qquad (*)$$

where $res_T : Hom(H, \mathbb{R}) \to Hom(T, \mathbb{R})$ is the restriction map.

2. In the case of a finitely generated Lie algebra L, an extension of abelian Lie algebras A by Q, the classification of finite presentability is in terms of the invariant $\Delta(Q, A)$. As already mentioned finite presentability is equivalent to $\Delta(Q, A)$ without non-zero antipodal elements. At first glance the definition of $\Delta(Q, A)$ looks fairly complicated but it becomes natural once it is explained that the above condition for non-antipodal points is equivalent to the finite generation of $A \otimes_K A$ over $K[Q]$ via the diagonal action i.e. $q \in Q$ acts via multiplication with $1 \otimes q + q \otimes 1$ (note that the diagonal action here is quite different from the group case). The latter result is proved using deep valuation theoretic methods [31, Prop.3.4]. In the Lie case there is a direct product formula too [31, Prop.3.3] and a Lie version of $(*)$ [31, Prop.3.1]. Furthermore the finite generation (over $K[Q]$) of the tensor square of A is equivalent to the finite generation of the exterior square of A (over $K[Q]$). As in the discrete case the finite presentability of L implies immediately that $H_2(A, K) \simeq A \wedge_K A$ is finite generated over $K[Q]$.

3. In the case of a finitely generated pro-p group G, an extension of abelian pro-p groups A by Q, we have by the results of J. King that the finite presentability of G is equivalent to the finite generation of the completed exterior square of A over the completed group algebra $\mathbb{Z}_p[[Q]]$. Again the invariant introduced by J. King has a fairly complicated definition that becomes natural once the following results are proved. By [52, Thm.D] every just-infinite compact local commutative ring of characteristic p embeds in $K[[T]]$, where K is the algebraic closure of $\mathbb{F}_p$ (remember by definition $\Delta_A(Q)$ contains the characters of Q with image in the multiplicative group of $K[[T]]$ that can be lifted to ring homomorphism $\mathbb{Z}_p[[Q]]/Ann(A) \to K[[T]]$). This easily implies A is (topologically) finitely generated if and only if $\Delta_A(Q) = \{1\}$. As before finite generation (over $\mathbb{Z}_p[[Q]]$ via the diagonal action given by multiplication with $q \otimes q$ for $q \in Q$) of completed exterior or tensor square of A are equivalent. There is pro-p version of $(*)$ given in [52, Cor.2.6]. And the completed exterior power of A over $\mathbb{Z}_p$ is isomorphic to $\hat{H}_2(A, \mathbb{Z}_p)$, thus explaining the link between finite presentability of G and finite generation of $A \wedge_{\mathbb{Z}_p} A$ over $\mathbb{Z}_p[[Q]]$.

By now we have discussed only similarities but in fact there are many differences, some are outlined below.

1. Principally what distinguishes the Lie from the discrete case (via the correspondence of the trivial module $\mathbb{Z}$ over $\mathbb{Z}[G]$ with the trivial module K over the universal enveloping algebra of L) is the type of action. In the Lie case the action of L on the exterior square of the abelian ideal A is given by multiplication with $1 \otimes q + q \otimes 1$ for $q \in L/A$. In the discrete case the action of G on the exterior square of the abelian normal subgroup $\widetilde{A}$ with abelian quotient $G/\widetilde{A}$ is given by multiplication with $q \otimes q$, for $q \in G/\widetilde{A}$. The more complicated nature of the action in the Lie case is the main culprit of the necessity to define the Lie invariant in its present form.

2. In the discrete case it is the decomposition of a finitely presented group as an HNN extension with finitely generated base and associated subgroups that led to the first definition of the geometric invariant. This decomposition gives the discrete part of the Bieri-Strebel-Neumann result that for finitely presented groups without free subgroups of rank two $\Sigma^1(G)^c$ does not contain antipodal points. Though the HNN construction has proved quite useful in the discrete case it has not given the expected results neither in the Lie nor in the pro-p cases. By now the best result obtained using the HNN equivalent for Lie algebras is that any ideal of codimension 1 in a finitely presented Lie algebra L, L without free subalgebras of rank 2, is itself finitely generated as a Lie algebra [65] (in fact in [65] this result is stated only for soluble Lie algebras but the proof uses only the absence of free Lie subalgebras of rank 2). In the discrete case the normal subgroups of a finitely presented group with quotients isomorphic to $\mathbb{Z}$ are not necessary finitely generated.

9 Related topics

In this section several questions are outlined, these are questions related to one of the main topics of this survey paper - the homological finiteness properties of groups.

1. One of the basic questions in the field is the characterisation of the finitely presented groups in the class of all finitely generated soluble groups. Unfortunately outside the class of metabelian groups not much is known. There is complete classification in the case of nilpotent-by-abelian S arithmetic groups defined over the rationals [1] that later motivated the results in [36]. The proof of the main result in [1] is long and complicated, and involves methods from the theory of algebraic groups plus results for extension of Lie groups over a local field (i.e. a finite extension of the p-adic field $\mathbb{Q}_p$) by discrete abelian group that acts on the Lie part via topological automorphisms. For locally compact groups there are compact versions of the homotopical and homological finiteness properties that are still not well understood [3]. For example it is not known (though it is probably true) whether a quotient of a CP_m group (CP_m is the compact version of FP_m) by a CP_{m-1} is again of type CP_m. The problem arises from the lack of homological criterion of the type of Lemma 5.1 for locally compact groups.

Going back to the class of discrete groups we note that recently a break through was done in [23] where the characterisation of the finitely presented abelian-by-nilpotent groups was reduced to the case of finitely presented abelian-by-nilpotent of class 2 groups. The latter question is still open.

Theorem 9.1 *[23, Cor.1] Let G be a finitely presented group. Either*

1. there is a prime p and a subgroup G_1 of finite index in G such that the pro-p completion of G_1 contains a non-abelian free pro-p group

or

2. if $\overline{G}$ is any image of G with $\overline{G}/Fitt(\overline{G})$ virtually nilpotent then $\overline{G}/Fitt(\overline{G})$ is virtually nilpotent of class at most 2.

As the following theorem shows considering finitely presented metanilpotent (=nilpotent-by-nilpotent) groups instead of finitely presented soluble ones is not very restrictive.

Theorem 9.2 *[24, Thm A] Every finitely presented abelian-by-polycyclic group has a metanilpotent subgroup of finite index.*

Theorem 9.2 gives a negative answer to a question raised by G. Baumslag [6] of whether every finitely generated abelian-by-polycyclic group can be embedded in a finitely presented abelian-by-polycyclic group.

2. The idea behind the geometric invariants is to study the behaviour of closed half subspaces of the universal coverings of $K(G,1)$-space with finite m-skeleton. There is an ∞ versions of this idea i.e. study the homological and homotopical pro-type of the neighbourhoods of ∞. It is interesting that in this case it is the cohomology $H^*(G, \mathbb{Z}G)$ that decides whether a group is m-acyclic at infinity i.e. $H^i(G, \mathbb{Z}[G]) = 0$ for $i \leq m-1$ and $H^m(G, \mathbb{Z}[G])$ is torsion-free if and only if G is m acyclic at infinity [40, Cor] . More results about connectivity at infinity can be found in [29], [8], [39].

3. As stated in the introduction there are few classes of groups where the Σ-invariants have been computed by now. One of them is the class of right angled Artin groups [48]. The work in [48] develops the ideas introduced in [7] where the first example of group that is of type FP_∞ but not finitely presented is constructed. More results about the geometric invariants in the case of graph product of groups can be found in [49] and [50].

4. The Σ theory for discrete groups considered in this paper is a special case of a more general recent study of groups acting on spaces with non-positive curvature [9], [10], [11]. Our special case comes from the action of G on $G/[G,G] \otimes_{\mathbb{Z}} \mathbb{R}$ via multiplication (either right or left depending on whether our group actions are right or left). The analogy between the flat case and the more general non-positive curvature case is in the good theory of geodesics i.e. two different geodesics that start at the same point never meet again, there is a well-defined boundary of a CAT(0)-space and for every geodesic there is an associated Busemann function [41]. In the general non-positive curvature case the boundary plays the role of the unit sphere $S(G)$.

5. In [47] the following direct product formula was suggested for groups G_1 and G_2 of type F_n : $[\chi_1, \chi_2] \in \Sigma^n(G_1 \times G_2)^c$ if and only if either one of the characters is 0 and for the other we have $[\chi_i] \in \Sigma^n(G_i)^c$ or there exist $n_1 + n_2 = n$, both n_1 and n_2 positive integers, $[\chi_i] \in \Sigma^{n_i}(G_i)^c$. This formula holds for $n = 1, 2$ [21], [38] but in general is wrong [48, Sect.6]. Still it is not known whether the homological counterpart holds and if not what other formula for the Σ-invariants of direct products of groups can be suggested. In the case when G_1 and G_2 are metabelian the above formula can be viewed as a corollary of the Σ^m-Conjecture, thus it is quite probably true.

References

[1] H. Abels, *Finite presentability of S–arithmetic groups. Compact presentability of Solvable groups. Lecture Notes in Mathematics 1261, Springer-Verlag, 1987*

[2] H. Abels, *An example of a finitely presented solvable group. Homological group theory (Proc. Sympos., Durham, 1977), pp. 205–211, LMS Lecture Note Ser., 36, Cambridge Univ. Press, Cambridge-New York, 1979.*

[3] H. Abels, A. Tiemeyer, *Compactness properties of locally compact groups*, Transform. Groups 2 (1997), no. 2, 119–135.

[4] H. Åberg, *Bieri-Strebel valuations (of finite rank)*, Proc. London Math. Soc. (3)52 (1986), 269–304

[5] J. M. Alonso, *Finiteness conditions on groups and quasi-isometries*, J. Pure Appl. Algebra 95 (1994), no. 2, 121–129.

[6] G. Baumslag, *Finitely presented metabelian groups, in Proc. second intern. conf. on the theory of groups(Springer-Verlag, 1974) 65-74*

[7] M. Bestvina, N. Brady, *Morse theory and finiteness properties of groups*, Invent. Math. 129, No.3(1997), 445-470

[8] N. Brady, J. Meier, *Connectivity at infinity for right angled Artin groups*, Trans. Amer. Math. Soc. 353 (2001), no. 1, 117–132

[9] R. Bieri, R. Geoghegan, *Kernels of actions on non-positively curved spaces. Geometry and cohomology in group theory (Durham, 1994)*, 24–38, London Math. Soc. Lecture Note Ser., 252, Cambridge Univ. Press, Cambridge, 1998

[10] R. Bieri, R. Geoghegan, Connectivity properties of group actions on non-positively curved spaces I : Controlled connectivity and openness results, submitted

[11] R. Bieri, R. Geoghegan, Connectivity properties of group actions on non-positively curved spaces II : the geometric invariants, submitted

[12] R. Bieri, *Homological dimension of discrete groups, Queen Mary College Mathematics Notes, London, 2nd ed. 1981.*

[13] R. Bieri, J.R.J. Groves, *Metabelian groups of type FP_∞ are virtually of type FP*, Proc London Math Soc (3)45, (1982), 365 - 384

[14] R. Bieri, J.R.J. Groves, *The geometry of the set of characters induced by valuations*, J. Reine Angew. Math. 347 (1984), 168–195

[15] R. Bieri, J.R.J. Groves, *Tensor powers of modules over finitely generated abelian groups*, J. Algebra 97 (1985), no. 1, 68–78

[16] R. Bieri, J. Harlander, *On the FP_3–Conjecture for metabelian groups*, J. London Math. Soc. (2) 64 (2001), no. 3, 595–610

[17] R. Bieri, J. Harlander, *A remark on the Polyhedrality Theorem for the Σ–invariants of modules over abelian groups*, Math. Proc. Cambridge Philos. Soc. 131 (2001), no. 1, 39–43

[18] R. Bieri, W. D. Neumann, R. Strebel, *A geometric invariant of discrete groups*, Invent. Math 90(1987), 451–477

[19] R.Bieri, B.Renz, *Valuations on free resolutions and higher geometric invariants of groups*, Comment. Math. Helv. 63(1988), 464-497

[20] R. Bieri, R. Strebel, *Valuations and finitely presented metabelian groups*, Proc. London Math. Soc. (3)41 (1980), 439–464

[21] R. Bieri, R. Strebel, *Geometric invariants for discrete groups*, preprint, 240 p.

[22] R. Bieri, R. Strebel, *Almost finitely presented soluble groups*, Comment. Math. Helv. 53 (1978), no. 2, 258–278

[23] C. J. B. Brookes, *Crossed products and finitely presented groups*, J. Group Theory 3 (2000), no. 4, 433–444

[24] C. J. B. Brookes, J. E. Roseblade, J. S. Wilson, *Exterior powers of modules for group rings of polycyclic groups*, J. London Math. Soc. (2) 56 (1997), no. 2, 231–244

[25] K. S. Brown, *Presentations for groups acting on simply–connected complexes*, J Pure Appl. Algebra 32 (1984), 1-10
[26] K. S. Brown, *Cohomology of Groups*, Springer-Verlag, 1982
[27] K. S. Brown, *Finiteness properties of groups*, J Pure Appl. Algebra 44 (1987), 45-75
[28] K.S. Brown, Trees, valuations and the Bieri-Newmann-Strebel invariant, Inv. Math. 90 (1987), 479-504
[29] K. S. Brown, J. Meier, *Improper actions and higher connectivity at infinity*, Comment. Math. Helv. 75 (2000), no. 1, 171–188
[30] R. Bryant, J.R.J. Groves, *Finite presentation of abelian-by-finite dimensional Lie algebras*, J London Math Soc (2), 60 (1999), no. 1, 45-57
[31] R. Bryant, J.R.J. Groves, *Finitely presented Lie algebras*, J Algebra 218(1999), no. 1, 1-25
[32] R. Bryant, J.R.J. Groves, *Finitely presented centre-by-metabelian Lie algebras*, Bull. Austral. Math. Soc. 60 (1999), no. 2, 221–226
[33] A. Brumer, *Pseudocompact algebras, profinite groups and class formations*, J. Algebra, 4, 442-470 (1966)
[34] K.U. Bux, *Finiteness properties of certain metabelian arithmetic groups in the function field case*, Proc. London Math. Soc., 75(1997),2, 308-322
[35] H. Cartan, *Algebres d'Eilenberg-MacLane*, Seminaire H. Cartan, Ecole Normale Superieure, included in Collected Works, Volume 3, Springer–Verlag, 1979.
[36] J.R.J. Groves, *Some finitely presented nilpotent- by-abelian groups*, Journal of Algebra 144(1991), 127-166
[37] J.R.J. Groves, *Metabelian groups with finitely generated integral homologies*, Quart. J. Math. Oxford (2), 33 (1982), 405–420.
[38] R. Gehrke, *The higher geometric invariants for groups with sufficient commutativity*, Comm. Algebra 26(1998), no. 4, 1097–1115
[39] R. Geoghegan, M. Mihalik, *The fundamental group at infinity. Topology* 35 (1996), no. 3, 655–669.
[40] R. Geoghegan, M. Mihalik, *A note on the vanishing of* $H^n(G, ZG)$ J. Pure Appl. Algebra 39 (1986), no. 3, 301–304
[41] P.K. Hotchkiss, The boundary of a Busemann space, Proc. Amer. Math. Soc. 125 (1997), 1903-1912
[42] J. Harlander, D. H. Kochloukova, *The* Σ^2*-Conjecture for metabelian groups : the general case*, to appear in J. Algebra
[43] J. Harlander, D. H. Kochloukova, *The* Σ^3*-Conjecture for metabelian groups*, to appear in J. London Math. Soc.
[44] H. Meinert, *Actions on 2–Complexes and the Homotopical Invariant* Σ^2 *of a Group*, J. Pure Appl. Algebra, 119(1997), 3, 297-317
[45] H. Meinert, *The homological invariants for metabelian groups of finite Prüfer rank: a proof of the* Σ^m*–Conjecture*, Proc London Math Soc, 72(1996) 2,385–424
[46] H. Meinert, *Iterated HNN-decomposition of constructible nilpotent-by-abelian groups*, Comm. Algebra 23(1995), 8, 3155-3164
[47] H. Meinert, The geometric invariants of direct products of virtually free groups, Comment. Math. Helvetici 69 (1994), 39-48
[48] J. Meier, H. Meinert, L. VanWyk, *Higher generation subgroup sets and the* Σ*–invariants of graph groups*, Comment. Math. Helv. 73(1998) no1. 22–44
[49] J. Meier, H. Meinert, L. Van Wyk, *The* Σ^2*-invariants for graph products of indicable groups*, Topology Appl. 99 (1999), no. 1, 41–65.
[50] J. Meier, H. Meinert, L. VanWyk, *On the* Σ*-invariants of Artin groups. Geometric topology and geometric group theory* (Milwaukee, WI, 1997). Topology Appl. 110 (2001), no. 1, 71–81.

[51] J. D. King, *Finite presentability of Lie algebras and pro-p groups*, PhD Thesis, University of Cambridge, 1995

[52] J. D. King, *A geometric invariant for metabelian pro-p groups*, J. London Math. Soc. (2) 60 (1999), no. 1, 83–94

[53] J. D. King, *Homological finiteness conditions for pro-p groups* Comm. Algebra 27 (1999), no. 10, 4969–4991

[54] D.H. Kochloukova, *The Σ^m–Conjecture for a class of metabelian groups*, Groups St Andrews'97 in Bath, LMS Lecture Note Series 261, Cambridge University Press,1999, 492–503

[55] D.H. Kochloukova, *The Σ^2–Conjecture for metabelian groups and some new conjectures: the split extension case*, J. Algebra, 222(1999), 357-375

[56] D.H. Kochloukova, *The FP_m–Conjecture for a class of metabelian groups*, J. Algebra, 184 (1996), 1175–1204

[57] D. H. Kochloukova, *On the homological finiteness properties of some modules over metabelian Lie algebras*, Israel J. Math., to appear

[58] D. H. Kochloukova, *Metabelian pro-p groups of type* FP_m, J. Group Theory 3 (2000), no. 4, 419–431.

[59] D. H. Kochloukova, *Groups with finitely generated integral homologies*, to appear in the Conference Proceedings of Groups St. Andrews at Oxford, 2001.

[60] D. H. Kochloukova, *Geometric invariants and modules of type FP_∞ over constructible nilpotent-by-abelian groups*, J. Pure Appl. Algebra, (2-3) 159 (2001), 187-202.

[61] D. H. Kochloukova, *A new characterisation of m-tame groups over finitely generated abelian groups*, J. London Math. Soc. (2) 60 (1999), no. 3, 802–816

[62] D. H. Kochloukova, *Subgroups of constructible nilpotent-by-abelian groups and a generalization of a result of Bieri-Neumann-Strebel*, J. Group Theory 5 (2002), no. 2, 219–231

[63] G. A. Noskov, *The Bieri-Strebel invariant and homological finiteness conditions for metabelian groups*, Algebra i Logika, 36 (1997), no 2, 194-218

[64] B. Renz, *Geometrische Invarianten und Endlichkeitseigenschaften von Gruppen*, Dissertation. Universität Frankfurt a.M. (1988)

[65] A. Wasserman, *A derivation HNN construction for Lie algebras*, Israel J. Math. 106(1998), 79–92

[66] J. Wilson, *Profinite groups*, Clarendon Press, Oxford, 1998

GROUPS WITH ALL NON-SUBNORMAL SUBGROUPS OF FINITE RANK

LEONID A. KURDACHENKO* and PANAGIOTIS SOULES†[1]

*Algebra Department, Dnepropetrovsk University, Vul. Naukova 13, Dnepropetrovsk 50, Ukraine 49000

†Mathematics Department, University of Athens, Panepistemiopolis, 15784 Athens, Greece

Abstract

In the present paper we start to study the soluble groups in which every non-subnormal subgroup has finite special rank.

1 Introduction

Let G be a group, $\mathcal{L}_{non-sn}(G)$ the set of all non-subnormal subgroups of G. If $\mathcal{L}_{non-sn}(G) = \emptyset$ then we obtain a group, every subgroup of which is subnormal. The study of these groups was very fruitful and has brought many interesting results (see the books [12], [13], [8]). The groups G in which the set $\mathcal{L}_{non-sn}(G)$ is "very small" in some sense is the natural next consideration. For many domains of Infinite Group Theory "to be very small" means to satisfy some finiteness conditions. The first natural finiteness conditions in Group Theory were the classical minimal and maximal conditions. The groups with minimal condition for non-subnormal subgroups have been considered by S. Franciosi and F. de Giovanni [1]. The groups with dual maximal condition for non-subnormal subgroups have been studied by L.A. Kurdachenko and H. Smith [4]. The groups with maximal condition for non-subnormal subgroups is nearly allied to the groups, in which the set $\mathcal{L}_{non-sn}(G)$ consists only of the finitely generated subgroups. Such groups have been considered. The results of papers [1], [4] have been extended on groups with weak minimal and maximal condition on non-subnormal subgroups [5], [6]. The weak minimal and maximal conditions are connected with the concept of special rank (or Mal'tsev-Prüfer rank). A group G is said to have finite special rank $\mathbf{r}$ if every finitely generated subgroup of G is at most $\mathbf{r}$-generator and $\mathbf{r}$ is the smallest integer with this property. We will denote the special rank of a group G by $r(G)$. In this paper we start the study of groups G with the following condition: if $H \in \mathcal{L}_{non-sn}(G)$ then H has finite (special) rank. These groups are studied in this paper under an additional condition of solubility. The solubility is here a natural condition, because the groups with all subgroups subnormal are soluble by a basic theorem of W. Möhres [11]. On the other hand, F. Menegazzo shows, that for each $d > 1$ there is a non-nilpotent group of derived length d, all of whose proper subgroups are subnormal [10].

[1]This work was supported by University of Athens (Research center, Grant No. 3403)

Lemma 1.1 *Let G be a group whose non-subnormal subgroups have finite rank.*
(1) If H is a subgroup of G then every non-subnormal subgroup of H has finite rank.
(2) If H is a normal subgroup of G then every non-subnormal subgroup of G/H has finite rank.

Proof The proof is straightforward. □

Lemma 1.2 *Let G be a group whose non-subnormal subgroups have finite rank, $g \in G$, $H = \mathbf{X}_{\lambda \in \Lambda} H_\lambda$, where each H_λ is a non-identity $< g >$-invariant subgroup of finite rank. If a set $\{r(H_\lambda) | \lambda \in \Lambda\}$ is not bounded then $< g >$ is subnormal in G.*

Proof Let $< x >=< g > \cap H$. Then $Supp\ x = \Delta$ is a finite subset of Λ and $(\mathbf{X}_{\lambda \in \mathbf{M}} H_\lambda) \cap < g >=< 1 >$ where $\mathbf{M} = \Lambda \backslash \Delta$ is infinite. There are two subsets $\mathbf{K}$ and $\mathbf{N}$ of $\mathbf{M}$ such that $\mathbf{K} \cup \mathbf{N} = \mathbf{M}$, $\mathbf{K} \cap \mathbf{N} = \emptyset$ and both sets $\{r(H_\lambda)|\lambda \in \mathbf{K}\}$ and $\{r(H_\lambda)|\lambda \in \mathbf{N}\}$ are unbounded. It follows that $< g > (\mathbf{X}_{\lambda \in \mathbf{K}} H_\lambda)$ and $< g > (\mathbf{X}_{\lambda \in \mathbf{N}} H_\lambda)$ have infinite rank, therefore these subgroups are subnormal. Thus their intersection $< g > (\mathbf{X}_{\lambda \in \mathbf{K}} H_\lambda) \cap < g > (\mathbf{X}_{\lambda \in \mathbf{N}} H_\lambda) =< g >$ is subnormal ([8], Proposition 1.1.2). □

If G is a group then denote by $B(G)$ **the Baer radical of G**, that is

$$B(G) = \{g \in G| < g > \text{ is subnormal in } G\}.$$

Corollary 1.3 *Let G be a group whose non-subnormal subgroups have finite rank, $H = \mathbf{X}_{\lambda \in \Lambda} H_\lambda$ where H_λ is a non-trivial $< g >$-invariant subgroup of finite rank. If the set $\{r(H_\lambda) \mid \lambda \in \Lambda\}$ is not bounded then $H \leq B(G)$.*

Lemma 1.4 *Let G be a group whose non-subnormal subgroups have finite rank, $g \in G$. Let A be an infinite $< g >$-invariant elementary abelian p-subgroup of G, p a prime. If $g^n \in C_G(A)$ for some $n \in \mathbf{N}$ then $g \in B(G)$.*

Proof Let $1 \neq a_1 \in A$, $A_1 =< a_1 >^{<g>}$. Since A is elementary abelian then $A = A_1 \times B_1$ for some subgroup B_1. The set $\{B_1^x \mid x \in < g >\}$ is finite, therefore $C_1 = \bigcap_{x \in <g>} B_1^x$ has finite index in A and $A_1 \cap C_1 =< 1 >$. Let $a_2, a_3, \ldots, a_k$ be elements of C_1 with the properties $< a_2, \ldots, a_k >=< a_2 > \times \cdots \times < a_k >$ and $k - 1 > r(A_1)$. Let $A_2 =< a_2, \ldots, a_k >^{<g>}$, then $A = A_1 A_2 \times B_2$ for some subgroup B_2 and $C_2 = \cap_{x \in <g>} B_2^x$ has finite index in A. With the help of similar arguments we construct the set $\{A_n | n \in \mathbf{N}\}$ of finite $< g >$-invariant subgroups such that $r(A_1) < r(A_2) < \ldots < r(A_n) < \ldots$. Using the Lemma 1.2 we obtain that $g \in B(G)$. □

Corollary 1.5 *Let G be a group whose non-subnormal subgroups have finite rank. If G contains an abelian periodic subgroup A of infinite rank, then $B(G)$ contains every element of finite order. In particular, the set T of all elements of finite order is a subgroup of G.*

Proof We have $A = \mathbf{X}_{p \in \Pi(G)} A_p$ where A_p is the Sylow p-subgroup of A. If there is a prime p such that $r(A_p)$ is infinite, then $A \leq B(G)$ by Lemma 1.4. If $r(A_p)$ is finite for each $p \in \Pi(A)$ then the set $\{r(A_p) | p \in \Pi(A)\}$ is not bounded. Then $A \leq B(G)$ by Lemma 1.2. In both cases $r(B(G))$ is infinite. The subgroup $B(G)$ is locally nilpotent ([8], Theorem 2.5.1). In particular, the set T of all elements of $B(G)$ having finite order is a (characteristic) subgroup of infinite rank. Let x be an element of finite order. By Zaitsev's Theorem [15], T contains an $< x >$-invariant abelian subgroup of infinite rank. Again using Lemmas 1.2, 1.4 we obtain that $x \in B(G)$, that is $x \in T$. □

Corollary 1.6 *Let G be a group whose non-subnormal subgroups have finite rank. If G is a locally finite group then either G is a Baer group or G has finite rank.*

Proof If G contains an abelian subgroup of infinite rank then $G = B(G)$ by Corollary 1.5. If every abelian subgroup of G has finite rank then G has finite rank by Shunkov's Theorem [14]. □

Let R be an integral domain, A an R - module. Put

$$\mathbf{t_R(A) = \{a \in A \mid \ Ann_R(a) \neq < 0 >\}}$$

Clearly $t_R(A)$ is an R-submodule of A. It is called the **R-periodic part of A**. If $t_R(A) = A$ then A is said to be an R-periodic module. If $t_R(A) =< 0 >$ then A is R-torsion free.

Let I be an ideal of R and put

$$\mathbf{A_I = \{a \in A \mid (a)I^n =< 0 >, \text{ for some } n \in N\}.}$$

Obviously A_I is an R-submodule of A. It is called the **I - component of A**. Let

$$\mathbf{\Omega_{I,n}(A) = \{a \in A \mid a(I^n) =< 0 >\}.}$$

$\Omega_{I,n}(A)$ is also an R-submodule of A, $\Omega_{I,n}(A) \leq \Omega_{I,n+1}(A)$ for any $n \in \mathbf{N}$ and $A_I = \bigcup_{n \in \mathbf{N}} \Omega_{I,n}(A)$.

Now let R be a principal ideal domain, $Spec(R) = \{P \mid P$ is a maximal ideal of $R\}$, $\Pi_R(A) = \{P \in Spec(R) \mid A_P \neq < 0 >\}$. It is not difficult to prove that

$$t_R(A) = \bigoplus_{P \in \Pi_R(A)} A_P.$$

Let C be a simple R-module, then $C \cong R/P$ for some $P \in Spec(R)$. Denote by $\mathbf{C_{P^\infty}}$ the R-injective envelope of C. The module $\mathbf{C_{P^\infty}}$ is said to be a **Prüfer P-module**. As in Abelian Group Theory we can show that

$$C_{P^\infty} \cong \varinjlim \{R/P^n, n \in \mathbf{N}\}.$$

Furthermore, every proper R - submodule of C_{P^∞} is isomorphic with R/P^n for some $n \in \mathbf{N}$, in particular, it is cyclic.

Let $P = yR$, then $\mathbf{C_{P^\infty}}$ has a subset $\{a_n | n \in \mathbf{N}\}$ of elements such that $a_1 y = 0$, $a_{n+1}y = a_n, n \in \mathbf{N}$, and $C_{p^\infty} \cong \bigcup_{n\in\mathbf{N}} a_n R$. As in Abelian Group Theory we can prove that if $\Omega_{P,1}(A) = C_1 \oplus \ldots \oplus C_n$ where C_i is a simple R-submodule, $1 \le i \le n$, then $A_P = E_1 \oplus \ldots \oplus E_n$ where either E_i is a Prüfer P-module or $E_i \cong R/P^{m_i}$ for some $m_i \in \mathbf{N}$, $1 \le i \le n$.

Lemma 1.7 *Let G be a group whose non-subnormal subgroups have finite rank, g an element of infinite order, A a $< g >$-invariant elementary abelian p-subgroup of G, p a prime. If $A =< a >^{<g>}$ for some element $a \in A$ then A is finite, and hence $g^n \in C_G(A)$ for some $n \in \mathbf{N}$.*

Proof We will consider A as $\mathbf{F_p} < g >$ - module. Then $A \cong_{\mathbf{F_p}<g>} \mathbf{F_p} < g > /I$ where $I = Ann_{\mathbf{F_p}<g>}(a)$. If we suppose that $I \ne < 0 >$ then $\mathbf{F_p} < g > /I$ is finite, and all is proved. Therefore assume that $I =< 0 >$. In this case $A \cong \mathbf{F_p} < g >$. Let q, r be the two distinct primes such that $q \ne p, r \ne p$. Then

$$\mathbf{F_p} < g >= \mathbf{F_p} < g^q > \oplus \mathbf{F_p} < g^q > g \oplus \ldots \oplus \mathbf{F_p} < g^q > g^{q-1}.$$

It follows that $A = A_q \times (A_q)^g \times \cdots \times (A_q)^{g^{q-1}}$ where $A_q =< a >^{<g^q>}$. The subgroups A_q and A_q^g are $< g^q >$-invariant and infinite, thus $< g^q > A_q$ and $< g^q > (A_q)^g$ are subgroups of infinite rank. Since $< g^q >=< g^q > A_q \cap < g^q > (A_q)^g$ then $< g^q >$ is subnormal in G ([8], Proposition 1.1.2). By the same reason $< g^r >$ is subnormal in G. The equation $< g >=< g^q >< g^r >$ yields that $< g >$ is subnormal ([8], Theorem 1.2.1). It follows that $A < g >$ is nilpotent ([2], Lemma 4). Since $A < g >=< a, g >$ then it is the periodic part of a finitely generated nilpotent subgroup. The periodic part of finitely generated nilpotent subgroup is finite, thus A is finite. □

Lemma 1.8 *Let G be a group whose non-subnormal subgroups have finite rank, g an element of infinite order, A an elementary abelian $< g >$-invariant p-subgroup of G, p a prime. If A is a Prüfer P-module for some $P \in Spec(\mathbf{F_p} < g >)$ then $g \in B(G)$.*

Proof There is an element y such that $P = y\mathbf{F_p} < g >$. Let $A =< a_n | n \in \mathbf{N} >$ where $a_1 y = 0$, $a_{n+1}y = a_n$, $n \in \mathbf{N}$. The subgroup $A_1 =< a_1, a_2 >^{<g>}$ is $< g >$-invariant and finite, thus the index $| < g >: C_{<g>}(A_1)|$ is finite. Put $< x >= C_{<g>}(A_1)$. We will consider A as $\mathbf{F_p} < x >$-module. Since A is $\mathbf{F_p} < g >$-divisible then A is also $\mathbf{F_p} < x >$-divisible. Let $Soc_{\mathbf{F_p}<\mathbf{x}>}(A) = C_1 \oplus \cdots \oplus C_k$. Since $a_1, a_2 \in C_A(x)$ then $k \ge 2$. As we noted above $A = E_1 \oplus \cdots \oplus E_k$ where E_i is a Prüfer Q_i-module for some $Q_i \in Spec(\mathbf{F_p} < x >)$. The subgroups E_1 and E_2 are $< x >$-invariant and infinite elementary abelian, hence $< x > E_1$ and $< x > E_2$ have infinite rank, so both these subgroups are subnormal. Then $< x >=< x > E_1 \cap < x > E_2$ is subnormal too ([8], Proposition 1.1.2]). It follows that $A < x >$ is nilpotent ([2], Lemma 4). But in this case there is a number $m \in \mathbf{N}$ such that $x^m \in C_G(A)$ ([2], Proposition 2). In other words, $g^t \in C_G(A)$ for some $t \in \mathbf{N}$. Lemma 1.4 yields that $g \in B(G)$. □

Lemma 1.9 *Let G be a group whose non-subnormal subgroups have finite rank, g an element of infinite order, A an infinite elementary abelian p-subgroup of G $< g >$-invariant, p a prime. Then $g \in B(G)$.*

Proof We will consider A as $\mathbf{F_p} < g >$-module. Lemma 1.7 shows that this module is periodic. Thus $A = \bigoplus_{P\in\pi} A_P$ where $\pi = \Pi_{\mathbf{F_p}<g>}(A)$. Assume first that π is infinite. There are infinite subsets ρ, σ of π with the following properties: $\rho \cap \sigma = \emptyset$, $\rho \cup \sigma = \pi$. The subgroups $B_1 = \bigoplus_{P\in\rho} A_P$, $B_2 = \bigoplus_{P\in\sigma} A_P$ are infinite, $< g >$-invariant and $B_1 \cap B_2 =< 1 >$. Hence $< g > B_1$ and $< g > B_2$ are subnormal and $< g >=< g > B_1 \cap < g > B_2$ is subnormal too ([8], Proposition 1.1.2).
Suppose now that π is finite. In this case there is $P \in \pi$ such that A_P is infinite. Put $C = \Omega_{P,1}(A) = \Omega_{P,1}(A_P)$, then $C = \oplus_{\lambda\in\Lambda} D_\lambda$ where D_λ is a simple $\mathbf{F_p} < g >$-submodule, $\lambda \in \Lambda$. If Λ is finite, then $A_P = \oplus_{\lambda\in\Lambda} E_\lambda$ where either E_λ is a Prüfer P-module or E_λ is a cyclic submodule, $\lambda \in \Lambda$. Since A_P is infinite there is an index k such that E_k is infinite. Lemma 1.8 yields that $g \in B(G)$. Finally suppose that Λ is infinite. Since every $D_\lambda \cong \mathbf{F_p} < g > /P$. In particular, $t = |\mathbf{F_p} < g > /P|$, $g^s \in C_G(D_\lambda)$ for some $s \leq t$. Since it is true for every $\lambda \in \Lambda$ then $g^s \in C_G(C)$. Again by Lemma 1.4 $g \in B(G)$. □

Lemma 1.10 *Let G be a group whose non-subnormal subgroups have finite rank, $g \in G$, A, B subgroups of $B(G)$ such that A is G-invariant and $|A| = p$, B is $< g >$ - invariant and B/A is an infinite elementary abelian p-group, p a prime. Then B contains a $< g >$-invariant infinite elementary abelian p-subgroup.*

Proof Lemmas 1.4, 1.9 yield that $gA \in B(G/A)$. Using the arguments of these Lemmas, also we see that $C_{B/A}(gA) = C/A$ is infinite. Note that C is $< g >$-invariant and $< C/A, gA >$ is abelian, that is $L =< C, g >$ has finite derived subgroup, in particular, L is an FC-group. Thus the centralizer of g in C has finite index in C, and we may assume that $[C, g] = 1$. Let E be a maximal abelian subgroup of C containing A. Then E is self-centralizing in C and therefore infinite. Now put $U = \Omega_1(E)$, then U is an infinite $< g >$-invariant elementary abelian p-subgroup of L. □

Corollary 1.11 *Let G be a group whose non-subnormal subgroups have finite rank, $g \in G$, A, B subgroups of $B(G)$ such that A is a finite G-invariant p-subgroup, B is $< g >$-invariant and B/A is an infinite elementary abelian p-group, p a prime. Then B contains a $< g >$-invariant infinite elementary abelian p-subgroup.*

Lemma 1.12 *Let G be a group whose non-subnormal subgroups have finite rank, $g \in G$, A, B subgroups of $B(G)$ such that A is a G-invariant divisible Chernikov p-subgroup, B is $< g >$-invariant and B/A is an infinite elementary abelian p-subgroup, p a prime. Then B contains an infinite $< g >$-invariant elementary abelian p-subgroup.*

Proof Since $B \leq B(G)$ then B is nilpotent ([7], Lemma 2.4). Since B is periodic then $A \leq \zeta(B)$ ([12], Lemma 3.13). Therefore, for every element $b \in B$ we have

$[b, B] \cong B/C_B(b)$. Since $A \leq \zeta(B)$ then $B/C_B(b)$ is an elementary abelian p-group. On the other hand, $[b, B] \leq A$, so that $[b, B]$ has finite rank $\leq r = r(A)$. From $[b, B] \cong B/C_B(b)$ we obtain that $|B/C_B(b)| \leq p^r$.
By B. H. Neumann's Theorem ([12], Theorem 4.35) $[B, B]$ is finite. Since $B/[B, B]$ is an abelian p-group, but not Chernikov, its lower layer $\Omega_1(B/[B, B]) = C/[B, B]$ is infinite. Clearly C is $< g >$-invariant. Now we can apply the Corollary 1.11. □

Corollary 1.13 *Let G be a group whose non-subnormal subgroups have finite rank, $g \in G$, A, B subgroups of $B(G)$ such that A is a G-invariant Chernikov p-subgroup, B is $< g >$-invariant and B/A is an infinite elementary abelian p-group, p a prime. Then B contains a $< g >$-invariant infinite elementary abelian p-subgroup.*

Proposition 1.14 *Let G be a soluble group whose non-subnormal subgroups have finite rank. If G contains an abelian periodic subgroup of infinite rank, then G is a Baer group.*

Proof By Corollary 1.5 the Baer radical $B(G)$ contains every element of finite order. Since $B(G)$ is locally nilpotent ([8], Theorem 2.5.1) then the set T of all elements of $B(G)$ having finite order is a characteristic subgroup of $B(G)$. If for every $p \in \Pi(T)$ the Sylow p-subgroup T_p of T has finite rank then the set $\{r(T_p)|p \in \Pi(T)\}$ is not bounded. Lemma 1.2 yields that in this case $G = B(G)$. Now assume that there is a prime p such that T_p has infinite rank. Since G is soluble then T_p contains G-invariant subgroups $A \leq B$ such that A has finite rank (i.e A is a Chernikov subgroup) and B/A is an abelian subgroup of infinite rank. In particular, $\Omega_1(B/A) = C/A$ is an infinite elementary abelian p-group. Let $g \in G$. By Corollary 1.13 G contains a $< g >$-invariant infinite elementary abelian p-subgroup. By Lemmas 1.4, 1.9 $g \in B(G)$. □

If $g \in G$ and A is a $< g >$-invariant abelian torsion-free subgroup then we can consider A as a $\mathbf{Z} < g >$-module.

Lemma 1.15 *Let G be a group whose non-subnormal subgroups have finite rank, g an element of infinite order, A a $< g >$-invariant torsion-free abelian subgroup of G. Then A is a $\mathbf{Z} < g >$-periodic module.*

Proof Let $1 \neq c \in A$, $C =< c >^{<g>}$. Then C is a cyclic $\mathbf{Z} < g >$-module. Suppose that $Ann_{\mathbf{Z}<g>}(c) =< 0 >$, that is $C \cong_{\mathbf{Z}<g>} \mathbf{Z} < g >$. Now we can repeat the arguments of Lemma 1.7 and obtain that $g \in B(G)$. It follows that $A < g >$ is nilpotent ([2], Lemma 1.7). In particular, $< C, g >=< c, g >$ is a finitely generated nilpotent subgroup. In this case $r(C)$ is finite, which is a contradiction to the isomorphism $C \cong_{\mathbf{Z}<g>} \mathbf{Z} < g >$. This contradiction shows that $Ann_{\mathbf{Z}<g>}(c) \neq< 0 >$. □

If A is a subgroup of a group G, $g \in G$, $n \in \mathbf{N}$, then put

$$[A,_n g] = [\underbrace{\ldots [[[A, g], g], \ldots, g]}_{n}.$$

Lemma 1.16 *Let G be a group whose non-subnormal subgroups have finite rank, g an element of infinite order, A a $< g >$-invariant torsion-free abelian subgroup of G. If A has infinite rank, then $< A, g >$ is hypercentral.*

Proof Let $E = A \otimes_{\mathbf{Z}} \mathbf{Q}$, then we can consider E as $\mathbf{Q} < g >$-module. Let $1 \neq c \in A$, $C =< c >^{<g>} = c\mathbf{Z} < g >$, $D = c\mathbf{Q} < g >$. By Lemma 1.15 $Ann_{\mathbf{Z}<g>}(c) \neq < 0 >$, that is $Ann_{\mathbf{Q}<g>}(c) \neq < 0 >$. Since $D \cong \mathbf{Q} < g > /Ann_{\mathbf{Q}<g>}(c)$ then $dim_{\mathbf{Q}} D$ is finite. In other words, C has finite rank. By P. Hall's Theorem ([13], Corollary 1 of Lemma 9.53]) C contains a free abelian subgroup L such that C/L is periodic with finite set $\Pi(C/L)$. This means that C is a minimax subgroup. It follows that for every finite subset $\{a_1, a_2, ..., a_n\}$ of elements of A the subgroup $< a_1, a_2, ..., a_n >^{<g>}$ is a minimax.
Let $a_1 \in A$, $A_1 =< a_1 >^{<g>}$. In A/A_1 we choose the elements a_2A_1, a_3A_1 of infinite order such that $< a_2A_1, a_3A_1 >=< a_2A_1 > \times < a_3A_1 >$. Put $A_2 =< A_1, a_2, a_3 >^{<g>}$, then $r(A_2/A_1) \geq 2$. Using the same arguments, we can construct an ascending chain of $< g >$-invariant subgroups

$$< 1 >= A_0 \leq A_1 \leq A_2 \leq ... \leq A_n \leq ...$$

satisfying the following conditions:

(i) A_n is a finitely generated $\mathbf{Z} < g >$ - module;

(ii) $r(A_{n+1}/A_n) \geq n + 1, n \in \mathbf{N}$.

We have already noted that every subgroup A_n is minimax. By P. Hall's Theorem ([12], Theorem 5.34) every factor A_{n+1}/A_n satisfies Max-$< g >$. It follows that its periodic part is finite. Choose in A_1 a free abelian subgroup C_1 such that A_1/C_1 is periodic (and hence A_1/C_1 is a Chernikov group). Let $p_1 \notin \Pi(A_1/C_1)$, then $r(C_1/(C_1)^{p_1}) = r(C_1)$ and $C_1/(C_1)^{p_1}$, is a Sylow p_1-subgroup of $A_1/(C_1)^{p_1}$. It follows that $A_1/(A_1)^{p_1} \cong C_1/(C_1)^{p_1}$, in particular, $r(A_1/(A_1)^{p_1}) = r(C_1/(C_1)^{p_1}) = r(C_1) = r(A_1)$. Put $B_1 = (A_1)^{p_1}$ and now consider A_2/B_1. Since the periodic part of A_2/A_1 is finite then the Sylow p_1-subgroup S/B_1 is also finite. It follows that $A_2/B_1 = S/B_1 \times U/B_1$. Let $s = |S/B_1|$ then $(A_2/B_1)^s$ is $< g >$ - invariant and lies in U/B_1, that is $(A_2/B_1)^s \cap S/B_1 =< 1 >$. Put $V_1/B_1 = (A_2/B_1)^s$. Using the previous arguments we can choose a prime $p_2 \neq p_1$ such that $V/B_1 \neq (V/B_1)^{p_2}$ and $r((V/B_1)/(V/B_1)^{p_2}) = r(V/B_1) = r(A_2/A_1)$. Put $B_2/B_1 = (V/B_1)^{p_2}$, then B_2 is a $< g >$-invariant subgroup, $r(V/B_2) \geq 2$, $B_2 \cap A_1 = B_1$. Furthermore A_2/B_2 is a finite $\{p_1, p_2\}$-group and $r(A_2/B_2) \geq 2$. With the help of similar arguments we can construct an ascending series

$$< 1 >= B_0 \leq B_1 \leq B_2 \leq ... \leq B_n \leq ...$$

of $< g >$-invariant subgroups satisfying the following conditions:

(iii) $B_n \leq A_n, A_n/B_n$ is finite;

(iv) $B_{n+1} \cap A_n = B_n$;

(v) $r(A_n/B_n) \geq n, n \in \mathbf{N}$.

Put $E = \bigcup_{n \in \mathbf{N}} A_n, B = \bigcup_{n \in \mathbf{N}} B_n$. Then $E/B = \bigcup_{n \in \mathbf{N}} A_nB/B$ and $A_nB/B \cong A_n/(A_n \cap B)$. It is not hard to show that (iv) implies $A_n \cap B = B_n$. This means that A_nB/B is finite and $r(A_nB/B) \geq n$. Consequently E/B is a periodic group of infinite rank. Lemma 1.2 yields that $< gB >$ is subnormal. In this case $< gB > E/B$ is nilpotent ([2], Lemma 4). In other words, there is a number t such that $[A_{1,t}\, g] \leq B$. Since A_1 is $< g >$-invariant then $[A_{1,t}\, g] \leq B \cap A_1 = B_1$. Let $m = r(A_1/B_1)$. Since A_1/B_1 is an elementary abelian p_1-subgroup then $[A_{1,m}\, g] \leq B_1$. This is valid for every prime $q \notin \Pi(A_1/C_1) = \sigma$. Let $L_1 = \bigcap_{q\notin\sigma}(A_1)^q$. We have already noted that $(A_1)^q \cap C_1 = C_1^q$, thus $L_1 \cap C_1 = \bigcap_{q\notin\sigma} C_1^q =< 1 >$. Since A_1/C_1 is periodic, it follows that $L_1 =< 1 >$.

Hence $[A_{1,m}\, g] =< 1 >$, i.e $< A_1, g >$ is nilpotent. In particular, $C_A(g) \neq< 1 >$. Using a transfinite induction, we can obtain that $< A, g >$ is hypercentral. □

Lemma 1.17 *Let G be a group whose non-subnormal subgroups have finite rank, g an element of infinite order, A a $< g >$-invariant torsion-free abelian subgroup of G. If A has infinite rank, then $g \in B(G)$.*

Proof By the previous Lemma the subgroup $< A, g >$ is hypercentral. Let

$$< 1 >= Z_0 \leq Z_1 \leq \ldots \leq Z_\omega \leq \ldots \leq Z_\gamma = A$$

be the upper $< g >$-central series of A. It is easy to prove that every subgroup Z_α is pure in A. If we assume that γ is finite, then $< A, g >$ is nilpotent. Every subgroup of a nilpotent group is subnormal. Since $< A, g >$ has infinite rank, it is subnormal and therefore $< g >$ is also subnormal. Assume now that γ is infinite. Put $Z = Z_\omega$. If $r(Z_n)$ is infinite for some $n \in \mathbf{N}$, then $< g > Z_n$ is subnormal. Since $< g > Z_n$ is nilpotent, $< g >$ is also subnormal. Therefore we may assume that $r(Z_n)$ is finite for every $n \in \mathbf{N}$. Choose in Z_2 a finitely generated subgroup K_2 such that Z_2/K_2 is periodic. Put $C_2 = (K_2)^{<g>}$. Since $< A, g >$ is hypercentral, $< K_2, g >$ is finitely generated, thus C_2 is finitely generated. If we assume that $[C_2, g] =< 1 >$ then $[Z_2, g] =< 1 >$, which is impossible. Thus $[C_2, g] \neq< 1 >$. Let $\pi_2 = \{p \mid p$ prime number such that $[C_2, g] \leq C_2^p\}$. If we suppose that π_2 is infinite, then $\bigcap_{p\in\pi_2} C_2^p =< 1 >$. But in this case $[C_2, g] \leq \bigcap_{p\in\pi_2} C_2^p =< 1 >$, and again we obtain a contradiction. Hence π_2 is finite. It follows that there is a prime p_1 such that $(C_2)^{p_1} = E_2$ does not contain $[C_2, g]$. In Z_5/Z_2 choose a maximal $\mathbf{Z}$-independent set of elements $\{z_1Z_2, \ldots, z_mZ_2\}$ and put $B_2/C_2 =< z_1C_2, \ldots, z_mC_2 >^{<g>}$. Since $< A, g >$ is hypercentral then B_2/C_2 is finitely generated. Hence there is a number k such that $C_3/C_2 = (B_2/C_2)^k$ is torsion free. Since Z_5/Z_2 is a pure envelope of C_3Z_2/Z_2 then $[C_3, g, g, g] \leq Z_2 \cap C_3 = C_2$, but $[C_3, g, g] \not\leq C_2$. Let $\pi_3 = \{p \mid p$ prime number such that $[C_3/C_2, g, g] \leq (C_3/C_2)^p\}$. Again the set π_3 is finite, therefore we may choose a prime $p_2 \notin \pi_2 \cup \{p_1\}$. Let $(C_3/C_2)^{p_2} = D/C_2$. Then C_2/E_2 is the periodic part of D/E_2, so that $D/E_2 = C_2/E_2 \times Q/E_2$. It follows that $E_3/E_2 = (D/E_2)^{p_1} \leq Q/E_2$, in particular, $E_3 \cap C_2 = E_2$. Moreover, E_3 is $< g >$-invariant which does not contain $[C_3, g, g]$ and $r(C_3/E_3) \geq 3$. With the help of similar arguments we can construct two ascending series

$$C_2 \leq C_3 \leq \ldots \leq C_n \leq \ldots \quad \text{and} \quad E_2 \leq E_3 \leq \ldots \leq E_n \leq \ldots$$

of $< g >$-invariant finitely generated subgroups and a set $\{p_n| \; n \in \mathbf{N}\}$ of primes satisfying the following conditions:

(i) $E_n \leq C_n, C_n/E_n$ is a finite $\{p_1, p_2, \ldots, p_n\}$-group;
(ii) $r(C_n/E_n) \geq n$;
(iii) E_n does not contain $[C_n,_n g]$;
(iv) $E_{n+1} \cap C_n = E_n, n \in \mathbf{N}$

Put $E = \bigcup_{n\in\mathbf{N}} E_n, C = \bigcup_{n\in\mathbf{N}} C_n$. Then $C/E = \bigcup_{n\in\mathbf{N}}(C_n)E/E$. By (iv) $C_n \cap E = E_n$, so that $(C_n)E/E \cong C_n/(C_n \cap E) = C_n/E_n$. It follows by (ii) that C/E is a periodic group of infinite rank and $\Pi(C/E)$ is infinite. Lemma 1.2 implies that $< gE >$ is subnormal in $< gG, C/E >$. Since C/E is abelian and $< g >$-invariant then $< C/E, gE >$ is nilpotent ([2], Lemma 1.7). However condition (iii) shows that $< C/E, gE >$ can not be nilpotent. This contradiction proved that γ is finite, and we have already noted that in this case $g \in B(G)$. □

Lemma 1.18 *Let G be a group whose non-subnormal subgroups have finite rank, g an element of finite order, A a $< g >$ - invariant torsion-free abelian subgroup of G. If A has infinite rank, then $g \in B(G)$.*

Proof Let $1 \neq a_1 \in A, A_1 =< a_1 >^{<g>}$. Since $|g|$ is finite then A_1 is finitely generated. Let B_1 be a subgroup of A maximal with respect to $A_1 \cap B_1 =< 1 >$. Then A/B_1 is a torsion - free group of finite rank. The set $\{B_1^x | x \in< g >\}$ is finite. Let $\{B_1^x | x \in< g >\} = \{B_1^{x_1}, \ldots, B_1^{x_m}\}$. Put $C_1 = \bigcap_{x\in<g>} B_1^x$, then $A/C_1 \leq A/B_1^{x_1} \times \cdots \times A/B_1^{x_m}$. In particular, C_1 is a $< g >$-invariant subgroup of A with the properties: $C_1 \cap A_1 =< 1 >$, A/C_1 is a torsion-free group of finite rank. So $C_1 \neq< 1 >$. Now we can repeat the arguments of Lemma 1.4 and obtain that $g \in B(G)$. □

Lemma 1.19 *Let G be a group whose non-subnormal subgroups have finite rank, $g \in G$, and let A, B be $< g >$-invariant subgroups of $B(G)$ satisfying the following conditions: $A \leq \zeta(B), A$ has finite rank, $[B, g] \leq A$ and B/A is an abelian group of infinite rank. Then $C_G(g)$ contains an abelian subgroup of infinite rank.*

Proof Consider the mapping $\phi_g : b \longrightarrow [b, g], b \in B$. We have

$$(bc)\phi_g = [bc, g] = [b, g]^c[c, g] = [b, g][c, g] = b\phi_g c\phi_g$$

because $A \leq \zeta(B)$. Furthermore $Ker\phi_g = C_B(g)$ and $Im\phi_g \leq A$, so that $B/C_B(g)$ has finite rank. This means that $C_B(g)$ has infinite rank. Since $C_B(g)$ is nilpotent then it contains an abelian subgroup of infinite rank ([13], Corollary 2 of Theorem 6.36). □

Corollary 1.20 *Let G be a group whose non-subnormal subgroup have finite rank, $g \in G$, and let A, B be $< g >$-invariant subgroups of $B(G)$ satisfying the following conditions: $A \leq \zeta(B), A$ has finite rank, B/A is an abelian group of infinite rank. Then $C_G(g)$ contains an abelian subgroup of infinite rank.*

Proof By Lemmas 1.17, 1.18 $< gA >$ is subnormal in $< gA, B/A >$. Then $< g, B > /A$ is nilpotent ([2], Lemma 4), that is B/A has a finite $< g >$-central series. Now we can use Lemma 1.19. □

Corollary 1.21 *Let G be a group whose non-subnormal subgroups have finite rank, $g \in G$, and let A, B be $< g >$-invariant subgroups of $B(G)$ satisfying the following conditions: A has finite rank, B/A is an abelian group of infinite rank, and B is nilpotent. Then $C_G(g)$ contains an abelian subgroup of infinite rank.*

Theorem 1.22 *Let G be a group whose non-subnormal subgroups have finite rank. If G is soluble then either G is a Baer group or G is a group of finite rank.*

Proof Suppose that G has infinite rank. By Kargapolov's Theorem [3] G contains an abelian subgroup of infinite rank. Let $B = B(G), T$ be the periodic part of B. If G contains an abelian periodic subgroup of infinite rank then $G = B(G)$ by Proposition 1.14. Therefore suppose that every periodic subgroup of G has finite rank, i.e. T is a subgroup of finite rank. Since B/T is locally nilpotent torsion-free and soluble, it has a finite series $T = B_0 \leq B_1 \leq \ldots \leq B_n = B$ of G-invariant subgroups with torsion-free abelian factors. By Lemma 1.2 every abelian subgroup of infinite rank lies in $B(G)$. So there is a number m such that B_m has finite rank but B_{m+1}/B_m has infinite rank. Then the nth term of the upper cent ral series of B/T, for some finite n contains B_m/T ([13], Lemma 6.37). In particular, B_{m+1}/T is nilpotent. Let $g \in G$. By Corollary 1.21 there is an abelian torsion-free subgroup U/T of infinite rank such that $[U/T, gT] =< 1 >$. Lemmas 1.17, 1.18 yield that $gT \in B(G/T)$, i.e G/T is Baer group. Let V/T be a free abelian subgroup of U/T having countable rank. Since G is soluble, T has a finite series of G - invariant subgroups $< 1 >= T_0 \leq T_1 \leq \ldots \leq T_s = T$ with abelian factors. From the proof of Theorem 5 of paper [9] follows that V/T_{s-1} contains a free abelian subgroup V_1/T_{s-1} of countable rank such that V/V_1T is periodic. Since V_1/T_{s-1} is subnormal then $< V_1/T_{s-1}, T/T_{s-1} >$ is nilpotent ([2], Lemma 4). By Corollary 1.20 there is a free abelian subgroup V_2/T_{s-1} of countable rank such that $[V_2/T_{s-1}, gT_{s-1}] =< 1 >$. Using the same arguments after finitely many steps we obtain that $C_G(g)$ contains a free abelian subgroup of infinite rank. Lemmas 1.17, 1.18 yield that $g \in B(G)$. □

We remark that there are Baer groups of finite rank in which not all subgroups are subnormal. Thus Baer groups, in which the non-subnormal subgroups have finite rank, can have non-subnormal subgroups, and their study requires further consideration.

Acknowledgements. We would like to thank the referee for his helpful comments and suggestions.

References

[1] S. Franciosi, F. de Giovanni, *Groups satisfying the minimal condition on non-subnormal subgroups* "Infinite groups 94", Proc. Interm. Conf. Ravello. Walter de Gruyter: 1996, 63-72.

[2] H. Heineken,L.A. Kurdachenko, *Groups with subnormality for all subgroups that are not finitely generated*, Annali Mat., **169**(1995), 203-232.

[3] M.I. Kargapolov, *On soluble groups of finite rank*, Algebra i logika., **1**(1962), no 5, 37-44.

[4] L.A. Kurdachenko, H. Smith, *Groups with the maximal condition on non-subnormal subgroups*, Bolletino Unione Mat. Ital., **10B**(1996), 441-460.

[5] L.A. Kurdachenko, H. Smith, *Groups with the weak minimal condition for non-subnormal subgroups*, Annali Mat., **173**(1997), 299-312.

[6] L.A. Kurdachenko, H. Smith, *Groups with the weak maximal condition for non-subnormal subgroups*, Ricerche Mat., **47**(1998), 29-49.

[7] L.A. Kurdachenko, H. Smith, *The nilpotency of some groups with all subgroups subnormal*, Publicationes Mat., **42**(1998), 411-421.

[8] J.C. Lennox, S.E. Stonehewer, *Subnormal subgroups of groups*, Clarendon Press: Oxford - 1987.

[9] A.I. Mal'tsev, *On certain classes of infinite soluble groups*, Mat. Sbornik, **28**(1951), 567-588; translated in Amer. Math. Soc. Translations, **2**(1956), 1-21.

[10] F. Menegazzo, *Groups of Heineken-Mohamed*, J. Algebra, **171**(1995),807-825.

[11] W. Möhres, *Auflösbarkeit von Gruppen, deren Untergruppen all subnormal sind*, Archiv Math., **54**(1990), 232-235.

[12] D.J.S. Robinson, *Finiteness conditions and generalized soluble groups*, Part 1. Springer Berlin - 1972.

[13] D.J.S. Robinson, *Finiteness conditions and generalized soluble groups*, Part 2. Springer Berlin - 1972.

[14] V.P. Shunkov, *On locally finite groups of finite rank*, Algebra i logika, **10**(1971), no 2, 199-225.

[15] D.I. Zaitsev, *On locally soluble groups of finite rank*, Doklady AN SSSR, **260**(1978), no 2, 257-260.

ON SOME INFINITE DIMENSIONAL LINEAR GROUPS

LEONID KURDACHENKO* and IGOR SUBBOTIN†

*Department of Mathematics, Dnepropetrovsk University, Vyl. Naukova 13, 49050 Dnepropetrovsk, Ukraine
†Mathematics Department, National University, 9920 S. La Cienega Blvd, Inglewood, CA 90301, USA

Abstract

This is a survey indicating the newest results on infinite-dimensional linear groups.

Let F be a field, A - a vector space over F. The group $\mathrm{GL}(F, A)$ of all automorphisms of A and its distinct subgroups (the linear groups) are the oldest subjects of investigation in Group Theory. Naturally, the first step here was the investigation of the case, when A has finite dimension over F. Under this condition every element of $\mathrm{GL}(F, A)$ (a non-singular linear transformation) defines some non-singular $n \times n$ - matrix over F where $n = \dim_F A$. Thus, for the finite-dimensional case the theory of linear groups coincides with the theory of matrix groups. That is why the theory of finite-dimensional linear groups is one of the best-developed theories in algebra.

The case when $\dim_F A$ is infinite is totally different. The study of this case always requires some additional essential restrictions. The situation here is similar to that one which appeared in the initial period of development of Infinite Group Theory. One of the very fruitful approaches is associated with the finiteness conditions. It seems very meaningful to apply it to the linear groups.

Take into consideration the following important aspect. Many problems related to groups with finiteness conditions require the study of modules over some group rings, frequently the group rings of the type FG where F is a field. Thus, the respectively finiteness conditions from groups could be transferred on the appropriate modules.

The investigation of modules always has the following two sides. The first one is the study of the internal structure of a module A; the second - the study of the structure of the group $G/C_G(A)$. The second side is especially important for many group-theoretical applications. It is crucial to admit that the group $G/C_G(A)$ is imbedded in $\mathrm{GL}(F, A)$.

A group $G \leq GL(F, A)$ is called a finitary linear group, if for each element $g \in G$ the factor-space $A/C_A(g)$ has finite dimension.

The investigation of such groups proves great effectiveness of applications of the finiteness conditions to the study of linear groups. We can consider these groups as linear analogies of FC-groups (the groups with finite conjugacy classes). The theory of finitary linear groups develops rather intensively and became rich with many interesting results (see, for example, a survey [10]). However, we would like to focus on some other types of infinite-dimensional linear groups.

The irreducible groups form the first important type of linear groups.

A group $G \leq GL(F, A)$ is called irreducible if every G - invariant subspace of A either coincides with $\langle 0 \rangle$ or with A.

In other words, the FG-module A is simple. Irreducible linear groups also play an essential role in the general group theory, where they arise as automorphism groups of abelian chief factors.

The finite-dimensional irreducible groups were studied rigorously from various points of view. Therefore we shall not consider problems of their structure, as they are sufficiently illuminated in the monographs [13, 14, 9]. One of the important research tools in this case is the classical Clifford's theorem. For the infinite-dimensional case its analogy is not always valid. Nevertheless, the following its weak form has been proved.

Theorem 1 *[2, 15]. Let F be a field, A a vector space over F, G an irreducible subgroup of $GL(F, A)$. If H is a normal subgroup of G having finite index, then A includes a simple FH-submodule B and $A = \oplus_{x \in S} Bx$ for some finite subset $S \subseteq G$.*

One of the first important results on irreducible groups is the following statement implied by the classical Schur's lemma.

Proposition 1 *[12]. Let F be a field, G a group, A a simple FG-module, $I = Ann_{FG}(A)$.*

(i) The center C/I of the factor-ring FG/I is an integral domain.

(ii) $\zeta(G/C_G(A))$ is embedded in a multiplicative group of some field.

(iii) The periodic part of $\zeta(G/C_G(A))$ is a locally cyclic p' - subgroup where $p =$ char F.

As usual, we denote the set of all primes by $0'$.

The structure of an irreducible group depends on the following its numerical characteristic.

A group G is said to have 0-rank $r_0(G) = r$ if it has a finite subnormal series with exactly r infinite cyclic factors, being the others periodic.

Note that every refinement of one of these series has only r factors which are infinite cyclic; any two finite subnormal series have isomorphic refinements. This allows us to claim that 0-rank of a group is independent of the series. This numerical invariant is also known as the *torsion-free rank of* G. For polycyclic-by-finite groups the 0-rank is exactly its *Hirsch number*.

Proposition 2 *[5]. Let F be a field, A be a vector space over F and G be a locally (polycyclic-by-finite) subgroup of $GL(F, A)$ of finite 0-rank. Suppose that* *the FG - module A is finitely generated and has only finitely many maximal FG-submodules. For each element $1 \neq z \in \zeta(G)$ there is a polynomial $f_z \in F[X]$ such that $Af_z(z) = \langle 0 \rangle$.*

Corollary *[5]. Let F be a field, A a vector space over F, G a locally (polycyclic-by-finite) subgroup of $GL(F, A)$ of finite 0-rank. Suppose that the FG-module A is finitely generated and has only finitely many maximal FG-submodules. Then the center $\zeta(G)$ is periodic. In particular, if G is irreducible, then $\zeta(G)$ is a locally cyclic p' - subgroup where $p =$ char F.*

The description of irreducible abelian groups is based on the following construction.

Let G be a group, F a field, $\bar{F}$ an algebraic field closure of $F, \theta : G \longrightarrow U(\bar{F})$ a group homomorphism, $C_\theta = F[G\theta]$. Since all elements of $G\theta$ are algebraic over F, C_θ is a subfield of $\bar{F}$. We can consider C_θ as a FG - module if we put $c\,g = c(g\theta), c \in C_\theta$. Also we shall designate this module by D_θ, in order to distinguish it from the subfield C_θ. It is easy to see that D_θ is a simple FG - module and $C_G(D_\theta) =$ Ker θ.

Theorem 2 *Let F be a field, A a vector space over F, G an abelian subgroup of $GL(F, A)$ having finite 0-rank. If G is irreducible, then the following assertion hold:*

(i) there is a group homomorphism $\theta : G \longrightarrow U(\bar{F})$ where $\bar{F}$ is an algebraic field closure of F such that an FG - module A is isomorphic with D_θ;

(ii) let $\theta : G \longrightarrow U(\bar{F})$ and $\phi\theta : G \longrightarrow U(\bar{F})$ are group homomorphisms; then $D_\theta \cong_{FG} D_\phi$ if and only if there is an F - isomorphism $\rho : C_\phi \longrightarrow C_\theta$ such that $\theta = \phi\rho$;

(iii) the periodic part T of the group G is a locally cyclic p'-subgroup where $p =$ char F, and if the field F is locally finite, then $G = T$.

For the case of a Chernikov group G this theorem has been proved in the paper [3]. However, basing on the mentioned above results, it is easy to extend the proof on the considered case. Note that the above construction shows that the conditions of Theorem 2 are also sufficient.

Theorem 3 *[12]. Let F be a field, A a vector space over F, G an abelian group of infinite 0-rank, T the periodic part of G. Then G is isomorphic with an irreducible subgroup of $GL(F, A)$ if and only if T is a locally cyclic p'-subgroup where $p =$ char F.*

As the following results of [6] show, the description of irreducible abelian groups mentioned above can be extended, practically without serious changes, on the case of the irreducible hypercentral groups.

Theorem 4 *[6]. Let F be a field, $p =$ char F, G a hypercentral group of finite 0-rank, $C = \zeta(G), T$ the periodic part of C.*

(i) If G is an irreducible subgroup of $GL(F, A)$, then T is a locally cyclic p'-group. Furthermore, if the field F is locally finite, then $C = T$.

(ii) Conversely, if the field F is locally finite, $C = T$ is a locally cyclic p'-group, then there exists a simple FG-module A such that $C_G(A) = \langle 1 \rangle$ (i.e. G is an irreducible subgroup of $GL(F, A)$).

(iii) If F is not locally finite and T is a locally cyclic p' - group, then there exists a simple FG-module A such that $C_G(A) = \langle 1 \rangle$ (i.e. G is an irreducible subgroup of $GL(F, A)$).

Theorem 5 *[6]. Let F be a field, A a vector space over F, G a hypercentral group of infinite 0-rank, $C = \zeta(G), T$ a periodic part of C. Then G is isomorphic with an irreducible subgroup of $GL(F, A)$ if and only if T is a locally cyclic p' - subgroup where $p =$ char F.*

The classes of irreducible groups, which we shall describe, are extensions of both the class of finite groups and the class of abelian groups. One of the effective ways for obtaining such generalizations is the following general method.

Let X be a class of groups. Then X is called a formation if it satisfies the following conditions:

(F 1) if $G \in X, H$ a normal subgroup of G, then $G/H \in X$;

(F 2) if G is a group, H_1, H_2 are its normal subgroups such that $G/H_1, G/H_2 \in X$, then $G/(H_1 \cap H_2) \in X$.

Let G be a group, $x \in G, x^G = \{x^g = g^{-1}xg | g \in G\}$. Clearly, $C_G(x^G)$ is a normal subgroup of G. Put $XC(G) = \{x \in G | G/C_G(x^G) \in X\}$. Since X is a formation, $XC(G)$ is a characteristic subgroup of G. This subgroup is called the *XC - center* of the group G. A group G is said to be an *XC - group or a group with X - conjugacy classes if $XC(G) = G$.*

If $X = I$ is the class of all identity groups, then the class of IC-groups coincides with the class A of all abelian groups.

If $X = F$ is the class of all finite groups, then we obtain the famous class of FC-groups.

And, finally, if $X = C$ is the class of all Chernikov groups, then we obtain the class of CC-groups, which have been introduced by Ya. D. Polovicky [11].

Our nearest aims are irreducible FC - and CC-groups.

Let G be a group. The subgroup $Soc(G)$ generated by all minimal normal subgroups of G is called *the socle* of G. If G has no minimal normal subgroups, then put $Soc(G) = \langle 1 \rangle$. Obviously, there is a family $\{M_\lambda | \lambda \in \Lambda\}$ of minimal normal subgroups of G such that $Soc(G) = X_{\lambda \in \Lambda} M_\lambda$. Put $\Lambda(ab) = \{\lambda \in \Lambda | M_\lambda$ is abelian$\}$ and $Soc_{ab}(G) = X_{\lambda \in \Lambda(ab)} M_\lambda$.

Let G be a CC-group, then it includes a normal subgroup S satisfying the following conditions:

(QS 1) $S = X_{\lambda \in \Lambda} S_\lambda$ where either S_λ is a minimal normal subgroup of G or S_λ is a normal infinite cyclic subgroup;

(QS 2) if H is a non-identity normal subgroup of G, then $H \cap S \neq \langle 1 \rangle$.

This subgroup S is said to be the *quasi-socle* of a CC - group G and denoted by $Qsoc(G)$. Note that $Soc(G) \leq Qsoc(G)$.

Theorem 6 *[4]. Let F be a field, A a vector space over F, G a CC-group of finite 0-rank.*

(i) Let G be an irreducible subgroup of $GL(F, A)$. Then $Socab(G)$ includes a subgroup H such that $Core_G(H) = \langle 1 \rangle$ and $Soc_{ab}(G)/H$ is a locally cyclic p'-group where $p =$ char F. Furthermore, if the field F is locally finite, then every quasi-socle of G coincides with its socle.

(ii) Conversely, let $Soc_{ab}(G)$ includes a subgroupH such that $Core_G(H) = \langle 1 \rangle$ and $Soc_{ab}(G)/H$ is a locally cyclic p'-group where $p =$ char F. Besides, if the field F is locally finite, then every quasi-socle of G is the socle. Then there exists a simple FG-module A such that $C_G(A) = \langle 1 \rangle$ (i.e. G is an irreducible subgroup of $GL(F, A)$).

Theorem 7 *[4]. Let F be a field, A a vector space over F, G a CC-group of infinite 0-rank. Then G is isomorphic with an irreducible subgroup of $GL(F, A)$ if and only if $Soc_{ab}(G)$ includes a subgroup H such that $Core_G(H) = \langle 1 \rangle$ and $Soc_{ab}(G)/H$ is a locally cyclic p' - group where $p =$ char F.*

Note that a special case (concerns with locally soluble irreducible linear FC-groups) of Theorems 6 and 7 has been obtained in [1].

Now we shall consider the linear groups, which are in some sense near to irreducible.

If $G \leq \mathrm{GL}(A)$, then we can consider A as a FG-module. If G is irreducible, then every proper FG-submodule of A is zero, in particular, it is finite-dimensional. We will consider the subgroups $G \leq \mathrm{GL}(F, A)$ with the following property:

every proper FG-submodule of A is finite-dimensional.

The following two cases appear here:

(i) A includes an FG-submodule B such that $\dim_F B$ is finite and A/B is a simple FG-module of infinite dimension;

(ii) every proper FG-submodule of A has finite dimension and A is the union of its finite-dimensional FG-submodules.

The first case is reduced to the cases of a finite-dimensional linear group and an irreducible group. Moreover, for some types of groups this case coincides with the irreducible case. For example, let G be a hypercentral group. If $C_G(B) = \langle 1 \rangle$, then $G \leq \mathrm{GL}_n(F)$ where $n = \dim_F(B)$. Remind that the finite-dimensional hypercentral groups have been already described (see, for example, [14], Chapter 8). Therefore we can assume that $C_G(B) = L \neq \langle 1 \rangle$. Let $1 \neq x \in L \cap \zeta(G)$; then the mapping $\phi : a \longrightarrow a(x-1)$, $a \in A$, is an FG - endomorphism. In particular, Kerϕ and $Im\phi$ are FG-submodules of A. By the selection of x, $B \leq$ Ker ϕ. Since Ker$\phi \neq A$, Ker$\phi = B$, thus $Im\phi \cong A/B$. In other words, $Im\phi$ is a simple FG - module of infinite dimension. This means that A is a simple FG-module.

This example shows that the second case promises to be much more interesting.

Let B be a proper FG-submodule of A. If $C_G(B) = \langle 1 \rangle$, then we can treat G as a subgroup of $\mathrm{GL}_n(F)$ where $n = \dim_F B$. We make the following definition.

Let F be a field, A a vector space over F, $G \leq GL(F, A)$. A group G is called a quasi-irreducible if A satisfies the following conditions:

(QI 1) if B is a proper FG-submodule of A, then $\dim_F B$ *is finite;*

(QI 2) A is the union of its proper FG-submodules.

(QI 3) $C_G(B) \neq \langle 1 \rangle$ for every proper FG - submodule B of A.

We will say also that in this case A is a quasifinite FG - module.

Clearly every quasifinite module is artinian.

The first natural step here is the description of hypercentral quasi-irreducible linear groups and some groups that are near to them. These groups have been considered in the paper [7]. We shall start with the main results of this paper.

Let F be a field, A a vector space over F of countable dimension, $\{a_n | n \in N\}$ a basis of A, $\langle x \rangle$ an infinite cyclic group. Define the action of x on A by the rule:

$$a_1x = a_1, a_{n+1}x = a_{n+1} + a_n \text{ or } a_1(x-1) = 0, a_{n+1}(x-1) = a_n \text{ , } n \in N.$$

Then we can consider A as $F\langle x \rangle$-module. It is easy to see that A is $F\langle x \rangle$ - hypercentral, $A = A(x-1)$ and every proper $F\langle x \rangle$-submodule of A coincides with some $a_1F + \ldots + a_nF$, $n \in N$. In particular, the $F\langle x \rangle$-module A is quasifinite.

We will called this module a *Prüfer* $(x-1)$ - *module* (by analogy with a Prüfer p-group).

Theorem 8 *[7]. Let F be a field, A a vector space over F, G a hypercentral subgroup of $GL(F, A)$. Suppose that G is a quasi-irreducible group. Then the following assertions hold:*

(i) G is abelian-by-finite;

(ii) the periodic part T of the group G is a p' - subgroup of finite (special) rank where $p =$ char F;

(iii) $T \cap \zeta(G)$ is a locally cyclic subgroup.

(iv) $\zeta(G)$ contains an element x of infinite order such that $A = C_1 \oplus \ldots \oplus C_n$ where C_i is a Prüfer $(x-1)$ - module, $1 \leq i \leq n$.

We can treat the FC - center as a generalization of the center of a group. Starting from the FC - center we can construct the *upper FC- central series of G*:

$$\langle 1 \rangle = G_0 \leq G_1 \leq \ldots G_\alpha \leq G_{\alpha+1} \leq \ldots G_\gamma$$

by the rule $G_1 = FC(G)$, $G_{\alpha+1}/G_\alpha = FC(G/G_\alpha), \alpha < \gamma, FC(G/G_\gamma) = \langle 1 \rangle$.

A group G is said to be FC - hypercentral if $G_\gamma = G$.

This definition implies that the class of FC - hypercentral groups is an extremely wide extension of the class of hypercentral groups.

Theorem 9 *[7]. Let F be a field, A a vector space over F, G an FC - hypercentral subgroup of $GL(F, A)$. Suppose that G is a quasi-irreducible group. Then the following assertions hold:*

(i) G includes a normal abelian subgroup U of finite index;

(ii) the periodic part of U is a subgroup of finite (special) rank where $p =$ char F;

(iii) if char $F = p > 0$, then $O_p(G) = \langle 1 \rangle$.

(iv) *U contains an element x of infinite order such that $A = C_1 \oplus \ldots \oplus C_n$ where C_i is a Prüfer $(x-1)$-module, $1 \leq i \leq n$.*

Now consider the dual situation. If G is an irreducible subgroup of $\mathrm{GL}(F, A)$, then every non-zero FG - submodule of A coincides with A, in particular, it has finite codimension. We consider the subgroups $G \leq \mathrm{GL}(F, A)$ with the following property: *every non-zero FG-submodule of A has finite codimension.*

The following two cases appear here:

(i) *A includes a simple FG-submodule B such that $\dim_F(A/B)$ is finite;*

(ii) *every non-zero FG-submodule of A has finite codimension and the intersection of all these submodules is zero.*

The first case is reduced to the cases of a finite-dimensional linear group and an irreducible group. Moreover, for some types of groups (as the quasiirreducible groups) the first case coincides with the irreducible case.

Let F be a field, G a group. An FG-module A is said to be a just infinite-dimensional, if it satisfies the following conditions:

(JI 1) *if B is a non-zero FG-submodule of A, then $\dim_F(A/B)$ is finite;*

(JI 2) *the intersection of all non-zero submodules of A is zero.*

In particular, every just infinite-dimensional FG-module A includes a non-zero proper submodule B. If $C_G(A/B) = \langle 1 \rangle$, then we can treat A as a subgroup of $\mathrm{GL}_n(F)$ where $n = \dim_F(A/B)$. We make the following natural definition.

Let F be a field, A a vector space over F. A group $G \leq GL(F, A)$ is called almost irreducible if A satisfies the following conditions:

(AI 1) *an FG-module A is just infinite-dimensional;*

(AI 2) *if B is a non-zero FG-submodule of A, then $C_G(A/B) \neq \langle 1 \rangle$.*

Theorem 10 *[8]. Let F be a field, A a vector space over F, G a hypercentral subgroup of $GL(F, A)$. If G is almost irreducible, then the following assertions hold:*

(i) *G is abelian-by-finite;*

(ii) *the periodic part T of the group G is a p'-subgroup of finite (special) rank where $p =$ char F;*

(iii) *$T \cap \zeta(G)$ is a locally cyclic subgroup.*

Theorem 11 *[8]. Let F be a field, A a vector space over F, G a locally soluble FC - hypercentral subgroup of $GL(F, A)$. If G is a almost irreducible then*

(i) *G includes a normal abelian subgroup U of finite index;*

(ii) *the periodic part of U is a subgroup of finite (special) rank;*

(iii) *if char $F = p > 0$, then $O_p(G) = \langle 1 \rangle$.*

The following two statements have played an essential role in the proofs of these two theorems above.

Proposition 3 *[8]. Let F be a field, G a group, H a normal subgroup having finite index in G, X a transversal to H in G, A a just infinite-dimensional FG-module, $C_G(A) = \langle 1 \rangle$. Then*

(i) A includes an FH-submodule B, such that A/Bx is a just infinite-dimensional FH-module for every $x \in X$;

(ii) $\cap_{x \in X} Bx = \langle 0 \rangle$, thus A is isomorphic with some FH-submodule of $\oplus_{x \in X}(A/Bx)$;

(iii) $\cap_{x \in X} x^{-1} C_H(A/B)x = \langle 1 \rangle$ thus H is isomorphic with some subgroup of $X_{x \in X}(H/x^{-1}C_H(A/B)x)$.

Proposition 4 *[8]. Let F be a field, G a group, A a just infinite-dimensional FG-module, $C_G(A) = \langle 1 \rangle, x$ an element of infinite order of $\zeta(G)$ such that A is $F\langle x \rangle$ - torsion-free as an $F\langle x \rangle$-module. Then F is imbedded into the field K and the FG-module A is imbedded into a KG-module R with the following properties:*

(i) $\dim_K R$ *is finite;*

(ii) R is a simple KG-module;

(iii) $C_G(R) = \langle 1 \rangle$.

References

[1] S. Franciosi, F. de Giovanni, and L.A. Kurdachenko, *Groups whose proper quotients are FC-groups*, Journal Algebra **186** (1996), 544 - 577.

[2] B. Hartley, *A class of modules over locally finite groups II*, Journal Austral. Math. Soc., series A 19 (4) (1975), 437 - 469.

[3] B. Hartley, and D. McDougall, *Injective modules and soluble groups satisfying the minimal conditions for normal subgroups*, Bull. Austral.Math. Soc. 4(1) (1971), 113 - 135.

[4] L.A. Kurdachenko and J. Otal, *Simple modules over CC-groups and monolithic just-non-CC-groups*, Bolletino Mat. Unione Italiana, to appear.

[5] L.A. Kurdachenko, B.V. Petrenko and I.Ya Subbotin, *On generalized hypercenters in artinian modules*, Comm. Algebra **25** (4) (1997), 1023 - 1046.

[6] L.A. Kurdachenko and I.Ya Subbotin, *Groups whose proper quotients are hypercentral*, Journal Austral. Math. Soc., series A 65 (1998), 224 - 237.

[7] L.A. Kurdachenko and I.Ya Subbotin, *On some infinite-dimensional linear groups I*, Comm. Algebra, to appear.

[8] L.A. Kurdachenko and I.Ya Subbotin, *On some infinite-dimensional linear groups II*, to appear

[9] Yu.I. Merzlyakov, Rational groups. Nayka: Moskow, 1980.

[10] R. Phillips, *Finitary linear groups: a survey, Finite and locally finite groups.* NATO ASI ser. C 471, Kluver: Dordrecht, 1995, 111 - 146.

[11] Ya.D. Polovicky, *Groups with extremal classes of conjugate elements.* Sibir. Math. J. **5** (1964), 891 - 895.

[12] D.J.S. Robinson and Z. Zhang, *Groups whose proper quotients have finite derived groups*, J. Algebra **118** (1988), 346 - 368.

[13] D.A. Suprunenko, The Matrix Groups. Nauka: Moskow, 1972.

[14] B.A.F. Werhfritz, Infinite Linear Groups. Springer: Berlin,1973.

[15] D. I. Zaitcev, *On the existence of direct complements in groups with operators*, Investigations in Group Theory, Math. Inst.: Kyiv, 1976, 26 - 44.

GROUPS AND SEMISYMMETRIC GRAPHS

SEYMOUR LIPSCHUTZ[1] and MING-YAO XU[2 1]

[1] Department of Mathematics, Temple University, Philadelphia, PA 19122-6094, USA
[2] Institute of Mathematics, Peking University, Beijing 100871, P. R. China China

Abstract

A simple undirected graph is said to be *semisymmetric* if it is regular and edge-transitive but not vertex-transitive. This paper uses finite groups to construct many infinite families of such graphs.

1 Introduction

There is an intimate relationship between groups and graphs. For example, any graph X gives rise to its automorphism group $A := \mathrm{Aut}(X)$. On the other hand, any group G with generating set S gives rise to its Cayley graph $\mathrm{Cay}(G, S)$. The main purpose of this paper is to show how groups can be used to construct examples of semisymmetric graphs, graphs which are regular and edge-transitive but not vertex-transitive. [Relevent definitions are given in Section 2.]

These semisymmetric graphs were first systematically studied in 1967 by J. Folkman [6]. For later works on semisymmetric graphs, the reader is referred to [1, 2, 3, 4, 5, 7, 8, 9].

Since the known semisymmetric graphs are not many and since they have very special symmetry properties, there is a common belief that semisymmetric graphs are rare in number. We do not share this belief since our contruction leads to infinite classes of such graphs.

This paper is organized as follows. First we give the necessary notation, definitions, and concepts needed for this paper, including that of semisymmetric graphs. In particular, we introduce the notion of co-neighbor blocks and noncontractable graphs, and we also give the definition of a bi-lexicographic product of a graph. Next we present a known construction of a graph from a group G and subgroups L and R with an example which illustrates our main result. Finally we give two types of group constructions of semisymmetric graphs.

2 Preliminaries

This section gives the necessary notation, definitions, and concepts needed for this paper.

[1]Supported in part by the National Natural Sciences Foundation of China and the Doctoral Program Foundation of Institutions of Higher Education of China. The work reported here was done during the second author visited Temple University in 2001, he is grateful to the university for its hospitality.

(a) Groups G acting on sets X:

Let X be a G-set. (Unless otherwise specified, X and G are finite.) We will let x^g denote the action of the group element $g \in G$ on $x \in X$. The G-set X is *transitive* if, for any $x, y \in X$, there exists $g \in G$ such that $x^g = y$. The *stabilizer* G_x of $x \in X$ consists of those elements in G which fix x, that is, $G_x = \{g \in G : x^g = x\}$.

A subset B of X is called a *G-block* if, for any $g \in G$, the sets B^g and B are either disjoint or equal, that is, $B^g \cap B = \emptyset$ or $B^g = B$. Note that $\emptyset$, X, and singleton sets $\{x\}$ are G-blocks. Any other G-block is said to be *nontrivial.*

Definition 2.1 A transitive G-set X is *primitive* if X contains no nontrivial G-blocks.

The following well-known characterization of primitive G-sets will be used in our main theorem.

Theorem 2.2 *Let X be a transitive G-set. Then X is primitive if and only if, for any $x \in X$, the stabilizer G_x is a maximal subgroup of G.*

(b) Graph notation

Let X denote a graph with vertex set $V(X)$ and edge set $E(X)$ and with automorphism group $A = Aut(X)$. We will sometimes denote such a graph by writing $X = X(V, E)$. Throughout this paper, X denotes a *simple* undirected graph, that is, X has no loops and no multiple edges.

Given a graph $X = X(V, E)$, we write uv for the edge in E with *endpoints* $u, v \in V$. Such vertices u and v are said to be *adjacent* or *neighbors.* Given any vertex $v \in V$, we write $N(v)$ to denote its *neighborhood*, that is, its set of neighbors. We also write deg (v) to denote the *degree* of v where deg $(v) = |N(v)|$ is the number of its neighbors.

X is called *complete* if any two vertices are adjacent. K_n is used to denote the complete graph with n vertices.

X is called *regular* or *k-regular* if all vertices have the same degree k. Clearly, the complete graph K_n is $(n-1)$-regular.

(c) Bipartite graphs

$X = X(V, E)$ is *bipartite* if V is the disjoint union of U and W where each edge uw has an endpoint in U and an endpoint in W. We denote such a bipartite graph by writing $X = [U, W]$. We also write $K_{m,n}$ to denote the *complete bipartite graph* $X = [U, W]$, that is, where $|U| = m$ and $|W| = n$ and where each vertex in U is adjacent to each vertex in W. Note that $K_{n,n}$ is n-regular, but $K_{m,n}$ is not regular when $m \neq n$.

Figure 1 gives examples of three bipartite graphs. Graph (a) is not complete. Graph (b) is the complete graph $K_{2,3}$ and (c) is the complete graph $K_{3,3}$.

Given a bipartite graph $X = [U, W]$, we let A^+ denote the following subgroup of $A = \text{Aut}(X)$:

$$A^+ = \langle g \in A : U^g = U \text{ and } W^g = W \rangle$$

That is, A^+ consists of the automorphism of X which map U into U and W into W. Moreover, if X is connected, then either $|A : A^+| = 2$ or $A = A^+$ according as there does or does not exist an automorphism which interchanges U and W.

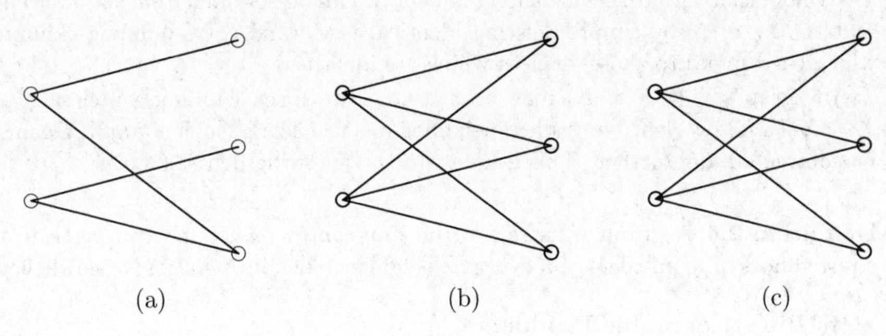

(a) (b) (c)

Figure 1

(d) Semisymmetric graphs

A graph $X = X(V, E)$ is *vertex-transitive* if, for any two vertices $u, v \in V$, there is an automorphism $\alpha \in \mathrm{Aut}(X)$ such that $u^\alpha = v$. Since an automorphism preserves the degree of a vertex, any vertex-transitive graph must be regular.

Analagously, $X = X(V, E)$ is *edge-transitive* if, for any two edges $e, e' \in E$, there is an automorphism $\alpha \in \mathrm{Aut}(X)$ such that $e^\alpha = e'$.

Observe that the graph $K_{2,3}$ in Figure 1(b) is edge-transitive, but it is not regular so it is not vertex-transitive. Generally speaking, the graphs K_n and $K_{n,n}$ are both vertex and edge transitive. On the other hand, the bipartite graph $K_{m,n}$, for $m \neq n$, is edge-transitive but it is neither regular nor vertex-transitive.

A bipartite graph $X = [U, W]$ is said to be *semitransitive* if A^+ acts (vertex) transitively on both U and W. More generally, $X = [U, W]$ is G-semitransitive if G is a subgroup of A^+ which acts transitively on U and W.

Definition 2.3 A graph $X = X(V, E)$ is *semisymmetric* if it is regular and edge-transitive but not vertex-transitive.

One can easily show that any semisymmetric graph must be bipartite with two parts of equal size and it must be semitransitive. Accordingly, any example of a semisymmetric graph must at least be a bipartite graph with two equal-size parts and be semitransitive.

(e) Co-neighbor blocks and non-contractable graphs

Consider any graph $X = X(V, E)$. Vertices $x, x' \in V$ will be called *co-neighbors* if they have the same set of neighbors, that is, if $N(x) = N(x')$. This is an equivalence relation on V, and its equivalence clases will be called *co-neighbor blocks* of X.

The graph $X = X(V, E)$ may be *contracted* to form a graph $X' = X'(V', E')$ by simply identifying co-neighbors. Specifically, consider any partition of each of the co-neighbor blocks of X into cells. Then we define $X' = X'(V', E')$ as follows:

$$\begin{aligned} V' &= \{[x] : x \in V \text{ and } [x] \text{ is the cell containing } x\} \\ E' &= \{[x][y] : xy \text{ is an edge in } E\} \end{aligned}$$

One can easily show that the graph X' is well defined. That is, if x and x' are co-neighbors and y and y' are co-neighbors, then xy is an edge if and only if $x'y'$ is an edge. The graph X' will be called a *contraction* of X.

We note that the adjacency matrix $M_{X'}$ of X' can be obtained from the adjacency matrix M_X of X by simply deleting identical rows (and corresponding columns) which correspond to those vertices which are identified.

Any graph $X = X(V, E)$ may be *contracted* to form a unique smallest graph $\bar{X} = \bar{X}(\bar{V}, \bar{E})$ by identifying all co-neighbors in X. Clearly, such a graph $\bar{X}$ cannot be contracted any further. This leads us to the following definition.

Definition 2.4 A graph X is said to be *non-contractable* if no two vertices are co-neighbors or, equivalently, if every co-neighbor block in X has only one element.

(f) Bi-lexicographic Products

The above graph operation of contraction may be reversed. Specifically, any graph $X = X(V, E)$ may be *expanded* to form a graph $X'' = X''(V'', E'')$ as follows. For each vertex $x \in X$, assign a set $\mathbb{N}_x = \{1, 2, ..., k_x\}$. Then we define $X'' = X''(V'', E'')$ as follows:

$$\begin{aligned} V'' &= \{(x,i) : x \in X, i \in \mathbb{N}_x\} = \bigcup (x \times \mathbb{N}_x : x \in X) \\ E'' &= \{(x,i)(y,j) : xy \in E, i \in \mathbb{N}_x, j \in \mathbb{N}_y\} \end{aligned}$$

The graph X'' will be called an *expansion* of X. Observe that if X'' is an expansion of X, then X is a contraction of X''.

Consider now a bipartite graph $X = [U, W]$. Given any two positive integers m and n, we define a particular expansion $Y = X(m, n)$ of X as follows. Let $\mathbb{N}_k = \{1, 2, .., k\}$. We assign $\mathbb{N}_m$ to each vertex $u \in U$ and we assign $\mathbb{N}_n$ to each vertex $w \in W$. Then $Y = X(m, n)$ is the expansion of X defined above. We note that Y is a bipartite graph $Y = [U \times \mathbb{N}_m, W \times \mathbb{N}_n]$, it will be called a *bipartite bi-lexicographic product* of X.

3 Bi-coset graph

This section gives the construction with an example of a bipartite graph from a group G and two subgroups L and R.

Let G be a finite group, and let L and R be two subgroups of G. Let D be a union of several double cosets of R and L in G, that is, $D = \bigcup_i Rg_iL$. [We emphasize that, in the double cosets of D, the subgroup R appears on the left and L appears on the right.] We define a bipartite graph $X = [U, W]$ as follows:

$$U = [G : L], \quad W = [G : R], \quad E(X) = \{(Lg, Rdg) : g \in G, d \in D\}$$

Here $[G : H]$ denotes the set of right cosets of a subgroup H of G. This graph, denoted by $\mathbf{B}(G, L, R; D)$, is called the *bi-coset graph* of G with respect to L, R, and D. [It is a generalization of the concept of a Sabidussi coset graph (see [10]).]

Next we give an example of such a bi-coset graph X where D is the single coset RL and $D \neq G$.

Example 3.1 Consider the following group G, subgroups L and R, and double coset $D = RL$:

$$G = A_4, L = A_3 = \langle(123)\rangle, R = \{1, (12)(34)\}, D = RL = L \cup (12)(34) \cdot L$$

Note $|G| = 12$, $|L| = 3$, $|R| = 2$, $|D| = 6$. Also, $L \cap R = 1$ and so $|L \cap R| = 1$. Furthermore, $D \neq G$.

Figure 2 shows the bi-coset graph $X = \mathbf{B}(G, L, R; D) = [U, W]$. We emphasize that there is an edge uw in X if $u = La$ and $w = Rda$ where $a \in G$ and $d \in D$. Note that X is non-contractable since no two vertices have the same neighborhood.

$|G : L| = 4$ $|G : R| = 6$

Cosets Lg: Cosets Rg:

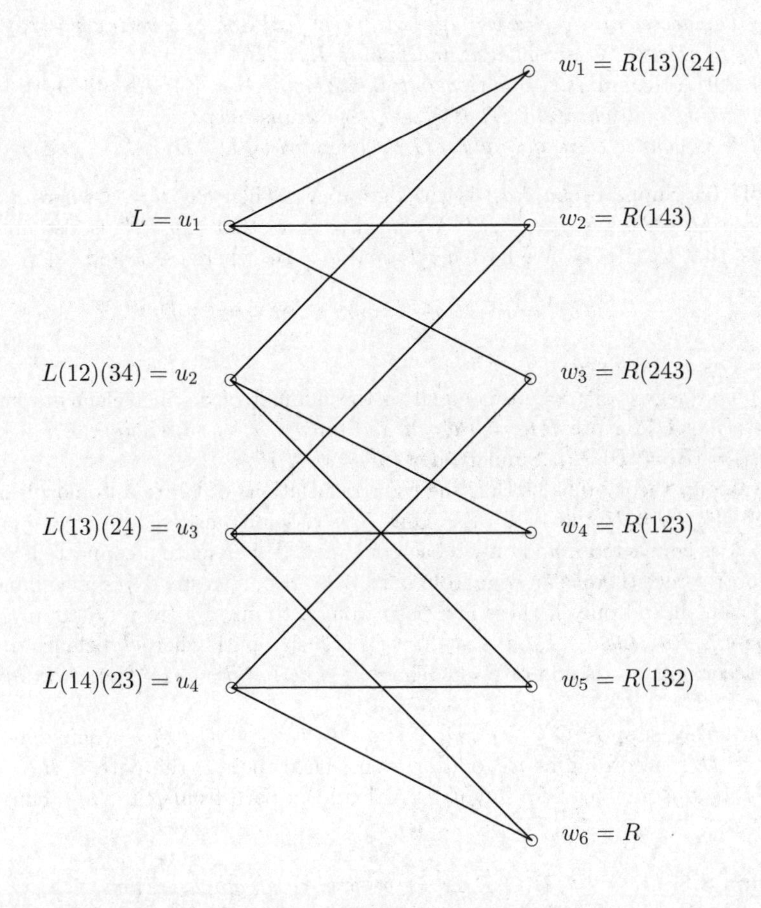

Figure 2

Bi-coset Graph $X = \mathbf{B}(G, R, L; D) = [U, W]$

The definition of the bi-coset graph $X = \mathbf{B}(G, R, L; D)$ does not assume that L and R are different subgroups of G. In the case that $L = R$, the cosets $[G : L]$ and $[G : R]$ are still distinct sets of vertices of X, and we will use Lg and Rg to denote different vertices in X. If $L = R = 1$, then D is simply a subset of G. In such a case, the bi-coset graph $\mathbf{B}(G, 1, 1; D)$ is called the bi-Cayley graph of G with respect to D and is simply denoted by $\mathbf{B}(G, D)$.

Well-known basic properties of the bi-coset graph are contained in the following two lemmas. For completeness we will prove them here.

Lemma 3.2 *Let $X = [U, W] = \mathbf{B}(G, L, R; D)$ be a bi-coset graph of G with respect to L, R, and D. Then:*

(a) X is a well-defined bipartite graph.

(b) The degree of a vertex $u = Lg$ in U is $|D|/|R|$ and of a vertex $w = Rg$ in W is $|D|/|L|$. Hence X is regular if and only if $|L| = |R|$.

(c) $\mathrm{Aut}(X)$ contains G and G acts transitively on $U = [G : R]$ and on $W = [G : L]$ by (right) multiplication. Thus X is G-semitransitive.

(d) X is connected if and only if G is generated by $D^{-1}D$.

Proof (a) Suppose (La, Ra') is an edge in X. Then $a'a^{-1} = d$ where $d \in D$. Suppose $La = Lb$ and $Ra' = Rb'$. We need to show that (Lb, Rb') is also an edge, that is, that $b'b^{-1} \in D$. We have $b = la$ and $b' = ra'$ where $l \in L$ and $r \in R$. Then

$$b'b^{-1} = (ra')(la)^{-1} = r(a'a^{-1})l = rdl \in D$$

Thus X is well defined.

(b) The degree of $u = Lg$ is equal to the number of distinct elements in $E = \{Rdg : d \in D\}$. Note $Rdg = Rd'g$ if and only if $d' = rd$ where $r \in R$. Hence $\deg(u) = |E| = |D|/|R|$. Similaly, $\deg(w) = |D|/|R|$.

(c) Group theory tells us that the right translations of G are automorphisms of X and also $U^g = U$ and $W^g = W$. Thus X is G-semitransitive.

(d) X is connected if and only if each of U and W belong to a connected component of X (since there is an edge from U to W). Also U (resp: W) is in a connected component if and only if there is a path from L to any Lg (resp: R to any Rg'). Suppose $L, Rd_1, Ld_2d_1, ..., Ld_{2s}...d_2d_1 = Lg$ is such a path where d_i belongs to D or D^{-1} according as i is odd or even. Then $g \in Ld_{2s}...d_2d_1 \subseteq \langle D^{-1}D \rangle$. Accordingly, $G = \langle D^{-1}D \rangle$.

Conversely, suppose $G \subseteq \langle D^{-1}D \rangle$. Let $g \in G$, say $g = d_{2s}...d_2d_1$ where d_i belongs to D or D^{-1} according as i is odd or even. Then there is the path $L, Rd_1, ..., Lg$ from L to Lg. Similarly, for any $g' \in G$, there is a path from R to Rg'. Thus X is connected. □

Lemma 3.3 *Let $X = \mathbf{B}(G, R, L; D)$ be a bi-coset graph. Then X is G-edge-transitive if and only if $D = RgL$ is a single double coset.*

Proof Suppose X is G-edge-transitive. Then, for any $d, d' \in D$, there is $g \in G$ such that

$$(La, Rda)^g = (Lb, Rd'b)$$

Hence $Lag = Lb$. Thus $agb^{-1} \in L$, say $agb^{-1} = l$. Also, $Rdag = Rd'b$, and so $Rdagb^{-1} = Rd'$. Hence $Rdl = Rd'$. Thus $dld'^{-1} \in R$, say $dld'^{-1} = r$. Then $r^{-1}dl = d'$. Therefore, d and d' belong to the same double coset RdL.

Conversely, suppose D is a single double coset RgL. Consider edges

$$(La, Rda) \text{ and } (Lb, Rd'b)$$

Say $d' = rdl$, $r \in R$ and $l \in L$. Hence $Rd'b = Rrdlb = Rdlb$. Choose $h = a^{-1}lb$. Then

$$(La)^h = Lah = La(a^{-1}lb) = Llb = Lb$$

and

$$(Rda)^h = Rdah = Rda(a^{-1}lb) = Rdlb = Rrdlb = Rd'b$$

Thus X is G-edge-transitive. □

Next we give an example to show how to construct a semisymmetric graph from the bi-coset graph in Example 3.1.

Example 3.4 Consider the bi-coset graph $X = [U, W] = \mathbf{B}(G, R, L; D)$ in Example 3.1. Choose $m = 3$ and $n = 2$ so $m|L| = n|R| = 12$. Form the bi-lexicographic product $Y = X(m, n) = [U \times \mathbb{N}_3, W \times \mathbb{N}_2]$ where each edge uw in X, corresponds to six edges $[u, i][w, j]$ in Y. Note first that Y is 6-regular. Also, Y is semitransitive and edge-transitive since X is semitransitive and edge-transitive. Furthermore, the co-neighbor block of each vertex $[u, i]$ has three elements, $[u, 1]$, $[u, 2]$, $[u, 3]$, and the co-neighbor block of each vertex $[w, j]$ has two elements, $[w, 1]$ and $[w, 2]$. Thus no automorphism can map a vertex $[u, i]$ into a vertex $[w, j]$, so Y is not vertex-transitive. Thus Y is a semisymmeytric graph.

4 Main results

Suppose G act on a graph X, that is, G is a subgroup of $\mathrm{Aut}(X)$. The next result is essentially the converse of Lemma 3.2. Here G_v denotes the *stabilizer* of v, that is, those elements $g \in G$ such that $v^g = v$. Also, recall $N(v)$ denotes the neighborhood of a vertex v in X.

Proposition 4.1 *Let $X = [U, W]$ be a bipartite G-semitransitive graph. Let $u \in U$ and $w \in W$. Set*

$$D = \{g \in G : w^g \in N(u)\}$$

Let $L = G_u$ and $R = G_w$. Then D is a union of double cosets of R and L in G, and $X \cong \mathbf{B}(G, L, R; D)$.

The following lemma is crucial for our purpose. The graph in Example 3.1 is an example of this result.

Lemma 4.2 *Let $X = [U, W]$ be a non-contractable bipartite graph which is edge-transitive and semitransitive. Assume $|U| \neq |W|$. Let $N = \mathrm{lcm}(|U|, |W|)$ and let $m = N/|U|$ and $n = N/|W|$. Then the bi-lexicographic product $Y = X(m, n)$ is a semisymmetric graph.*

Proof Y is edge-transitive since X is edge-transitive. By hypothesis, X is non-contractable, that is, its co-neighbor blocks have size one. Thus the co-neighbor blocks of Y are of the form

$$\{[u,i] : i \in \mathbb{N}_m\} \text{ and } \{[w,i] : i \in \mathbb{N}_n\}$$

which have different sizes. Thus X is not vertex-transitive. On the other hand,

$$\deg([u,i]) = N(\deg(u))/|W| = N(\deg(w))/|U| = \deg([w,j])$$

Thus Y is regular. □

Theorem 4.3 *Let $X = \mathbf{B}(G, L, R; D) = [U, W]$ be a bipartite bi-coset graph of G with respect to L, R, and D. Assume $D = RgL$ is a single double coset. Then:*

(1) *X is a bipartite complete graph if and only if $D = G$.*

(2) *X is non-contractable if and only if the following conditions hold for any $x, y \in G$:*

(a) *For any $x, y \in G$, $Dx = Dy$ implies $Lx = Ly$.*

(b) *For any $x, y \in G$, $D^{-1}x = D^{-1}y$ implies $Rx = Ry$.*

Proof (1) The degree of a vertex of the form Lx is $|D : R|$ and that of the form Ry is $|D : L|$. Suppose $D = G$. Then the degree of Lx is $|G : R| = |W|$ and that of the form Ry is $|G : L| = |U|$. Thus Lx joins every right coset in W, and every Ry joins every left coset in U. Thus X is a complete bipartite graph. Each step is reversable, so the converse is also true.

(2a) The neighborhood of Lx follows:

$$N(Lx) = \{Rdx : d \in D\} = \{Rdlx : l \in L\}$$

Since D is a single double coset, the union of the neighbors of Lx is Dx. Similarly, the union of the neighbors of Ly is Dy. Accordingly, if $Lx \neq Ly$ then $Dx \neq Dy$ and hence Lx and Ly have different neighborhoods.

(2b) Similarly, the union of the neighbors of Rx and Ry are, respectively, $D^{-1}x$ and $D^{-1}y$. Thus, if $Rx \neq Ry$ then Rx and Ry have different neighborhoods.

Both conditions (2a) and (2b) imply X is non-contractable. The steps are easily reversed, so the converse is also true. □

Theorem 4.4 *Let G be a finite group, and let L and R be two maximal subgroups of G with different orders. Let $D = RL$ and suppose $D \neq G$. Let $X = [U, W] = \mathbf{B}(G, L, R; D)$ be the bipartite bi-coset graph. [Then G may be viewed as a group acting on the set X.]*

(a) *X is not a complete bipartite graph.*

(b) *The bi-coset graph $X = \mathbf{B}(G, L, R; D)$ is noncontractable.*

(c) *Let $N = \operatorname{lcm}(|U|, |W|)$, and let $m = N/|U|$ and $n = N/|W|$. Then the bi-lexicographic product $Y = X(m, n)$ is a semisymmetric graph.*

Proof (a) Since $D \neq G$, X is not a complete bipartite graph by Theorem 4.3.

(b) Consider a vertex $u = Lx \in U$. Let

$$B = \{b \in U : N(b) = N(u)\}$$

Then B is the co-neighbor block containing $Lx = u$. Then, for any $g \in G$, B^g is also a co-neighbor block. Since the co-neighbor blocks form a partition, either $B^g \cap B = \emptyset$ or $B^g = B$. Thus the co-neighbor blocks are G-blocks where we view G as a group acting on X. [See Section 2(a).]

Furthermore, the stabilizier G_u of any vertex $u = Lx$ is $x^{-1}Lx$ and the stabilizer G_w of any vertex $w = Ry$ is $y^{-1}Ry$. Since L and R are maximal subgroups of G, so are the stabilizers G_u and G_w. Since X is a transitive G-set, Theorem 2.1 tells us that X is primitive. Thus the G-blocks or, in other words, the co-neighbor blocks of X are trivial. Thus X is noncontractable.

(c) Part (b) shows that the hypothesis of Lemma 4.2 is satisfied. Thus $Y = X(m, n)$ is a semisymmetric graph. □

Example 4.5 Let $G = S_n$, the symmetric group acting on $\{1, 2, ..., n\}$ with $n \geq 5$. Let $L = G_n$, the stabilizer of n. Thus $L \cong S_{n-1}$. Let $R = G_{\{n-1,n\}}$, the setwise stabilizer of $n-1, n$. Thus $R \cong S_{n-2} \times S_2$. It is known that L and R are maximal subgroups of G. Let $D = RL$. Then $|D| = |R||L|/|R \cap L| = 2(n-1)!$. Thus $D \neq G$. By Theorem 4.4, $X = \mathbf{B}(G, L, R; D) = [U, W]$ is a semisymmetric graph.

(a) For $n = 5$, $|U| = 5$ and $|W| = 10$. Then X has $2 \times 10 = 20$ vertices. This graph was first discovered by Folkman who proved [6] that it is the smallest semisymmetric graph, that is, the one with the minimal number of vertices.

(b) Generally speaking, the above semisymmetric graphs X will have $2n(n-1)$ vertices or $n(n-1)$ vertices according as n is even or odd.

(c) Letting $n = 4$ in the above example, we get $R \cong S_2 \times S_2$ which is not maximal in $G = S_4$. Hence we cannot use Theorem 4.4. However, $X = \mathbf{B}(S_4, L, R; D)$ is still a semisymmetric graph. In fact, it is the graph in Example 3.1.

Example 4.6 This is a generalization of the graph in Example 4.1. Let $G = S_n$, and suppose $1 < i < j \leq n/2$. Let $L = G_{\{1,...,i\}}$, $R = G_{\{1,...,j\}}$, $D = RL$. Again L and R are maximal subgroups of G, and $D \neq G$. Thus by Theorem 4.4, $X = \mathbf{B}(G, L, R; D)$ is a semisymmetric graph.

Example 4.7 Let $q = p^s$, an odd prime power. Let $G = PGL(2, q)$. Then G has maximal subgroups $L = D_{2(q-1)}$ and $R = D_{2(q+1)}$, where D_k is the dihedral group of order k. Choose an L and R such that $|L \cap R| = 4$. Let $D = RL$. Then $|D| = (q-1)(q+1)$ whereas $|G| = q(q-1)(q+1)$. Thus $D \neq G$, and $X = \mathbf{B}(G, L, R; D)$ gives another infinite family of semisymmetric graphs.

5 Problems

We now state a number of open problems.

Example 4.5(c) shows that the maximality of L and R in G in Theorem 4.4 is sufficient, but not necessary, for $X = \mathbf{B}(G, L, R; D)$ to be non-contractable.

Problem 5.1 Find weaker conditons or simple necessary conditions on L and R so that $X = \mathbf{B}$(G,L,R;D) is non-contractable.

Let Graph(n) denote the set of graphs with n given vertices. Note Graph(n) has $2^{C(n,2)} = 2^{n(n-1)/2}$ elements. We assume that Graph(n) is given the uniform probability distribution.

Problem 5.2 Find the probability that a graph with n vertices is semisymmetric.

The property of non-contractability does not seem to have been investigated in any depth.

Problem 5.3 Find non-contractable semisymmetric graphs.

Problem 5.4 Find the probability that a graph with n vertices is non-contractable.

References

[1] I. Z. Bouwer, On edge but not vertex transitive cubic graphs, *Canad. Math. Bull.* **11** (1968) 533–535.

[2] I. Z. Bouwer, On edge but not vertex transitive regular graphs, *J. Combin. Theory Ser. B*, **12** (1972) 32–40.

[3] S. F. Du, Construction of Semisymmetric Graphs, *Graph Theory Notes of New York* **XXIX**, 1995.

[4] S.F. Du and D. Marušič, An infinite family of biprimitive semisymmetric grpahs, *J. Graph Theory* **32**(3)(1999), 217-228.

[5] Shao-Fei Du and Ming-Yao Xu, Semisymmetric Graphs of Order $2pq$, *Communications in Algebra*, **28**(2000), 2685–2715.

[6] J. Folkman, Regular line-symmetric graphs, *J. Combin. Theory Ser. B*, **3** (1967), 215–232.

[7] M. E. Iofinova and A. A. Ivanov, Biprimitive cubic graphs (Russian), in *Investigation in Algebraic Theory of Combinatorial Objects* Proceedings of the Seminar, Institute for System Studies, Moscow, 1985, pp. 124–134.

[8] I. V. Ivanov, On edge but not vertex transitive regular graphs. *Comb. Annals of Discrete Mathematices* **34** (1987) 273–286.

[9] M. H. Klin, On edge but not vertex transitive regular graphs, *Colloquia Mathematica Societatis Janos Bolyai, 25. Algebraic Methods in Graph Theory, Szeged (Hungary)*, 1978 Budapest, 1981, pp. 399–403.

[10] G.O. Sabidussi, Vertex-transitive graphs, *Monatsh. Math.* **68**(1964), 426-438.

ON THE COVERS OF FINITE GROUPS

MARIA SILVIA LUCIDO

Dipartimento di Matematica e Informatica, Università di Udine
via delle Scienze 200, I-33100 Udine, Italy
E-mail: lucido@dimi.uniud.it

Abstract

A *cover* for a group is a collection of proper subgroups whose union is the whole group. We report some recent results concerning the different covers of a finite group.

1 Minimal covers

Let G be a group. A *cover* of G is a collection $\mathcal{A} = \{A_i | 1 \leq i \leq n\}$ of proper subgroups of G whose union is G. The subgroups in $\mathcal{A}$ are called the *components* of the cover. The cover is *irredundant* if no proper sub-collection is also a cover. The cover is *minimal* if no cover of G has fewer than n members. In this case, J. H. E. Cohn [8] defined $\sigma(G)$ to be this minimal number of subgroups. It is clear that to study $\sigma(G)$, it is enough to consider covers consisting of maximal subgroups.

A number of results were proved for soluble groups. In particular it was proved in 1997 by Tomkinson in [14] that if G is a finite noncyclic soluble group, then $\sigma(G) = p^a + 1$, where p^a is the order of a particular chief factor of G. In fact, he proves that

Theorem 1.1 *[14] Let G be a finite soluble group and let H/K be the smallest chief factor of G having more than one complement in G. Then $\sigma(G) = |H/K| + 1$.*

In his paper he suggested that it might be of interest to investigate $\sigma(G)$ for families of simple groups.

In 1999, Bryce, Fedri and Serena calculated $\sigma(G)$ for the groups $\mathrm{PSL}(2,q)$, $\mathrm{PGL}(2,q)$, $\mathrm{SL}(2,q)$ and $\mathrm{GL}(2,q)$. They prove the following:

Theorem 1.2 *[4] Let G be $PSL(2,q)$, $PGL(2,q)$, $SL(2,q)$ and $GL(2,q)$ and $q = 4, 8$ or $q \geq 11$, then*

$$\sigma(G) = \frac{q(q+1)}{2} \quad \textit{if } q \textit{ is even}$$

$$\sigma(G) = \frac{q(q+1)}{2} + 1 \quad \textit{if } q \textit{ is odd.}$$

A similar result can be obtained for other groups with a partition. In fact if G has a partition with n subgroups, we have $\sigma(G) \leq n$, because a partition is a cover.

In particular we can calculate $\sigma(G)$ for the family of Suzuki groups.

Proposition 1.3 *Let $G = Sz(q)$ be a Suzuki group, where $q = 2^{2m+1}$, then we have*

$$\sigma(G) = \frac{q^2(q^2+1)}{2}.$$

Proof Let $G = Sz(q) \leq \mathrm{GL}(4,q)$ be a Suzuki group, where $q = 2^{2m+1}$, m a positive integer and let $r = 2^{m+1}$. Then $|G| = q^2(q-1)(q^2+1)$ and $(q-r+1)(q+r+1) = q^2+1$. Let U be the subgroup of the lower unitriangular matrices of G: then U is a 2-Sylow subgroup of exponent 4 and of order q^2. Let H be the subgroup of the diagonal matrices of G. Then H is isomorphic to the multiplicative group of the field, and therefore has order $q-1$; it is a $\pi(q-1)$-Hall subgroup of G and it normalizes U. By Theorem 3.10, chapter XI of [10] we get that the set $\Psi = \{U^g, H^g, T_1^g, T_2^g\}$ is a partition of G, where T_1 is a cyclic maximal torus of order $q+r+1$ and T_2 is a cyclic maximal torus of order $q-r+1$. By [15] the only maximal subgroup of G containing T_i is $N_i = N_G(T_i) = T_i < t_i >$, with t_i an element of order 4 and $|N_i : T_i| = 4$, for $i = 1, 2$.

Let now $\mathcal{A}$ be a minimal cover of G with maximal components. Since T_i is cyclic and N_i is the only maximal subgroup containing it (for $i = 1, 2$), all the conjugates of N_1 and N_2 must be among the components of $\mathcal{A}$. Moreover N_1 is maximal, and therefore self-normalizing, so the number of distinct conjugates of N_1 is $q^2(q-1)(q-r+1)/4$, while the number of distinct conjugates of N_2 is $q^2(q-1)(q+r+1)/4$. We observe that the 2-elements are contained in the union of the conjugates of N_1. In fact there exists two conjugacy classes of elements of order 4 and one class of elements of order 2. Moreover the elements of order 4 are not real, that is if $|x| = 4$, then x^{-1} is not conjugate to x (see [15]).

Since t_1 is of order 4 and $t_1 \in N_1$, then t_1^{-1} and t_1^2 belongs to N_1. Therefore the conjugacy classes $[t_1], [t_1^{-1}]$ and $[t_1^2]$ are contained in the union $\bigcup_g N_1^g$. Since Ψ is a partition, the $\pi(q-1)$-elements are contained in some conjugate of H, and none of them can belong to $\mathcal{N} = \bigcup_g N_1^g \cup \bigcup_g N_2^g$. Therefore, since H^g is cyclic, $\mathcal{A}$ must contain a maximal subgroup of G containing H^g, for any $g \in G$. By [15], there are two kinds of maximal subgroups containing H^g and these are

$N_G(U)^g = (UH)^g = B^g$, the Borel subgroups of G

$N_G(H)^g = (H < t >)^g$, where $t \notin U$ is an involution.

The only elements of G not belonging to $\mathcal{N}$ are the $\pi(q-1)$-elements, then we can substitute the components $N_G(H)^g$ with B^g. We therefore consider the subgroups B^g. Since B is a Frobenius group, there are q^2 conjugates of H in B and there are $q^2(q^2+1)/2$ conjugates of H in G, since $|N_G(H)| = 2|H|$.

We now want to examine the intersection of the Borel subgroup B with its conjugates. To do this we need the following two facts (see [10] or [7])

i) $B \cap B^t = H$

ii) any element of $G \setminus B$ can be written uniquely in the form $g = btu$, with $b \in B$ and $u \in U$.

Therefore, if $g \in G \setminus B$, we have $B \cap B^g = H^u$, where u is the unique element of U such that $g = btu$. It follows that if $B^{g_1} \neq B^{g_2}$, then $B \cap B^{g_1} \neq B \cap B^{g_2}$. Therefore there are $q^2 + (q^2 - 1)$ conjugates of H in $B \cup B^g$, if $g \in G \setminus B$, and so on. If we consider the union of n conjugates of B, this union contains $\sum_{i=0}^{n-1}(q^2 - i)$ conjugates of H. Hence to get all the conjugates of H, we need at least q^2 conjugates of B.

In fact
$$\sum_{i=0}^{q^2-1}(q^2 - i) = \sum_{i=1}^{q^2} i = \frac{q^2(q^2+1)}{2}.$$

It follows that

$$\sigma(G) \geq q^2 + q^2(q-1)(q-r+1)/4 + q^2(q-1)(q+r+1)/4 = q^2(q^2+1)/2.$$

On the other hand, the elements of G are either contained in the conjugates of the subgroups H, T_1 and T_2 or are 2-elements, and therefore contained in $\mathcal{N}$. Therefore
$$\mathcal{A} = \{N_1^g, N_2^g, B^x | g \in G, x \in G \setminus B\}$$
is a minimal cover of G.

We finally observe that any minimal cover of G is of this type: it contains $\mathcal{N}$ and all except one conjugate of B. □

2 Covers with conjugate subgroups

It is well known that a finite group G is never the set-theoretical union of the G-conjugates of a proper subgroup H, that is the family

$$\mathcal{I} = \{H^{i_g} \mid H < G,\ i_g \in Inn(G)\}$$

can never be a cover.

Brandl in [2] considers the covers

$$\delta = \{H^\alpha :\ H \leq G,\ \ \alpha \in Aut(G)\}$$

and shows how some properties of H extends to G.

Praeger in [12] and [13] explores the more general covers

$$\Phi = \{H^\alpha : \alpha \in A\} \ \text{ where } \ Inn(G) \leq A \leq Aut(G)$$

to study the covers of the Galois group of certain extensions of fields. She proves that H and G must have the same composition factors. In particular there is no such a cover, if G is a simple group.

Another way to extend the family is to consider the G-conjugates of H and another subgroup K or the set of the G-conjugates of K. In some cases it is possible to cover the group G. This happens, for example, in a Frobenius group G if H and K are respectively a complement and the kernel. In [5] two kinds of collections of subgroups for a group G were introduced:

$$(*) \quad \delta = \{H^g, K : g \in G\},$$

$$(**) \quad \delta = \{H^g, K^g : g \in G\},$$

where H and K are fixed proper subgroups of G. If a group G is the set-theoretical union of subgroups of the type $(*)$ or $(**)$, then G is said respectively $(*)$-*coverable* or $(**)$-*coverable* and the subgroups in δ are called the *components* of the cover. The characterization of the $(*)$-coverable groups in [5] shows that no simple group is $(*)$-coverable. But it is easily seen that there exist simple $(**)$-coverable groups. In fact it can be proved the following

Theorem 2.1 *i)[5] G is $(*)$-coverable if and only if G is a Frobenius-Wielandt group.*

ii) [5] The alternating group A_n is $(**)$-coverable if and only if $4 \leq n \leq 8$ and the symmetric group S_n is $(**)$-coverable if and only if $3 \leq n \leq 6$.

iii) [6] The groups $PSL(n,q)$, $PGL(n,q)$, $SL(n,q)$ and $GL(n,q)$ are $(**)$-coverable if and only if $2 \leq n \leq 4$.

In [3] the $(**)$-covers of the Symmetric groups are applied to the algebraic number theory.

For a finite simple group of Lie type G, it is reasonable to expect that G is $(**)$-coverable if and only if G has small rank.

3 Hall covers

We define a *Hall cover* of a finite group G as a set $\mathcal{H} = \{H_1, H_2, \ldots, H_r\}$ of proper Hall subgroups of G such that:

a) $\bigcup_{i=1}^{r} H_i = G$

b) either $|H_i| = |H_j|$ or $(|H_i|, |H_j|) = 1$ for $i, j = 1, ..., r$.

If the elements of $\mathcal{H}$ all have order a prime power, then $\mathcal{H}$ is called a *Sylow cover of G*. The finite groups G with a Sylow cover have been studied independently by G. Higman ([9]) and G. Zacher ([16], [17]) in the case in which G is soluble, by M. Suzuki ([15]) in the case of a simple group G and by R. Brandl ([1]) in the general situation.

Theorem 3.1 *Let G be a finite group with a Sylow cover. Then G is one of the following*

i) G is soluble and let p be a prime such that $O_p(G) \neq 1$. Then if p is odd, $G/O_p(G)$ is cyclic of prime power order or a generalized quaternion group, otherwise $G/O_p(G)$ is a group of order $p^a q^b$ with cyclic Sylow subgroups, where $q = kp^a + 1$ a prime.

ii) G is one of the following finite (almost) simple groups:

$$PSL(2,4), PSL(2,7), PSL(2,8), PSL(2,9), PSL(2,17), PSL(3,4), Sz(8), Sz(32),$$

$$\textit{the non split extension of } PSL(2,9).$$

iii) G has a nontrivial normal 2-subgroup P and G/P is isomorphic with $PSL(2,4)$, $PSL(2,8)$, $Sz(8)$, $Sz(32)$. Moreover P is elementary abelian and isomorphic with a direct sum of natural modules for G/P.

In the paper [11], Jabara and Lucido study the groups which admit a Hall cover. They prove

Theorem 3.2 *[11] Let G be a simple non abelian group, such that G admits a Hall cover. Then G is one of the following:*

$$PSL(2,q),\ PSL(3,4),\ PSL(3,q) \text{ with } (3,q-1)=1,\ \ {}^2B_2(2^{2n+1}),\ A_7,\ M_{22}.$$

Acknowledgements. We wish to thank D. Bubboloni and S. Dolfi for their valuable suggestions on the topic of this paper.

References

[1] R. Brandl, *Finite groups all of whose elements are of prime power order*, Boll. U.M.I. (5) **18 - A**, 1981, 491 - 493.
[2] R. Brandl, *A cover property of finite groups*, Bull. Austral. Math. Soc. **23** (1981), 227–235.
[3] R. Brandl, D. Bubboloni and I. Hupp, *Polynomials with roots mod p for all primes p*, J. Group Theory **4** (2001), 233-239.
[4] R. Bryce, V. Fedri, L. Serena, *Subgroup covers of some linear groups*, Bull. Austral. Math. Soc. **60** (1999), 227–238.
[5] D. Bubboloni, *Covers of the Symmetric and Alternating groups*, Dipartimento di matematica "U. Dini" - Universita' di Firenze **7**, (1998).
[6] D. Bubboloni and M. S. Lucido, *Covers of Linear groups*, to appear in Communications in Algebra.
[7] R. W. Carter, *Simple groups of Lie type*, J. Wiley and sons - London 1972.
[8] J. H. E. Cohn, *On n-sum groups*, Math. Scand. **75** (1994), 44-58.
[9] G. Higman, *Finite groups in which every element has prime power order*, J. London Math. Soc **32**, 1957, 335 - 342.
[10] B. Huppert and N. Blackburn, *Finite Groups III* - Springer-Verlag, Berlin - Heidelberg - New York, 1982.
[11] E. Jabara and M. S. Lucido, *Finite Groups with Hall coverings*, to appear.
[12] C.E. Praeger, *Cover subgroups of groups and Kronecker classes of fields*, J. Algebra **118** (1988), 455–463.
[13] C.E. Praeger, *Kronecker classes of fields and cover subgroups of finite groups*, J. Austral. Math. Soc. (Series A) **57** (1994), 17–34.
[14] M. J. Tomkinson, *Groups as the union of proper subgroups*, Math. Scand. **81** (1997), 191-198.
[15] M. Suzuki, *On a class of doubly transitive groups*, Ann. Math., **75**, 1962, 105 - 145.
[16] G. Zacher, *Sull' ordine di un gruppo finito risolubile somma dei sottogruppi di Sylow*, Atti Acc. Naz. Lincei (8) **20**, 1956, 171 - 174.
[17] G. Zacher, *Sui gruppi finiti somma dei loro sottogruppi di Sylow*, Rend. Sem. Mat. Padova, **27**, 1957, 267 - 275.

GROUPLAND

OLGA MACEDOŃSKA

Institute of Mathematics, Silesian University of Technology, Gliwice, 44-100, Poland

Abstract

Groupland is a flat planet where groups live in regions defined by their properties, e.g. residually finite groups, groups of different growth types, locally graded groups, etc. This visualizes relations of properties, and formulation of problems, such as, for example, where do Engel groups live?

A law $u(x_1, ..., x_n) = v(x_1, ..., x_n)$ is called positive if u, v are written without inverses of variables. The simplest examples are $xy = yx$ and $x^{n+1} = x$, $n > 0$. If a group G satisfies a positive law, then the variety $var(G)$ has a basis of positive laws [11]. We note that groups satisfying positive laws can not contain a free non-abelian subsemigroup, denoted by $\mathcal{F}$.

In Fig.1 of Groupland we indicate the regions: of groups which have no laws, groups with only non-positive laws, and groups satisfying positive laws.

All groups containing a free non-abelian subsemigroup $\mathcal{F}$ are in the right half of Groupland. Groups in the left half do not contain $\mathcal{F}$.

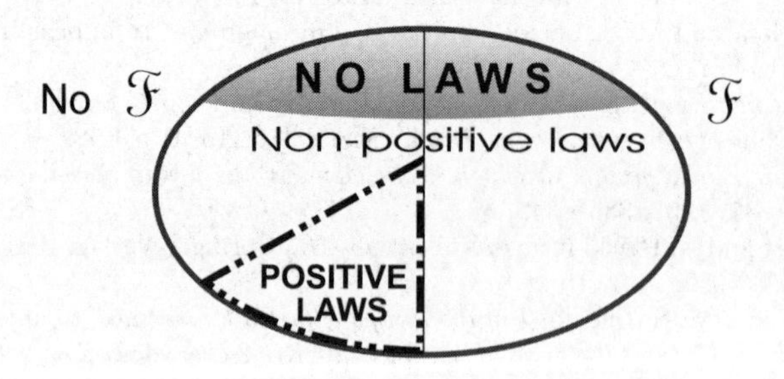

Fig.1, Groupland

We need the following results.

Lemma 1

(a) If a group G is nilpotent-by-(finite exponent), then G satisfies a positive law.

(b) If a finitely generated group G is soluble-by-finite, then either G is nilpotent-by-finite or G contains $\mathcal{F}$.

(c) If G is a finitely generated periodic residually finite group which satisfies a nontrivial law, then G is finite.

Proof *(a)* It was shown independently in [13] and [16], that nilpotent groups satisfy positive laws. So if a variety $\mathfrak{N}_c$ of c-nilpotent groups satisfies a positive law $u(x_1, ..., x_c) = v(x_1, ..., x_c)$, then a nilpotent-by-Burnside variety $\mathfrak{N}_c\mathcal{B}_e$ satisfies the positive law $u(x_1^e, ..., x_c^e) = v(x_1^e, ..., x_c^e)$.

(b) It is shown in [19] 4.7, that a finitely generated soluble group with no free subsemigroup on two generators must be polycyclic. Also by [19] 4.12, if G is a polycyclic group, then G either has a nilpotent subgroup of finite index or G contains a free subsemigroup on two generators. These imply (b).
(c) This result can be found in [27] IV.2.6. $\square$

Since 1953 it was conjectured that a group satisfies positive laws if and only if it is an extension of a nilpotent group by a group of finite exponent. A counterexample was found only in 1996 [17], where it was shown that there exist groups satisfying positive laws, which are not even (locally soluble)-by-(finite exponent). By [1], relatively free groups of prime exponent ≥ 665 and of finite rank > 1 have exponential growth. In Fig.2 of Groupland these groups are in the region with the small star. We add now four more regions to the picture.

Statement 1 *Fig.2 of Groupland shows the regions of:*
(i) groups of polynomial growth (hatched pattern),
(ii) torsion groups of intermediate growth (a small white circle),
(iii) polycyclic groups (a small inner grey ellipse),
(iv) residually finite groups (a large dotted ellipse).

Proof (i) By results of Gromov [6], Milnor [14] and Wolf [26], a group G has polynomial growth if and only if it is nilpotent-by-finite. So by Lemma 1 (a), groups of polynomial growth satisfy positive laws.
(ii) All known finitely generated infinite torsion groups of exponent zero, in particular the groups of intermediate growth, are residually finite and hence by Lemma 1 (c), they do not satisfy any law (see e.g. [1], ([15], 34.63), [5], [7], [23]).
(iii) The region of polycyclic groups is defined by Lemma 1 (b). As an example of a polycyclic group containing $\mathcal{F}$ we have the group $G = \langle a, b, c \mid [a, b] = 1, a^c = ab, b^c = a^2 b \rangle$, which by ([15] 32.35) generates the whole metabelian variety $\mathfrak{A}^2$. So G is not nilpotent-by-finite and hence by Lemma 1 (b), it contains $\mathcal{F}$.
(iv) By P. Hall, nilpotent-by-finite groups are residually finite [8], and polycyclic groups are also residually finite ([15] 32.1). $\square$

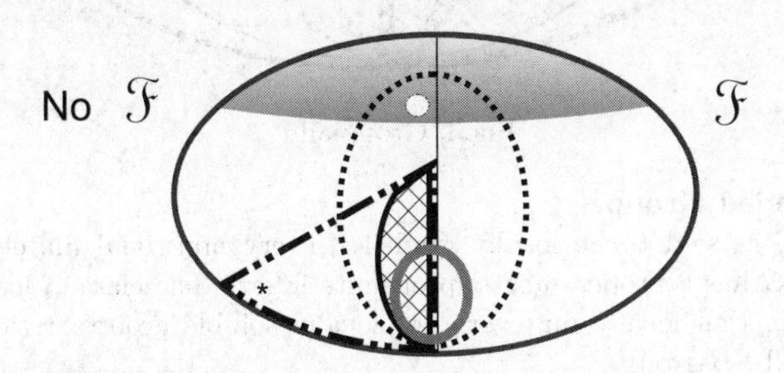

Fig.2, Groupland

We denote by $\mathfrak{B}_e$ the Restricted Burnside variety of exponent e, i.e. the variety generated by all finite groups of exponent e. All groups in $\mathfrak{B}_e$ are locally finite of exponent e. The existence of such varieties for each positive integer e follows from the positive solution of the Restricted Burnside Problem and relies on the classification of finite simple groups (see [25]). Clearly $\mathfrak{S}_n\mathfrak{B}_e \subseteq \mathfrak{S}_n\mathcal{B}_e$.

Definition [2] We define an *SB-group* to be one lying in some product of finitely many varieties each of which is either soluble or a $\mathfrak{B}_e$ (for varying e).

In particular the class of SB-groups contains all groups G which are soluble-by-(locally finite of finite exponent), that is $G \in \mathfrak{S}_n\mathfrak{B}_e$ for some n, e.

Statement 2 *Fig.3 of Groupland contains new regions of:*
(i) SB-groups (inner horizontal ellipse),
(ii) groups which are nilpotent-by-(locally finite of finite exponent), $G \in \mathfrak{N}_c\mathfrak{B}_k$ for some c, k (lined and hatched regions).

Proof (i) As an example of an SB-group containing $\mathcal{F}$, we have a free metabelian group of rank > 1 [13]. SB-groups without $\mathcal{F}$ may satisfy positive laws as, for example, nilpotent-by-finite groups. However a cartesian product of metabelian nilpotent groups of increasing nilpotency class need not satisfy positive laws.
(ii) It follows from [2] Thm.B, that an SB-group G satisfies a positive law if and only if G is nilpotent-by-(locally finite of finite exponent). So the intersection of regions of SB-groups and groups satisfying positive laws coincides with the region of groups which are nilpotent-by-(locally finite of finite exponent). □

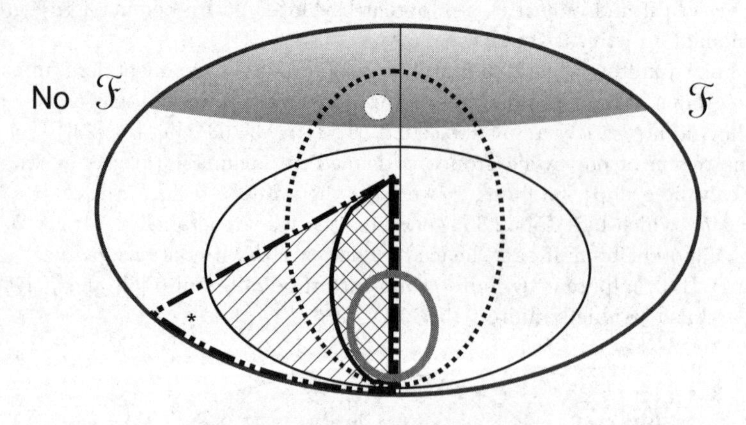

Fig.3, Groupland

Locally Graded Groups

A group G is said to be locally graded if every nontrivial finitely generated subgroup of G has a proper subgroup of finite index. The class of locally graded groups contains all locally finite groups, locally soluble groups, residually finite groups and all SB-groups.

Statement 3 *Fig.4 of Groupland shows the region of the class of locally graded groups (the region inside the dashed ellipse, including the solid ellipse.*

Proof We have to check that locally graded groups which satisfy a positive law are nilpotent-by-(locally finite of finite exponent. For finitely generated groups this follows from Theorem A [10], and then for any group, from Theorem B (ii) [4]. □

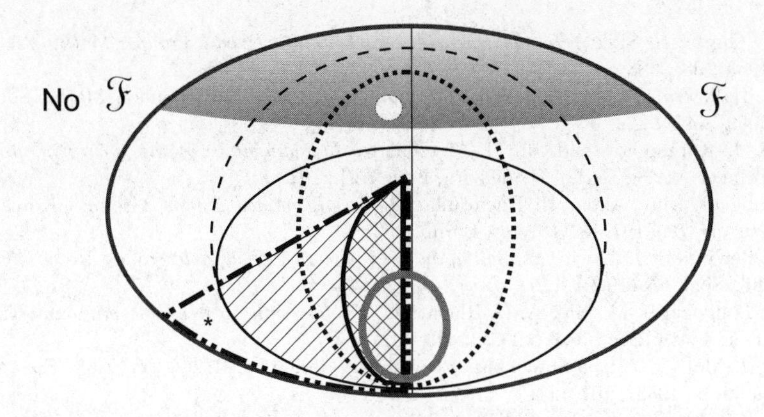

Fig.4, Groupland

Question In which region of Groupland do Engel groups live?

The question **whether every n-Engel group is locally nilpotent** dates from 1936, when Zorn proved that a finite n-Engel group is nilpotent. Positive answers are known for soluble Engel groups (Gruenberg, 1953), residually finite Engel groups (Wilson, 1991), profinite Engel groups (Wilson and Zelmanov, 1992), see [3]. The problem is still open in general.

The question **whether n-Engel group satisfies a positive law** was posed by Shirshov in 1963 [22]. He proved that 2-Engel law $[[x, y], y]$ is equivalent to $xy^2x = yx^2y$, and 3-Engel law is equivalent to two positive laws. Then in [12] Maj et al. proved that every torsion-free 4-Engel group satisfies positive laws. In 1998 Traustason [24] proved that every 4-Engel group satisfies positive laws. For $n > 4$ the problem is open.

However in the class of locally graded groups the answer is known. It was shown by Kim and Rhemtulla 1995 [10], that every locally graded n-Engel group is locally nilpotent. Then by result of Burns and Medvedev 1998 [3], every locally graded n-Engel group is contained in a variety $\mathfrak{N}_{c(n)}B_{e(n)} \cap \mathfrak{B}_{e(n)}\mathfrak{N}_{c(n)}$, and hence satisfies a positive law.

References

[1] S. V. Aleshin, *Finite automata and the Burnside problem for periodic groups*, Mat. Zametki, **11** (1972), 319–328.

[2] R. G. Burns, O. Macedońska, Yu. Medvedev, *Groups satisfying semigroup laws, and nilpotent-by-Burnside varieties*, J. Algebra, **195** (1997), 510–525.

[3] R. G. Burns and Yuri Medvedev, *A note on Engel groups and local nilpotence*, J. Austral. Math. Soc., (Series A) **64** (1998), 92–100.

[4] R. G. Burns and Yuri Medvedev, *Group Laws Implying Virtual Nilpotence*, to appear.

[5] R. I. Grigorchuk, *Degrees of growth of finitely generated groups and the theory of invariant means*, Izv. Akad. Nauk SSSR Ser. Mat., **48** (1984), 939–985.

[6] M. Gromov, *Groups of polynomial growth and expanding maps*, Publs. Math. Inst. hautes étud. sci., **53** (1981), 53–73.

[7] N. Gupta, S. Sidki, *On the Burnside problem for periodic groups*, Math. Z., **182** (1983), 385–388.

[8] P. Hall, *On the finiteness of certain soluble groups*, Proc. London Math. Soc., **9** (1959), 595–622.

[9] M. I. Kargapolov and Ju. I. Merzljakov, *Fundamentals of the theory of groups*, Springer–Verlag Berlin, Heidelberg, New York, 1979.

[10] Yangkok Kim, Akbar H. Rhemtulla, *On locally graded groups*, Groups—Korea '94 (Pusan), 189–197, de Gruyter, Berlin, 1995.

[11] J. Lewin and T. Lewin, *Semigroup laws in varieties of solvable groups*, Proc. Cambr. Phil. Soc., **65** (1969), 1–9.

[12] P. Longobardi, M. Maj, A. H. Rhemtulla, *Groups with no free subsemigroups*, Trans. Amer. Math. Soc., **347** (1995), 1419–1427.

[13] A. I. Malcev, *Nilpotent semigroups*, Ivanov. Gos. Ped. Inst. Uc. Zap. Fiz. Mat. Nauki, **4** (1953), 107–111.

[14] J. Milnor, *Growth of finitely generated solvable groups*, J. Differential Geometry, **2** (1968), 447–449.

[15] H. Neumann, *Varieties of groups*, Springer-Verlag, Berlin, Heidelberg, New York, 1967.

[16] B. H. Neumann, T. Taylor, *Subsemigroups of nilpotent groups*, Proc. Roy. Soc., (Series A) **274** (1963), 1–4.

[17] A. Yu. Olshanskii, A. Storozhev, *A group variety defined by a semigroup law*, J. Austral. Math. Soc. (Series A) **60** (1996), 255–259.

[18] D.J.S. Robinson, *Finiteness conditions and generalized soluble groups, Part I*, Springer-Verlag Berlin, Heidelberg, New York, 1972.

[19] J. M. Rosenblatt, *Invariant measures and growth conditions*, Trans. Amer. Math. Soc., **193** (1974), 33–53.

[20] J. F. Semple, A.Shalev, *Combinatorial conditions in residually finite groups I*, J. Algebra, **157** (1993), 43–50.

[21] A. Shalev, *Combinatorial conditions in residually finite groups II*, J. Algebra, **157** (1993), 51–62.

[22] A. I. Shirshov, *On certain near-Engel groups*, Algebra i Logika Sem., **2**(5) (1963), 5–18.

[23] V. I. Sushchanskii, *Periodic p-groups of permutations and the unrestricted Burnside problem*, Dokl. Akad. Nauk SSSR, **247** (1979), 557–561.

[24] G. Traustason, *Semigroup identities in 4-Engel groups*, J. Group Theory, **2** (1999), 39–46.

[25] M. Vaughan-Lee, *On Zelmanov's solution of the restricted Burnside problem*, J. Group Theory, **1** (1998), 65–94.

[26] J.A.Wolf, *Growth of finitely generated solvable groups and curvature of Riemannian manifolds*, J. Differential Geometry, **2** (1968), 421–446.

[27] E. I. Zelmanov, *Nil Rings and Periodic Groups*, The Korean Mathematical Society Lecture Notes in Mathematics, (1992), 1-79.

ON MAXIMAL NILPOTENT π-SUBGROUPS

JUAN MEDINA[1]

Universidad Politécnica de Cartagena,
Departamento de Matemática Aplicada y Estadística, Paseo Alfonso XIII 52,
30203 Cartagena (Murcia), Spain

Abstract

In this paper, we prove the existence of injectors in any group for the class of nilpotent π-groups. We also prove that in a soluble group G, the number of conjugacy classes of $G/\mathbf{O}_\pi(\mathbf{F}(G))$ is bounded by the index of an injector of this class.

1 Introduction

All groups considered are finite and if π is a set of primes, we will denote by $\mathcal{N}_\pi$ the class of all nilpotent π-groups. Notice that if $\pi = \{p\}$, it coincides with the class of p-groups and if $\pi = \mathbb{P}$, it coincides with the class of nilpotent groups. Some properties of this class were analyzed by H. Bender in [1]. In this paper, H. Bender affirms that if G is a π-constrained group ($\mathbf{C}_{\bar{G}}(\mathbf{F}(\bar{G})) \leq \mathbf{F}(\bar{G})$ where $\bar{G} = G/\mathbf{O}_{\pi'}(G)$), $\mathcal{F}_\pi(G)$ denotes the set of (nilpotent) π-subgroups X of G such that $\bar{X}\mathbf{F}(\bar{G})$ is nilpotent and $\mathcal{F}^*_\pi(G)$ are the maximal elements of this set, these families of subgroups behave like p-subgroups, since they satisfy the following:

1. If $X \leq Y \in \mathcal{F}_\pi(G)$ then $X \in \mathcal{F}_\pi(G)$.
2. If $X, Y \in \mathcal{F}_\pi(G)$ and $Y \leq \mathbf{N}_G(X)$ then $XY \in \mathcal{F}_\pi(G)$.
3. If $N \trianglelefteq G$ then N is a π-constrained group, $\mathcal{F}_\pi(N) = \mathcal{F}_\pi(G) \cap N$ and $\mathcal{F}^*_\pi(N) = \mathcal{F}_\pi(G) \cap N$.
4. $\mathcal{F}^*_\pi(G)$ is a conjugacy class of subgroups of G.

It is not difficult to prove that if G is a π-constrained group then $\mathbf{Inj}_{\mathcal{N}_\pi}(G) = \mathcal{F}^*_\pi(G)$ and hence we have the existence and conjugation of the $\mathcal{N}_\pi$-injectors in π-constrained groups. Moreover we have that if G is π-constrained then

$$\mathbf{Inj}_{\mathcal{N}_\pi}(G) = \{H \mid H \in \mathbf{Hall}_\pi(I) \text{ where } \bar{I} \in \mathbf{Inj}_{\mathcal{N}}(\bar{G})\}$$

and if H is an $\mathcal{N}_\pi$-injector of G then $\bar{I}$ is an $\mathcal{N}$-injector of $\bar{G}$. Recall that if $\mathcal{F}$ is a Fitting class, a subgroup I of G is a $\mathcal{F}$-injector of G whenever $I \cap N$ is a maximal $\mathcal{F}$-subgroup of N for every subnormal subgroup N of G.

An important problem in Fitting classes is to prove the existence of injectors in any group with respect to a Fitting class $\mathcal{F}$. We find several results which solve this problem for certain families of Fitting classes. For instance, in [2] it is

[1]The research is supported by DGICYT, MEC of Spain. Proyecto PB-97-0674-C02-02.

proved that every group has nilpotent injectors and in [6], it is proved that if all $\mathcal{F}$-constrained groups have $\mathcal{F}$-injectors, then all groups have $\mathcal{F}$-injectors, where $\mathcal{F}$ is an homomorph and Fitting class such that $\mathrm{E}_Z\mathcal{F} = \mathcal{F}$. In particular, if $\mathcal{F}$ is a class of quasinilpotent groups containing the class of nilpotent groups, then every group has $\mathcal{F}$-injectors. In the first section we will prove the existence of $\mathcal{N}_\pi$-injectors in any group. Now, it is clearly true that $\mathcal{N}_\pi$ does not satisfy the property $\mathrm{E}_Z\mathcal{F} = \mathcal{F}$, so the techniques and the results of [6] do not apply.

If G is a group, $\mathbf{k}(G)$ will denote the number of conjugacy classes of G. In the second section we will analize a question introduced in [5] about the bound in a soluble group G of $\mathbf{k}(G/G_\mathcal{F})$ by $|G : I|$ where $\mathcal{F}$ is a Fitting class and I is an $\mathcal{F}$-injector of G. This paper shows that the answer to this question is affirmative for the classes of p-groups and nilpotent groups in Theorems A and B respectively. In the second section we will prove that the answer is affirmative for class $\mathcal{N}_\pi$.

2 The existence of $\mathcal{N}_\pi$-injectors

In this section we prove the existence of the $\mathcal{N}_\pi$-injectors in any finite group. For this purpose, the following results will be useful:

Proposition 2.1 *Let G be a group and π a set of primes.*
If $I/\mathbf{O}_{\pi'}(G) \in \mathbf{Inj}_{\mathcal{N}_\pi}(G/\mathbf{O}_{\pi'}(G))$ and $H \in \mathbf{Hall}_\pi(I)$ then $H \in \mathbf{Inj}_{\mathcal{N}_\pi}(G)$.

Proof Since $I/\mathbf{O}_{\pi'}(G)$ is a π-group, we have that $H\mathbf{O}_{\pi'}(G) = I$.

Suppose that $N \trianglelefteq\trianglelefteq G$ satisfying $H \cap N \leq L \leq N$ where $L \in \mathcal{N}_\pi$. Then:

$$(H \cap N)\mathbf{O}_{\pi'}(G)/\mathbf{O}_{\pi'}(G) \leq L\mathbf{O}_{\pi'}(G)/\mathbf{O}_{\pi'}(G) \leq N\mathbf{O}_{\pi'}(G)/\mathbf{O}_{\pi'}(G).$$

We will show that $(H \cap N)\mathbf{O}_{\pi'}(G) = I \cap N\mathbf{O}_{\pi'}(G)$.

Since $N \cap I \trianglelefteq\trianglelefteq I \in \mathcal{S}_{\pi'}\mathcal{S}_\pi$, $N \cap I \in \mathcal{S}_{\pi'}\mathcal{S}_\pi$. Moreover, as $H \in \mathbf{Hall}_\pi(I)$ and $N \cap I \trianglelefteq\trianglelefteq I$, $H \cap N \in \mathbf{Hall}_\pi(N \cap I)$. From this, it follows that $N \cap I = (H \cap N)\mathbf{O}_{\pi'}(N \cap I)$.

Therefore $I \cap N\mathbf{O}_{\pi'}(G) = \mathbf{O}_{\pi'}(G)(N \cap I) = \mathbf{O}_{\pi'}(G)(H \cap N)\mathbf{O}_{\pi'}(N \cap I) = \mathbf{O}_{\pi'}(G)(H \cap N)$.

Thus $\mathbf{O}_{\pi'}(G)(H \cap N) = L\mathbf{O}_{\pi'}(G)$ and then $|H \cap N| = |L|$ and $H \cap N = L$.

From this, we can conclude that $H \in \mathbf{Inj}_{\mathcal{N}_\pi}(G)$. □

Notice that subgroup I in the above result has π-Hall subgroups, since I is a π-soluble group.

Lemma *For any Fitting class $\mathcal{F}$ of groups and any group G, let $E_1, \ldots, E_n$ be a set of $\mathcal{F}$-components such that it is invariant under conjugation by elements of G. Let $F_i \in \mathbf{Inj}_\mathcal{F}(E_i)$ and put $F = \langle F_1, \ldots, F_n \rangle$. Then $\mathbf{Inj}_\mathcal{F}(\mathbf{N}_G(F)) \subseteq \mathbf{Inj}_\mathcal{F}(G)$.*
(Theorem 2.1 of [7])

Theorem 2.2 *If G is a group then $\mathbf{Inj}_{\mathcal{N}_\pi}(G) \neq \emptyset$.*

Proof Suppose the theorem is false, and let G be a minimal counterexample.

If $\mathbf{O}_{\pi'}(G) \neq 1$, there exists $I/\mathbf{O}_{\pi'}(G) \in \mathbf{Inj}_{\mathcal{N}_\pi}(G/\mathbf{O}_{\pi'}(G))$ and $H \in \mathbf{Hall}_\pi(I)$. Using proposition 1, we obtain that $H \in \mathbf{Inj}_{\mathcal{N}_\pi}(G)$.

Therefore we can suppose that $\mathbf{O}_{\pi'}(G) = 1$.

Let $E_1, E_2, \ldots, E_r$ be the components of G and let $I = I_1 I_2 \ldots I_r$ where I_i is a maximal $\mathcal{N}_\pi$-subgroup of E_i containing $\mathbf{O}_\pi(\mathbf{F}(E_i))$.

If $\mathbf{N}_G(I) \neq G$ then, by minimality of G, we have that $\mathbf{Inj}_{\mathcal{N}_\pi}(\mathbf{N}_G(I)) \neq \emptyset$. Then, using the above lemma, we obtain that $\emptyset \neq \mathbf{Inj}_{\mathcal{N}_\pi}(\mathbf{N}_G(I)) \subseteq \mathbf{Inj}_{\mathcal{N}_\pi}(G)$.

Hence we can suppose that $I \trianglelefteq G$ and as $I_i \leq I \cap E_i \leq E_i$, we have that $I_i = I \cap E_i \trianglelefteq E_i$. Thus, as $\mathbf{O}_{\pi'}(E_i) = 1$, $\mathbf{Z}(E_i)$ is a π-group so $\mathbf{Z}(E_i) \leq I_i$. From this, $I_i/\mathbf{Z}(E_i) \trianglelefteq E_i/\mathbf{Z}(E_i)$ which is simple. Thus $I_i \leq \mathbf{Z}(E_i)$ and then $\mathbf{Z}(E_i)$ is an $\mathcal{N}_\pi$-maximal subgroup of E_i. Therefore $\mathbf{Z}(E_i)$ is a Hall π-subgroup of E_i and $E_i = \mathbf{Z}(E_i) \times H_i$ where H_i is a Hall π'-subgroup of E_i. Since E_i es perfect group we can conclude that $E_i = 1$, $1 \leq i \leq r$. Hence G is an $\mathcal{N}$-constrained group satisfying $\mathbf{O}_{\pi'}(G) = 1$, therefore G is π-constrained and then $\mathbf{Inj}_{\mathcal{N}_\pi}(G) \neq \emptyset$, which is a contradiction.

□

It is known that if G possesses a single conjugacy class of nilpotent injectors then G is an $\mathcal{N}$-constrained group ([6]). The analogous result for class $\mathcal{N}_\pi$ does not generally hold, that is, there exist groups with a single conjugacy class of $\mathcal{N}_\pi$-injectors that are not π-constrained, because if we consider $G = \mathbf{A}_5$ and $\pi = \{2\}$, by Sylow's theorem, the $\mathcal{N}_\pi$-injectors of G are the Sylow 2-subgroups which are conjugate, but $G/\mathbf{O}_{\pi'}(G) = G/\mathbf{O}_{\{3,5\}}(G) \cong G$ which is not an $\mathcal{N}$-constrained group.

Recall that if G is a group, $\mathcal{A}(2, G)$ denotes the set of subgroups with maximum order among the nilpotent subgroups of G with nilpotency class less or equal than 2 and $\mathcal{NI}(G)$ denotes the set of the maximal nilpotent subgroups of G containing an element of $\mathcal{A}(2, G)$. It is known that if G is an $\mathcal{N}$-constrained group then $\mathbf{Inj}_{\mathcal{N}}(G) = \mathcal{NI}(G)$.

Definition 2.3 If G is a group, we define $\mathcal{A}_\pi(2, G)$ as the set of the $\mathcal{N}_\pi$-subgroups of G with nilpotency class less or equal than 2.

Proposition 2.4 *Let G be a π-constrained group and let A be a π-subgroup of G. Then, the following statements are equivalent:*

1. $A \in \mathcal{A}_\pi(2, G)$.
2. $\bar{A} \in \mathcal{A}(2, \bar{G})$.

Proof If $A \in \mathcal{A}_\pi(2, G)$ then $\bar{A} \cong A$ has a nilpotency class less or equal than 2.

If $\bar{R} \in \mathcal{A}(2, \bar{G})$ then, by Theorem B in [4] we obtain that $\bar{R}\mathbf{F}(\bar{G})$ is nilpotent and since $\bar{G}$ is $\mathcal{N}$-constrained, we have that $\bar{R}$ is a π-group. Consequently, if $H \in \mathbf{Hall}_\pi(R)$, $R = H\mathbf{O}_{\pi'}(G)$ and $H \cong \bar{R} \in \mathcal{N}_\pi$ with a nilpotency class less or equal than 2, then $|H| \leq |A|$. Hence $|\bar{R}| = |\bar{H}| \leq |\bar{A}|$ and $\bar{A} \in \mathcal{A}(2, \bar{G})$.

Reciprocally, if $\bar{A} \in \mathcal{A}(2, \bar{G})$ then $A \in \mathcal{N}_\pi$ with nilpotency class less or equal than 2 and if $B \in \mathcal{A}_\pi(2, G)$ we have $\bar{B}$ has a nilpotency class less or equal than 2. Hence $|\bar{B}| \leq |\bar{A}|$ and thus $|B| \leq |A|$ and $A \in \mathcal{A}_\pi(2, G)$. □

Definition 2.5 If G is a group, we define $\mathcal{NI}_\pi(G)$ as the set of maximal nilpotent π-subgroups of G containing an element in $\mathcal{A}_\pi(2, G)$.

Theorem 2.6 *If G is a π-constrained group then $\mathbf{Inj}_{\mathcal{N}_\pi}(G) = \mathcal{NI}_\pi(G)$.*

Proof If $I \in \mathbf{Inj}_{\mathcal{N}_\pi}(G)$ then $\bar{I} \in \mathbf{Inj}_{\mathcal{N}}(\bar{G})$ and thus there exists $\bar{R} \in \mathcal{A}(2, \bar{G})$ such that $\bar{R} \leq \bar{I}$. If $R_\pi \in \mathbf{Hall}_\pi(R)$ then, using the above proposition, $R_\pi \in \mathcal{A}_\pi(2, G)$ and $R_\pi \leq I\mathbf{O}_{\pi'}(G)$. Then, there exists $g \in G$ such that $R_\pi^g \leq I$ and then $R_\pi^g \in \mathcal{A}_\pi(2, G)$.

Reciprocally, let $I \in \mathcal{NI}_\pi(G)$. Then, if $A \in \mathcal{A}_\pi(2, G)$ such that $A \leq I$ we have that $\bar{A} \in \mathcal{A}(2, \bar{G})$ and $\bar{A} \leq \bar{I}$.

We will show that $\bar{I}$ is a maximal nilpotent subgroup of $\bar{G}$. If $\bar{J}$ is a maximal nilpotent subgroup of $\bar{G}$ containing $\bar{I}$, since $\mathbf{C}_{\bar{G}}(\bar{A}) \leq \bar{A}$ we have that $\bar{J}$ is a π-group.

Therefore, if $J_\pi \in \mathbf{Hall}_\pi(J)$, $I\mathbf{O}_{\pi'}(G) \leq J = J_\pi \mathbf{O}_{\pi'}(G)$, so there exists $g \in J$ such that $I \leq J_\pi^g \in \mathcal{N}_\pi$ and hence, by $\mathcal{N}_\pi$-maximality of I, we obtain that $I = J_\pi^g$ and thus $|I| = |J_\pi|$ and $\bar{I} \leq \bar{J}$. Hence, $\bar{I} = \bar{J}$ and $\bar{I}$ contains an element of $\mathcal{A}(2, \bar{G})$. Consequently, $\bar{I}$ is an $\mathcal{N}$-injector of $\bar{G}$ and then I is an $\mathcal{N}_\pi$-injector of G.

□

Returning to the conjugacy of maximal $\mathcal{N}_\pi$-subgroups in any group we obtain:

Proposition 2.7 *Let $N \leq G$ satisfying $\mathbf{C}_G(N) \leq N\mathbf{O}_{\pi'}(G)$. Suppose X_1, X_2 are $\mathcal{N}_\pi$-maximal subgroups of G such that $N \leq X_1 \cap X_2$. Then X_1 and X_2 are conjugate.*

Proof

We have that $\mathbf{C}_{\bar{G}}(\bar{N}) = \overline{\mathbf{C}_G(N)} \leq \bar{N}$.

We shall see that $\bar{X}_i$ is a maximal nilpotent subgroup of $\bar{G}$, $i = 1, 2$.

Let $i \in \{1, 2\}$ and suppose $\bar{X}_i \leq \bar{J}$, where J is a nilpotent subgroup of $\bar{G}$. Then, since $\bar{J}$ contains $\bar{N}$ satisfying $\mathbf{C}_{\bar{G}}(\bar{N}) \leq \bar{N}$, it follows that $\bar{J}$ is a π-group. Thus there exists a π-subgroup R of J such that $J = R\mathbf{O}_{\pi'}(G)$.

Hence, since $X_i\mathbf{O}_{\pi'}(G) \leq R\mathbf{O}_{\pi'}(G)$, X_i is a π-subgroup of $R\mathbf{O}_{\pi'}(G)$. Thus there exists $g \in R\mathbf{O}_{\pi'}(G)$ such that $X_i \leq R^g$ and since $R \cong J/\mathbf{O}_{\pi'}(G)$, which is a nilpotent π-group, we have that $X_i = R^g$.

Thus we obtain that $\bar{X}_i = \overline{R^g} = (R\mathbf{O}_{\pi'}(G))^g/\mathbf{O}_{\pi'}(G) = \bar{R} = \bar{J}$.

Now $\mathbf{C}_{\bar{G}}(\bar{N}) \leq \bar{N} \leq \bar{X}_1 \cap \bar{X}_2$ and using Proposition 2 in [9] there exists $g \in G$ such that $\bar{X_1}^g = \bar{X}_2$. Hence $X_1{}^g\mathbf{O}_{\pi'}(G) = X_2\mathbf{O}_{\pi'}(G)$ and since $X_1{}^g$ and X_2 are Hall π-subgroups of $X_2\mathbf{O}_{\pi'}(G)$, there exists $h \in X_2\mathbf{O}_{\pi'}(G)$ such that $X_1{}^{gh} = X_2$ and finally X_1 and X_2 are conjugate in G.

□

3 A bound of the number of conjugacy classes relatively to the class $\mathcal{N}_\pi$

The following result is due to Gallagher [3].

Lemma *If G is a finite group and $N \trianglelefteq G$ then:*

$$\mathbf{k}(G) \leq \mathbf{k}(G/N)\mathbf{k}(N)$$

We shall see in this section that in a soluble group G, the number of conjugacy classes of $G/\mathbf{O}_\pi(\mathbf{F}(G))$ is bounded by the index in G of an $\mathcal{N}_\pi$-injector.

We will denote by $\mathcal{F} = \mathcal{S}_{\pi'} \diamond \mathcal{N}_\pi = \{G \mid G/\mathbf{O}_{\pi'}(G) \in \mathcal{N}_\pi\}$ which is a Fitting homomorph. First of all, we will prove the following result:

Lemma 3.1 *If $G \in \mathcal{F}$ and $\pi \subseteq \pi(G)$ then:*

$$\mathbf{k}(G/\mathbf{O}_\pi(G)) \leq |G|_{\pi'}.$$

Proof

We argue by induction on $|G|$.

If $\mathbf{O}_\pi(G) \neq 1$ then $G/\mathbf{O}_\pi(G) \in \mathcal{F}$ and by induction:

$$\mathbf{k}(G/\mathbf{O}_\pi(G)) \leq |G/\mathbf{O}_\pi(G)|_{\pi'} = |G|_{\pi'}.$$

Thus we can suppose that $\mathbf{O}_\pi(G) = 1$.

Let N be a minimal normal subgroup of G. Since G is soluble, N is an elementary abelian p-group and $p \notin \pi$.

Let $M/N = \mathbf{O}_\pi(G/N)$. By induction:

$$\mathbf{k}((G/N)/\mathbf{O}_\pi(G/N)) \leq |G/N|_{\pi'}$$

and thus:

$$\mathbf{k}(G/M) \leq |G/N|_{\pi'}.$$

If $M < G$, since $M \in \mathcal{F}$, by induction:

$$\mathbf{k}(M/\mathbf{O}_\pi(M)) \leq |M|_{\pi'}$$

but since $\mathbf{O}_\pi(M) = 1$ and M/N is a π-group, we obtain:

$$\mathbf{k}(M) \leq |N|_{\pi'}$$

and then, applying the above lemma, we obtain:

$$\mathbf{k}(G) \leq \mathbf{k}(G/M)\mathbf{k}(M) \leq |G/N|_{\pi'}|N|_{\pi'} = |G|_{\pi'}.$$

Hence, we can suppose that $M = G$ and then $\mathbf{O}_\pi(G/N) = G/N$. Thus G/N is a π-group and $G = NR$, where N is an elementary abelian normal p-subgroup of G and R is a nilpotent p'-subgroup of G. Therefore N is a faithful irreducible $\mathbf{GF}(p)[R]$-module and since R is supersoluble, applying a theorem of Knörr (7.4 of [8]) we obtain:

$$\mathbf{k}(G) = \mathbf{k}(NR) \leq |N| = |G|_{\pi'}.$$

□

Theorem 3.2 *If G is a soluble group then:*

$$\mathbf{k}(G/\mathbf{O}_\pi(\mathbf{F}(G))) \leq |G : J|$$

where $J \in \mathbf{Inj}_{\mathcal{N}_\pi}(G)$.

Proof

Let $\mathbf{F}(G/\mathbf{O}_{\pi'}(G)) = R\mathbf{O}_{\pi'}(G)/\mathbf{O}_{\pi'}(G)$ where R is a π-group. Since $\bar{J}$ is a nilpotent injector of $\bar{G}$, applying Theorem B in [5], it follows:

$$\mathbf{k}((\bar{G}/\mathbf{F}(\bar{G})) \leq |\bar{G} : \bar{J}|$$

and thus:

$$\mathbf{k}(G/(R\mathbf{O}_{\pi'}(G))) \leq |G : J\mathbf{O}_{\pi'}(G)|.$$

Now, we consider $M = R\mathbf{O}_{\pi'}(G) \trianglelefteq G$. Therefore we have that $\mathbf{O}_\pi(M) \leq R$ which is nilpotent and since $\mathbf{O}_\pi(\mathbf{F}(G)) \leq M$, it follows that $\mathbf{O}_\pi(M) = \mathbf{O}_\pi(\mathbf{F}(G))$.

Thus, since $M \in \mathcal{S}_{\pi'} \diamond \mathcal{N}_\pi$, applying the above lemma we obtain:

$$\mathbf{k}(M/\mathbf{O}_\pi(M)) \leq |M|_{\pi'}$$

and:

$$\mathbf{k}(R\mathbf{O}_{\pi'}(G)/\mathbf{O}_\pi(\mathbf{F}(G))) \leq |\mathbf{O}_{\pi'}(G)|.$$

Since $R\mathbf{O}_{\pi'}(G)/\mathbf{O}_\pi(\mathbf{F}(G)) \trianglelefteq G/\mathbf{O}_\pi(\mathbf{F}(G))$, if we denote $\widetilde{G} = G/\mathbf{O}_\pi(\mathbf{F}(G))$, applying previous lemma:

$$\mathbf{k}(\widetilde{G}) \leq \mathbf{k}(\widetilde{G}/(\widetilde{R\mathbf{O}_{\pi'}(G)}))\mathbf{k}((\widetilde{R\mathbf{O}_{\pi'}(G)})) \leq |G : J\mathbf{O}_{\pi'}(G)||\mathbf{O}_{\pi'}(G)| = |G : J|.$$

□

An argument like that of [5] yields:

Corollary 3.3 *If G is a soluble group and $\theta \in \mathbf{Irr}(\mathbf{O}_\pi(\mathbf{F}(G)))$ is G-invariant then $|\mathbf{Irr}(G|\theta)| \leq |G : J|$ where $J \in \mathbf{Inj}_{\mathcal{N}_\pi}(G)$.*

Acknowledgement The author wishes to express his thanks to M. J. Iranzo and F. Pérez-Monasor for their kind help in the preparation of this paper.

References

[1] H. Bender, *Nilpotent π-subgroups behaving like p-subgroups*, Proc. Rutgers group theory year 1983/84 (1984), 119–125.

[2] P. Förster, *Nilpotent injectors in finite groups*, Bull. Austral. Math. Soc. 32 (1985), no. 2, 293–297.

[3] P. X. Gallagher, *The number of conjugacy classes in a finite group*, Math. Z. 118 (1970), 175–179.

[4] G. Glauberman, *On Burnside's other p^aq^b theorem*, Pacific J. Math. 56 (1975), no. 2, 469–476.

[5] M. J. Iranzo, G. Navarro, F. Pérez-Monasor, *A conjecture on the number of conjugacy classes in a p-solvable group*, Israel J. Math. 93 (1996), 185–188.

[6] M. J. Iranzo, F. Pérez-Monasor, *Fitting classes $\mathcal{F}$ such that all finite groups have $\mathcal{F}$-injectors*, Israel J. Math. 56 (1986), 97–101.

[7] M. J. Iranzo, F. Pérez-Monasor, M. Torres, *A criterion for the existence of injectors in finite groups*, Proceedings of the Second International Group Theory Conference (Bressanone, 1989).

[8] R. Knörr, *On the number of characters in a p-block of a p-solvable group*, Illinois J. Math. 28 (1984), no. 2, 181–210.

[9] H. Lausch, *Conjugacy classes of maximal nilpotent subgroups*, Israel J. Math. 47 (1984), no. 1, 29–31.

[10] A. Mann, *Injectors and normal subgroups of finite groups*, Israel J. Math. 9 (1971), 554–558.

CHARACTERS OF p-GROUPS AND SYLOW p-SUBGROUPS

ALEXANDER MORETÓ[1]

Departamento de Matemáticas, Universidad del País Vasco, Apartado 644, 48080 Bilbao (Spain)
e-mail: mtbmoqua@lg.ehu.es

1 Introduction

The aim of this note is to present some problems and also partial results in some cases, mainly on characters of p-groups. (In the last section we deal with a problem that consists in obtaining information about characters of a Sylow p-subgroup of an arbitrary group from information about the characters of the whole group.) This survey is far from being exhaustive. The topics included are strongly influenced by the author's interests in the last few years. There seems to be an increasing interest in the character theory of p-groups and we hope that this expository paper will encourage more research in the area. In the sixties I. M. Isaacs and D. S. Passman [17, 18] wrote two important papers that initiated the study of the degrees of the irreducible complex characters of finite groups (henceforth referred to as character degrees). The study of the influence of the set of character degrees on the structure of a group was taken up again in the eighties, in large part due to B. Huppert and his school. In particular, this has led to several papers dealing with the character degrees of important families of p-groups since the nineties (see [6, 8, 12, 28, 30, 32, 33, 34, 35, 36, 37]). Here we are mostly concerned with character degrees, but instead of studying particular families of p-groups, we intend to obtain general structural properties of groups according to their character degrees. Other problems on characters of p-groups appear in [25].

The notation is standard. All the groups considered are finite. We write $\mathrm{cd}(G)$ to denote the set of character degrees of a group G, $b(G)$ the maximum of the character degrees of G, $\mathrm{cs}(G)$ the set of conjugacy class sizes, and $c(G)$ and $\mathrm{dl}(G)$ the nilpotence class and derived length of G, respectively. The terms of the ascending Fitting series of a group G will be denoted $F_i(G)$ and the Fitting subgroup $F(G)$. If P is a p-group we write $\Omega_i(P)$ to denote the subgroup of P generated by the elements of order $\leq p^i$.

2 Bounding the derived length and the nilpotence class

Taketa proved that if G is a monomial group (in particular, a p-group) then $\mathrm{dl}(G) \leq |\,\mathrm{cd}(G)|$. An important problem in character theory of finite solvable groups is the Isaacs-Seitz conjecture, which asserts that the derived length of any solvable group G is bounded by $|\,\mathrm{cd}(G)|$. Despite the fact that this conjecture is not proved yet, it is widely believed that the "right" bound for $\mathrm{dl}(G)$ in terms of $|\,\mathrm{cd}(G)|$ is logarithmic. This is the case for several families of p-groups, like the Sylow

[1]Research supported by the Basque Government and the University of the Basque Country.

subgroups of the symmetric groups or the Sylow p-subgroups of the general linear groups in characteristic p.

A related problem was studied by Isaacs and Knutson in [15]. If N is a normal subgroup of G, we write $\mathrm{cd}(G|N)$ to denote the set of character degrees of G whose kernel does not contain N, i.e., the set of degrees of the irreducible characters of G that "'say something about N". With this notation, they obtained bounds for $\mathrm{dl}(N)$ in terms of $|\,\mathrm{cd}(G|N)|$.

In an impressive series of papers, T. M. Keller (see [21, 22, 23]) has reduced the problem of finding a logarithmic bound for the derived length of a solvable group in terms of the cardinality of the set of character degrees to the following conjecture.

Conjecture 1 Let G be a solvable group. Then there exist constants C_1 and C_2 such that

$$\mathrm{dl}(F(G)) \leq C_1 \log|\,\mathrm{cd}(G|F(G))| + C_2.$$

The reason for the inclusion of this conjecture in this survey is that its proof should not be much harder than that of the following conjecture.

Conjecture 2 Let P be a p-group. Then $\mathrm{dl}(P)$ is bounded logarithmically in terms of $|\,\mathrm{cd}(P)|$.

In other words, Keller's work comes close to reducing to p-groups the problem of replacing Taketa's bound by a logarithmic one. Unfortunately, the p-group case seems to be extremely hard. For instance, it is not known the answer to the following well-known question.

Question 2.1 Does there exist a p-group of derived length 4 with 4 character degrees?

M. C. Slattery has proved that the set of degrees of such a p-group cannot be $\{1, p, p^2, p^3\}$ [38].

Of course, one cannot hope to obtain any bounds for the nilpotence class of a p-group in terms of the number of character degrees. (It is well-known that there exist p-groups P of maximal class of arbitrarily large order with an abelian subgroup of index p and, therefore, $\mathrm{cd}(P) = \{1, p\}$.) However, if we fix the set $\mathcal{S}$ of character degrees then, sometimes, we can obtain bounds for the nilpotence class. In the following, $\mathcal{S}$ denotes a finite set of powers of p containing 1. We say that $\mathcal{S}$ is class bounding if there exists a constant C (depending on $\mathcal{S}$) such that $c(P) \leq C$ for any p-group P with $\mathrm{cd}(P) = \mathcal{S}$. In 1968 Isaacs and Passman [18] proved that if $|\mathcal{S}| = 2$ then $\mathcal{S}$ is class bounding if and only if p does not belong to $\mathcal{S}$. Later, in 1994, Slattery [39] found sets of arbitrarily large size that are class bounding within the class of metabelian groups. However, apart from the result of Isaacs and Passman on sets of cardinality 2, there were no more theorems asserting that a given set $\mathcal{S}$ is class bounding (or non-class-bounding) until 2001. In [16] Isaacs and the present author proved, among other results, that if $\mathcal{S} \subseteq \{1, p^a, \ldots, p^{2a}\}$ then $\mathcal{S}$ is class bounding. All the class bounding sets $\mathcal{S}$ found in that paper have the property that $p \notin \mathcal{S}$. In [19] Isaacs and Slattery prove that this a necessary condition for a set $\mathcal{S}$ to be class bounding. However, an example constructed in [20] shows that this is not a sufficient condition. In [20] Jaikin-Zapirain and the author also find more class bounding sets. We refer the reader to [16, 19] and [20] for the detailed results and to [20] for some specific questions related to this problem.

Question 2.2 Which are the class bounding sets?

A complete answer to this question seems to be out of the scope of the known methods. We do not know even what to conjecture. I am inclined to think that the probability that a set is class bounding is 0, in the sense that

$$\lim_{n\to\infty} \frac{\#\{\mathcal{S} \mid \mathcal{S} \text{ is class bounding and } \max(\mathcal{S}) \leq p^n\}}{\#\{\mathcal{S} \mid \max(\mathcal{S}) \leq p^n\}} = 0,$$

but there is little evidence for this.

We close this section with a question that relates the two problems discussed here.

Question 2.3 Is it true that if $\mathrm{cd}(P) = \mathcal{S}$, $|\mathcal{S}| = 3$ and $p \notin \mathcal{S}$, then P is metabelian?

On the one hand, an affirmative answer to this question would give a new proof of part of Theorem D of [16]. On the other hand, it would provide another situation where Taketa's bound is not best possible.

3 Minimal characters and normal subgroups

We write $m(P)$ to denote the minimal degree of the non-linear irreducible characters of P (for convenience, we write $m(P) = 1$ if P is abelian). It is well-known that this number has a strong influence on the structure of the group P (see, for instance, Problem 5.14 of [11] and [26]). In particular, we want to stress the influence of $m(P)$ in the last problem discussed in the previous section (see [20]). We introduce a new invariant associated to any p-group. Write $|G : Z(G)| = p^{2n+e}$, where $e \in \{0,1\}$. We define $m_1(P) = n - \log_p m(P)$, i.e., $m(P) = p^{n-m_1(P)}$.

With this notation, the groups studied in [4] are exactly those that satisfy $m_1(P) = 0$. One of the main results of that paper characterizes such groups in terms of their normal subgroups. More precisely, we prove the following. (See Theorems B and C of [4].)

Theorem 3.1 1. *Assume that $e = 0$. Then $m_1(P) = 0$ if and only if P satisfies the strong condition on normal subgroups.*

2. *Assume that $e = 1$. Then $m_1(P) = 0$ if and only if P satisfies the weak condition on normal subgroups.*

We recall from [4] the definition of the strong (weak) condition. We say that a p-group P satisfies the strong (weak) condition on normal subgroups if for every $N \trianglelefteq P$, either $P' \leq N$ or $N \leq Z(P)$ (either $P' \leq N$ or $|NZ(P) : Z(P)| \leq p$).

The proof of Theorem 3.1 requires a careful study of the groups satisfying these properties. This analysis leads G. A. Fernández-Alcober and the author to group-theoretical properties of the groups under consideration which, we think, have some interest by themselves. For instance, in Theorem F of [4] we obtain a bound for the index of the center of the groups with any of these properties and class ≥ 3.

These definitions and part of Theorem F of [4] have been generalized by Isaacs [14]. Isaacs defines an invariant $a(P)$ associated to any finite p-group P as follows: $a(P) = a$ is the minimum integer such that if N is a normal subgroup of P and $|NZ(P) : Z(P)| \geq p^a$ then $P' \leq N$. Observe that $a(P) = 0$ if and only if P is abelian, $a(P) = 1$ if and only if P satisfies the strong condition and is not abelian and $a(P) = 2$ if and only if P satisfies the weak condition but not the strong condition. Therefore, Theorem 3.1 says that if $e = 0$, then $m_1(P) = 0$ if and only if $a(P) \leq 1$, while if $e = 1$, then $m_1(P) = 0$ if and only if $a(P) \leq 2$. We think that it should be possible to generalize this result too. Our aim now is to present some results that suggest this.

We begin by proving that if $m_1(P)$ is small then $a(P)$ is also small.

Proposition 3.2 *With the notation above, if $m_1(P) \leq m$ for some $a \in \mathbb{N}$, then $a(G) \leq 2m + 1 + e$.*

Proof Let N be a normal subgroup of P such that $P' \not\leq N$. Since P/N is not abelian and $\mathrm{cd}(P/N) \subseteq \mathrm{cd}(P)$, we have that

$$p^{2(n-m)} \leq |P/N : Z(P/N)| \leq |P : NZ(P)| \leq |P : Z(P)| = p^{2n+e},$$

and it follows that $|NZ(P) : Z(P)| \leq p^{2m+e}$. Therefore $a(P) \leq 2m + 1 + e$. □

The bound obtained here is best possible, as direct products of p-groups of maximal class with non-trivial abelian p-groups show. This result generalizes the easy part of Theorem 3.1. However, in this general setting the converse is not true, as the following example shows.

Example 3.3 Let $P = \langle x, y \mid x^{p^n} = y^{p^{2n}} = 1, y^x = y^{1+p^n} \rangle$ for $n \in \mathbb{N}$. Then $P' = Z(P) = \langle y^{p^n} \rangle$, $|P : Z(P)| = p^{2n}$ and $N = \langle x^p, y^{p^{n+1}} \rangle$ is a normal subgroup of P such that $P' \not\leq N$ and $|NZ(P) : Z(P)| = p^{n-1}$. Now it is possible to prove that $a(P) = n$ and $m_1(P) = n - 1$.

Now we prove that if P is a group of class 2 and $a(P)$ is small, then $m_1(P)$ is small. We first need a lemma, which is a generalization of [4, Theorem D]. In order to prove it it is enough to mimic the proof of Theorem D of [4], so we will just give a sketch of it.

Lemma 3.4 *Let P be a p-group of class 2. If $a(P) > 1$, then $\exp P/Z(P) \leq p^{a(P)}$. Furthermore, if $\exp P/Z(P) = p^{a(P)}$ then $P/Z(P) \cong C_{p^{a(P)}} \times C_{p^{a(P)}}$ and $P' \cong C_{p^{a(P)}}$.*

Proof Let M be a maximal subgroup of P' and let N be a normal subgroup of P maximal with respect to the property $P' \cap N = M$. Let $K/N = Z(P/N)$. Since the derived subgroup of P/N is of order p, P/K is elementary abelian. By the choice of N, we have that K/N is cyclic, so $K/NZ(P)$ is cyclic. Since $P' \not\leq N$, $|NZ(P) : Z(P)| \leq p^{a(P)-1}$. Now the bound for the exponent of $P/Z(P)$ follows from the fact that this abelian group has at least two cyclic factors of maximal order.

If $\exp P/Z(P) = p^{a(P)}$ the previous argument shows that $P/Z(P)$ is a direct product of two cyclic groups of order $p^{a(P)}$ and, perhaps, some cyclic factors of smaller order. We want to prove that in fact all the cyclic factors are of order $p^{a(P)}$. In this case, we can choose $M = \Omega_{a(P)-1}(P')$. Let $T/Z(P) = \Omega_{a(P)-1}(P/Z(P))$. Then $\exp[T,P] = \exp T/Z(P) = p^{a(P)-1}$ and consequently $[T,P] \leq M$. Hence $T \leq K$. We know that $K/NZ(P)$ is cyclic and $|NZ(P) : Z(P)| \leq p^{a(P)-1}$. Therefore,

$$|\Omega_{a(P)-1}(P/Z(P))| \leq p^{2a(P)-2}.$$

It follows that $P/Z(P) \cong C_{p^{a(P)}} \times C_{p^{a(P)}}$ and $P' \cong C_{p^{a(P)}}$. □

The previous example shows that the bound obtained in this lemma cannot be improved. Of course, the order of $P/Z(P)$ cannot be bounded in terms of $a(P)$ when $c(P) = 2$. This lemma proves that, at least, it is possible to bound the exponent.

Theorem 3.5 *Let P be a p-group of class 2. If $a(P) = a > 1$ then $m_1(P) \leq a - 1 - e$.*

Proof Since $c(P) = 2$, $\chi(1)^2 = |P : Z(\chi)|$ for any $\chi \in \mathrm{Irr}(P)$, by Theorem 2.31 of [11]. Let $K = \operatorname{Ker}\chi$. If χ is non-linear, then $|KZ(P) : Z(P)| \leq p^{a-1}$. We also know that $Z(\chi)/K$ is cyclic and $\exp P/Z(P) \leq p^{a-e}$ by the previous lemma. It follows that $|Z(\chi) : Z(P)| \leq p^{2a-1-e}$, so

$$|P : Z(\chi)| \geq p^{2n-2a+1+2e}.$$

Since this index is a square, we have that $|P : Z(\chi)| \geq p^{2(n-a+1+e)}$, and we deduce that $\chi(1) \geq p^{n-a+1+e}$, as we wanted to prove. □

The bound in this theorem cannot be improved, as Example 3.3 shows. However, I think that the hypothesis on the class can be removed.

Conjecture 3 Let P be a p-group. Then $m_1(P) \leq a(P)-1-e$ whenever $a(P) > 1$.

4 Miscellaneous questions

In the last years a number of similar results for character degrees and conjugacy class sizes have been obtained. However, the reason for the existence of this parallelism, if any, is still unknown. Isaacs [10] proved that any set of powers of p containing 1 occurs as the set of character degrees of a p-group of class ≤ 2.

The analog result for conjugacy class sizes has been obtained by J. Cossey and T. Hawkes [2]. Although it is easy to find sets of character degrees (resp. conjugacy class sizes) that impose restrictions on the class sizes (resp. character degrees), Fernández-Alcober and the author [5] proved that there is not any relation between the cardinalities of these sets, i.e., given any two integers m and n greater than 1, there exists a p-group with m character degrees and n conjugacy class sizes. A complete answer to the following question, which was first raised in [5], seems to be very difficult.

Question 4.1 Determine the pairs (A, B) of sets of powers of p containing 1 such that there exists a p-group P with $\mathrm{cd}(P) = A$ and $\mathrm{cs}(P) = B$.

The particular case $|A| = |B| = 2$ is being studied by Cossey and Hawkes [7].

We recall that a group G is normally monomial if any irreducible character of G is induced from a linear character of a normal subgroup. It was thought that, perhaps, the derived length of normally monomial groups was bounded by some constant. However, in [24] L. Kovacs and C. R. Leedham-Green constructed normally monomial p-groups of derived length the integer part of $\log_2(p + 1)$. As far as I know, the answer to the following question is still unknown.

Question 4.2 Does there exist a function f such that if P is a normally monomial p-group then $\mathrm{dl}(P) \leq f(p)$?

A classical problem in character theory is to determine what kind of group theoretical information can be determined from the knowledge of the character table. For instance, R. Brauer [1] asked whether or not the derived length of a solvable group can be read off from the character table. This question was answered negatively even for p-groups by S. Mattarei [29], but the following question remains open.

Question 4.3 Do there exist p-groups P_1 and P_2 with the same character table and $\mathrm{dl}(P_1) \geq \mathrm{dl}(P_2) + 2$?

5 Character degrees of Sylow p-subgroups

There are some known results that give information on the structure of a finite group in terms of the p-parts of the irreducible complex (or Brauer) characters (see, for instance, [27]). In this section we deal with the following problem.

Conjecture 4 Let G be a finite group and write $e_p(G)$ to denote the exponent of the largest p-part of the irreducible complex characters of G. Then $e_p(P)$ is bounded by some function of $e_p(G)$, where $P \in \mathrm{Syl}_p(G)$.

If $e_p(G) = 0$ then P is abelian (by Ito-Michler's theorem), so we will assume $e_p(G) > 0$ in the remaining. This might lead one to think that, in fact, $e_p(P) \leq e_p(G)$ but this is false even for solvable groups, as the following example of Isaacs shows.

Example 5.1 Let V be a row vector space of dimension 3 over the finite field with two elements. Consider the natural action of the subgroup H of $\mathrm{GL}(3,2)$ consisting of the matrices

$$\begin{pmatrix} \mathrm{GL}(2,2) & \mathbf{0} \\ * \; * & 1 \end{pmatrix}$$

on V. If $G = VH$ then $e_2(G) = 1$ and $e_2(P) = 2$, where P is a Sylow 2-subgroup of G. There are similar examples for $p = 3$.

As far as I know, it might be true that $e_p(P) \leq e_p(G)$ for $p \geq 5$ and G solvable and that $e_p(P) \leq 2e_p(G)$ for any prime and any group. This would imply that if G is solvable then $\mathrm{dl}(P) \leq 2e_p(G) + 1$, using Taketa's theorem. This last inequality is the main theorem of [9].

We are able to prove the bound $e_p(P) \leq e_p(G)$ for metanilpotent groups (in particular for supersolvable groups). However this bound does not hold for groups of Fitting height 3 as Isaacs' example shows. For solvable groups we obtain a bound for $e_p(P)$ in terms of $e_p(G)$ and the Fitting height of G.

We begin with an elementary lemma.

Lemma 5.2 *Let N be a normal subgroup of G. Then $e_p(N) \leq e_p(G)$. In particular, if G has a normal Sylow p-subgroup P, then $e_p(P) \leq e_p(G)$.*

Proof This is an immediate consequence of Clifford theory. □

Theorem 5.3 *Let G be a solvable group and R the smallest normal subgroup of G such that G/R is nilpotent. Assume that R is p-nilpotent. Then $e_p(P) \leq e_p(G)$ for any $P \in \mathrm{Syl}_p(G)$.*

Proof We may assume that $PR > R$ (otherwise the result follows by induction and Lemma 5.2). Since G/R is nilpotent and PR/R is a Sylow p-subgroup of G/R, we have that $PR \trianglelefteq G$ and, again by induction and Lemma 5.2, we may assume that $G = PR$. Then G is p-nilpotent, and the result follows. □

Corollary 5.4 *Let G be a metanilpotent group and $P \in \mathrm{Syl}_p(G)$. Then $e_p(P) \leq e_p(G)$.*

Our next aim is to bound $e_p(P)$ in terms of the Fitting height $h(G)$ and $e_p(G)$. The key is the following lemma, which gives a bound for the p-part of the order of the quotient of two consecutive terms of the ascending Fitting series $F_{i+1}(G)/F_i(G)$ when $F_i(G)/F_{i-1}(G)$ is a p'-group.

Lemma 5.5 *Let G be a solvable group and assume that $O_p(G) = 1$. Then*

$$|F_2(G)/F(G)|_p \leq p^{2e_p(G)-1}.$$

Proof By Gaschutz's theorem (see [27, Theorem 1.12]) $G/F(G)$ acts faithfully on the abelian group $F(G)/\Phi(G)$. Therefore it also acts faithfully on $\mathrm{Irr}(F(G)/\Phi(G))$. Let $P_1/F(G) \in \mathrm{Syl}_p(F_2(G)/F(G))$ and consider the faithful action of the p-group

$P_1/F(G)$ on the p'-group $\mathrm{Irr}(F(G)/\Phi(G))$. We may assume that $|P_1/F(G)| > 1$. By [13], there exists $\lambda \in \mathrm{Irr}(F(G)/\Phi(G))$ such that $|I_{P_1/F(G)}(\lambda)| < |P_1/F(G)|^{1/2}$. Since $P_1 \trianglelefteq G$,

$$p^{e_p(G)} \geq |P_1/F(G) : I_{P_1/F(G)}(\lambda)| > |P_1/F(G)|^{1/2}$$

and the result follows. □

Let G be a group such that $O_p(G) = 1$. We define the following series

$$1 = P_0 \lhd N_1 \lhd P_1 \lhd N_2 \lhd P_2 \lhd \cdots \lhd G,$$

where N_{i+1}/P_i is the largest nilpotent subgroup of G/P_i and P_i/N_i is the largest normal p-subgroup of G/N_i. Let $s_p(G)$ be the minimum integer such that $G = N_s$ or $G = P_s$. It is clear that $l_p(G) \leq s_p(G) \leq h(G) \leq 2s_p(G)$. Observe that if there are just two primes dividing $|G|$, then this series is exactly the ascending p', p-series.

Corollary 5.6 *Let G be a solvable group and $P \in \mathrm{Syl}_p(G)$. Then $|P : O_p(G)| \leq p^{s_p(G/O_p(G))(2e_p(G)-1)}$.*

Proof It is enough to apply repeatedly the previous lemma. □

Now we are ready to bound $e_p(P)$ in terms of $e_p(G)$ and $s_p(G)$.

Theorem 5.7 *Let G be a solvable group and $P \in \mathrm{Syl}_p(G)$. Then*

$$e_p(P) \leq (2s_p(G/O_p(G)) + 1)e_p(G) - s_p(G).$$

Proof By Lemma 5.2, $e_p(O_p(G)) \leq e_p(G)$ and, by the previous corollary, $|P : O_p(G)| \leq p^{s_p(G)(2e_p(G)-1)}$. Therefore

$$b(P) \leq |P : O_p(G)| b(O_p(G)) \leq p^{s_p(G)(2e_p(G)-1)+e_p(G)}$$

and the result follows. □

Corollary 5.8 *Let G be a group of order $p^a q^b$ and $P \in \mathrm{Syl}_p(G)$. Then*

$$e_p(P) \leq C_1 e_p(G) \log e_p(G) + C_2.$$

Proof It is enough to observe that $s_p(G/O_p(G)) = l_p(G/O_p(G))$ and apply [27, p. 194] to the bound in Theorem 5.7. □

It seems hard to find a counterexample to the conjecture, since it would be necessary to construct an infinite family of groups with Fitting height going to infinity and at least 3 primes dividing the order of each of the groups.

There is a similar conjecture, due to G. Navarro [31].

Conjecture 5 Let G be a finite group. Then

$$\prod_{p\in\pi(G)} b(G_p) \leq b(G),$$

where G_p is a Sylow p-subgroup of G.

Despite the similar flavor of these conjectures, it does not seem easy to apply results obtained for one of them to the other.

We close by remarking that it is not difficult to prove the conjugacy-class analog of Conjecture 4 for solvable groups. (We believe that the solvability hypothesis is not necessary, however.) On the other hand, the analog of Conjecture 5 is completely false even for solvable groups (see [3]).

References

[1] R. BRAUER, *Representations of finite groups*, Lectures on modern mathematics I, Wiley, New York, 1963, 133–175.
[2] J. COSSEY, T. HAWKES, *Sets of p-powers as conjugacy class sizes*, Proc. Amer. Math. Soc. **128** (2000), 49–51.
[3] J. COSSEY, T. HAWKES, *On the largest conjugacy class size in a finite group*, Rend. Sem. Mat. Univ. Padova **103** (2000), 171–179.
[4] G. A. FERNÁNDEZ-ALCOBER, A. MORETÓ, *Groups with extreme character degrees and their normal subgroups*, Trans. Amer. Math. Soc. **353** (2001), 2271–2292.
[5] G. A. FERNÁNDEZ-ALCOBER, A. MORETÓ, *On the number of conjugacy class sizes and character degrees in finite p-groups*, Proc. Amer. Math. Soc. **129** (2001).
[6] R. GOW, M. MARJORAM, A. PREVITALI, *On the irreducible characters of a Sylow* 2*-subgroup of the finite symplectic group in characteristic* 2, J. Algebra **241** (2001), 393–409.
[7] T. HAWKES, *Character degrees and class sizes -a tantalizing duality*, Papers on group theory, International conference on group theory, Doerk's 60th birthday in Calpe, Calpe, 1999.
[8] B. HUPPERT, *A remark on the character-degrees of some p-groups*, Arch. Math. **59** (1992), 313–318.
[9] I. M. ISAACS, *The p-parts of character degrees in p-solvable groups*, Pacific J. Math **36** (1971), 677–691.
[10] I. M. ISAACS, *Sets of p-powers as irreducible character degrees*, Proc. Amer. Math. Soc. **96** (1986), 551–552.
[11] I. M. ISAACS, "Character Theory of Finite Groups", Dover, New York, 1994.
[12] I. M. ISAACS, *Characters of groups associated with finite algebras*, J. Algebra **177** (1995), 708–730.
[13] I. M. ISAACS, *Large orbits in actions of nilpotent groups*, Proc. Amer. Math. Soc. **127** (1999), 45–50.
[14] I. M. ISAACS, private communication.
[15] I. M. ISAACS, G. KNUTSON, *Irreducible character degrees and normal subgroups*, J. Algebra **199** (1998), 302–326.
[16] I. M. ISAACS, A. MORETÓ, *The character degrees and nilpotence class of a p-group*, J. Algebra **238** (2001), 827–842.
[17] I. M. ISAACS, D. S. PASSMAN, *A characterization of groups in terms of the degrees of their characters*, Pacific J. Math **15** (1965), 877–903.
[18] I. M. ISAACS, D. S. PASSMAN, *A characterization of groups in terms of the degrees of their characters II*, Pacific J. Math. **24** (1968), 467–510.

[19] I. M. Isaacs, M. C. Slattery, *Character degree sets that do not bound the class of a p-group*, to appear in Proc. Amer. Math. Soc.
[20] A. Jaikin-Zapirain, A. Moretó, *Character degrees and nilpotence class of finite p-groups: an approach via pro-p groups*, preprint.
[21] T. M. Keller, *Orbit sizes and character degrees*, Pacific J. Math. **187** (1999), 317–332.
[22] T. M. Keller, *Orbit sizes and character degrees II*, J. reine angew. Math. **516** (1999), 27–114.
[23] T. M. Keller, *Orbit sizes and character degrees III*, to appear in J. reine angew. Math.
[24] L. Kovacs, C. R. Leedham-Green, *Some normally monomial p-groups of maximal class and large derived length*, Quart. J. Math. Oxford Ser. (2) **87** (1986), 49–54.
[25] A. Mann, *Some questions about p-groups*, J. Austral Math. Soc. (Series A) **67** (1999), 356–379.
[26] A. Mann, *Minimal characters of p-groups*, J. Group Theory **2** (1999), 225–250.
[27] O. Manz, T. R. Wolf, "Representations of Solvable Groups", Cambridge University Press, 1993.
[28] M. Marjoram, *Irreducible characters of a Sylow p-subgroup of the orthogonal group*, Comm. Algebra **27** (1999), 1171–1196.
[29] S. Mattarei, *An example of p-groups with identical character tables and different derived lengths* , Arch. Math. **62** (1994), 12–20.
[30] A. Moretó, *An elementary method for the calculation of the set of character degrees of some p-groups*, to appear in Rend. Circ. Mat. Palermo.
[31] G. Navarro, private communication.
[32] A. Previtali, *Orbit lengths and character degrees in p-Sylow subgroups of some classical Lie groups*, J. Algebra **177** (1995), 658–675.
[33] A. Previtali, *On a conjecture concerning character degrees of some p-groups*, Arch. Math. **65** (1995), 375–378.
[34] J. M. Riedl, *Character degrees, class sizes, and normal subgroups of a certain class of p-groups*, J. Algebra **218** (1999), 190–215.
[35] I. A. Sagirov, *Degrees of irreducible characters of the Suzuki 2-groups*, Math. Notes **66** (1999), 203–207.
[36] J. Sangroniz, *Conjugacy classes and characters in some quotients of the Nottingham group*, J. Algebra **211** (1999), 26–41.
[37] J. Sangroniz, *Characters of the Sylow p-subgroups of classical groups*, preprint.
[38] M. C. Slattery, *Character degrees and derived length in p-groups*, Glasgow Math. J. **30** (1988), 221–230.
[39] M. C. Slattery, *Character degrees and nilpotence class in p-groups*, J. Austral Math. Soc. (Series A) **57** (1994), 76–80.

ON THE RELATION BETWEEN GROUP THEORY AND LOOP THEORY

MARKKU NIEMENMAA

Department of Mathematical Sciences, University of Oulu,
PL 3000, 90014 Oulu, Finland

Abstract

Loops are nonassociative algebras (and are sometimes known as 'nonassociative groups') which can be investigated by using their multiplication groups. This connection to group theory is a source of several interesting group theoretical investigations and in the following four sections we try to cover some of the major problems in this area.

1 Introduction

Let Q be a groupoid with a neutral element e. If for any $a, b \in Q$ each of the equations $ax = b$ and $ya = b$ has a unique solution, then we say that Q is a *loop*. For each $a \in Q$ we have two permutations L_a *(left translation)* and R_a *(right translation)* on Q defined by $L_a(x) = ax$ and $R_a(x) = xa$ for every $x \in Q$. The permutation group $M(Q)$ generated by the set of all left and right translations is called the *multiplication group* of Q. It is easy to see that $M(Q)$ is transitive on Q and the stabilizers of elements of Q are conjugated in $M(Q)$. The stabilizer of $e \in Q$ is denoted by $I(Q)$ and this stabilizer is called the *inner mapping group* of Q (it is interesting to observe that if Q is a group, then $I(Q)$ is just the group of inner automorphisms of Q). The concepts of the multiplication group and the inner mapping group of a loop were defined by Bruck [3] in 1946 in an article where he laid the foundation of loop theory. In the present article we shall consider various problems linked with the structure of $M(Q)$ and $I(Q)$ and we shall also consider the relation between the structure of the loop Q and the corresponding multiplication group and inner mapping group.

The first two major problems are:

- 1) Which permutation groups are multiplication groups of loops?
- 2) Which groups are isomorphic to multiplication groups of loops?

When we investigate these two problems (and the structure of $M(Q)$ in general), then it is very useful to see what is happening with the inner mapping group $I(Q)$. When we answer the question: "Which groups can (can not) be in the role of $I(Q)$?", we simultaneously get tools that can be used to provide answers to problems 1) and 2).

In loop theory solvability and (central) nilpotency of loops can be defined in a way which is analogous to the way we define these notions in group theory. We immediately have the following interesting problems here: How does the structure of Q influence the structure of $M(Q)$ and $I(Q)$ and conversely: if we assume that

$M(Q)$ (or $I(Q)$) has a special structure, then what do we know about the structure of Q?

The following sections are devoted to the problems mentioned before. We wish to point out that there are several other links between loops and groups. Conway [4], for example, has used a special loop in the construction of the Fischer-Griess Monster and Foguel and Ungar [10], [11] have found interesting links to hyperbolic geometries and relativistic physics. For those who are interested in the history of loop theory we recommend the article of Pflugfelder [22]. This article is based on the talk that she gave during the international loop theory conference held in Prague in 1999.

2 Loops, groups and transversals

Loops appear in a natural way as algebraic structures on transversals of a subgroup in a group. This observation can be traced back to the work by Baer [1]. Let Q be a loop and denote $A = \{L_a : a \in Q\}$ and $B = \{R_a : a \in Q\}$. Now the two sets A and B are left (also right) transversals to $I(Q)$ in $M(Q)$. It is easy to see that the commutator subgroup $[A, B]$ is a subgroup of $I(Q)$ and we say that the transversals A and B are $I(Q)$-*connected* in $M(Q)$. In general, if G is a group, H is a subgroup of G and there exist two left transversals A and B to H in G such that $[A, B] \leq H$, then we say that A and B are H- connected transversals in G. The following theorem that was proved by Kepka and Niemenmaa [16] describes the relation between multiplication groups of loops and connected transversals.

Theorem 2.1 *A group G is isomorphic to the multiplication group of a loop if and only if there exist a subgroup H satisfying $H_G = 1$ (the core of H in G is trivial) and H-connected transversals A and B such that $G = \langle A, B\rangle$.*

This theorem gives us a purely group theoretical characterization of multiplication groups of loops and in recent papers it has very often been the starting point of the investigation. From Theorem 2.1 it follows that if $I(Q)$ is cyclic, then $M(Q)$ is an abelian group and, in fact, our loop Q must be an abelian group (then, of course, $I(Q) = 1$). It also follows that $I(Q)$ can not be the Prüfer group C_{p^∞} for a prime p. Furthermore, we know that the normalizer of $I(Q)$ in $M(Q)$ equals the direct product of $I(Q)$ and $Z(M(Q))$. By combining these properties of $I(Q)$ we can show that S_3, S_4 and A_4 are not isomorphic to the multiplication group of a loop. This is also true for hamiltonian groups, dihedral groups and Blackburn groups (these groups are neither hamiltonian nor abelian and the intersection of all nonnormal subgroups is not trivial [2]). As these examples indicate, it seems easier to prove non-existence results than to actually construct loops with given multiplication groups. To further demonstrate this we shall now consider the situation in finite simple groups. Thus assume that Q is a finite loop such that $M(Q)$ is simple (and here 'simple' means noncommutative simple). The first crucial observation is that $I(Q)$ is a maximal subgroup in $M(Q)$ (see [16]). When we decide whether a given simple group G is isomorphic to the multiplication group of a loop, we first have to list all the maximal subgroups of G and then we try to find the corresponding connected transversals such that the conditions of Theorem 2.1 are satisfied. By using representation theory and the centralizer ring of $M(Q)$ (for the details, see [13] and [25]), we get the following inequality:

$$| M(Q) |\leq| I(Q) | ((C-1) | I(Q) | +1),$$

where C is the number of conjugacy classes in G. This result can be used to rule out maximal subgroups of small order.

We can now list some nonexistence results. Vesanen ([23] and [25]) was able to show that $PSL(2,q)$ is isomorphic to some $M(Q)$ if and only if $q=9$. If G is one of the groups

1) $U_n(q)$, where $n \geq 6$,

2) $O_n(q)$, where n is odd and $n \geq 7$, or

3) $O_n^{\epsilon}(q)$, where n is even and $n \geq 7-\epsilon$

acting in the set Ω of the isotropic points of the corresponding projective space, then there exists no loop Q defined in the set Ω such that $M(Q) \leq G$ (for the details, see Vesanen [26].

Finally, we list some results which are positive in the sense that some finite simple groups are (isomorphic to) multiplication groups of loops. Drapal and Kepka [6] showed that the alternating group A_5 is not a loop group but if $n \geq 6$ then there exists a loop of order n such that $M(Q) \cong A_n$. Here we have an interesting example: there exists a loop Q of order 15 such that $M(Q) \cong A_8$. This is the only known example of a finite simple group being the multiplication group of two loops of different orders. A loop is called a *Moufang loop* if it satisfies the law $x(y.zy) = (xy.z)y$. If Q is a finite simple nonassociative Moufang loop, then $M(Q)$ is a nonabelian finite simple group with triality . Liebeck proved in [14] that the only simple groups with triality are the groups $O_8^+(q)$ and by using this result he was able to classify all finite simple Moufang loops: they are the finite simple groups and the loops defined by Paige [21].

Doubly transitive groups in which only the identity fixes three points are called Zassenhaus groups. In 2001 Drapal [7] managed to show that no finite loop has a multiplication group that is a Zassenhaus group.

3 Nilpotent and solvable loops

If Q is a loop, then the centre of Q is denoted by $C(Q)$ and $a \in C(Q)$ provided that $ax = xa$ for every $x \in Q$ and, in addition, $a(xy) = (ax)y$, $(xy)a = x(ya)$ and $(xa)y = x(ay)$ for all $x, y \in Q$. Clearly, the centre is an abelian group and it is not difficult to show that $C(Q)$ and $Z(M(Q))$ are isomorphic. If we put $C_0 = 1$, $C_1 = C(Q)$ and $C_i/C_{i-1} = C(Q/C_{i-1})$, then we obtain a series of normal subloops of Q. If C_{n-1} is a proper subloop of Q but $C_n = Q$, then we say that Q is (centrally) nilpotent of class n.

In 1946 Bruck [3] showed that if Q is a nilpotent loop, then $M(Q)$ is a solvable group. He also showed that if $M(Q)$ is a nilpotent group, then Q is a nilpotent loop. The latter result was investigated more closely by Vesanen [24] and he was able to prove

Theorem 3.1 *If $M(Q)$ is a finite nilpotent group, then Q is a direct product of nilpotent p-loops.*

Thus a finite nilpotent group G is isomorphic to the multiplication group of a loop if and only if its Sylow subgroups are isomorphic to multiplication groups of loops. Here one should remember that there exist nilpotent loops which are not direct products of p-loops and which thus have nonnilpotent (but solvable) multiplication groups (see Bruck [3] , Chapter III).

As we pointed out in the introduction, it is also interesting to see how the structure of $I(Q)$ influences the structure of Q. In [17] Kepka and Niemenmaa considered the situation that H is an abelian subgroup of a finite group G such that there exist H-connected transversals A and B in G. If $G = \langle A, B \rangle$, then H is subnormal in G. From this it follows that if Q is a finite loop such that $I(Q)$ is abelian, then Q is nilpotent. In [18] Niemenmaa investigated a similar situation with a subgroup that is a dihedral 2-group. Again, it turned out that the subgroup is subnormal and as a loop theoretical application we get: If Q is a loop such that $I(Q)$ is a dihedral 2-group, then Q is nilpotent.

Bruck defined the solvability of loops as follows: A loop Q is solvable if it has a series $1 = Q_0 \subseteq ... \subseteq Q_n = Q$, where Q_{i-1} is a normal subloop of Q_i and Q_i/Q_{i-1} is an abelian group (normal subloops are the kernels of loop homomorphisms). Of course, nilpotent loops are solvable. In 1968, Glauberman [12] managed to show that every finite Moufang loop of odd order is solvable (since the class of Moufang loops includes the class of groups, this result generalizes the theorem of Feit and Thompson [9]).

In 1996, Vesanen [27] investigated the relation between solvable loops and solvable groups and he was able to prove

Theorem 3.2 *Let Q be a finite loop. If $M(Q)$ is a solvable group, then Q is a solvable loop.*

This result is very fundamental and deep and opens a large variety of possibilities to create solvability criteria for finite loops in terms of the properties of the inner mapping group. Hence, we are interested in those properties of $I(Q)$ which guarantee the solvability of $M(Q)$.

The results by Kepka and Niemenmaa in [17] and [18] indicate that the solvability of a loop Q follows provided that $I(Q)$ is a finite abelian group or $I(Q)$ is a dihedral 2-group. Therefore it is quite natural that we next turn our attention to the case where $I(Q)$ is a nonnilpotent dihedral group. In [5] Csörgö and Niemenmaa showed that if Q is a loop such that $\mid I(Q) \mid = 2p$, where p is an odd prime, then $M(Q)$ is solvable. If Q is finite, then Vesanen's result implies that Q is a solvable loop. In 2000, Myllylä [15] was able to show that a finite loop Q is solvable provided that $I(Q)$ is a dihedral group of order $2k$, where k is an odd number.

Quite recently, Drapal [8] has written an interesting paper where he considers the case that $I(Q)$ is a nonabelian group of order pq, where $q < p$ are prime numbers. He shows that $Q/C(Q)$ is a loop of order pk, where $2 \leq k \leq q$ and $M(Q)/Z(M(Q))$ is a solvable group of order p^2qk. Now it is clear that both $M(Q)$ and Q are solvable.

We put an end to this section by introducing the following open problems:

- *Is it true that the solvability of $M(Q)$ implies the solvability of Q (without the condition that Q is finite)?*
- *If $I(Q)$ is a nilpotent group, does it then follow that $M(Q)$ is a solvable group and Q is a solvable loop?*
- *If $I(Q)$ is a dihedral group, does it then follow that $M(Q)$ is a solvable group and Q is a solvable loop?*

4 Abelian inner mapping groups

The starting point here is the observation that $I(Q) = 1$ if and only if Q is an abelian group. We also know that $I(Q)$ can neither be (nontrivial) cyclic nor the Prüfer group. In [19] we managed to show that if Q is a finite loop, then $I(Q)$ can not be isomorphic to $C_n \times D$, where C_n is a cyclic group of order $n > 1$, D is an abelian group and $gcd(n, \mid D \mid) = 1$. We continued our investigations of finite loops in [20] and it turned out that $I(Q)$ can not be isomorphic to $C_{p^2} \times C_p$, where p is a prime number. Furthermore, if D is a finite abelian group such that D has a Sylow p-subgroup S which is isomorphic to $C_{p^2} \times C_p$, then $I(Q)$ can not be isomorphic to D. By using this result we could show that certain solvable groups can not occur as multiplication groups of loops (for the details, see section 3 in [20]). To conclude this survey, we introduce the following problem:

- *Classify those finite abelian groups which are isomorphic to inner mapping groups of loops.*

References

[1] R.Baer, *Nets and groups*, Trans. Amer. Math. Soc. **54** (1939), 110–141.

[2] N.Blackburn, *Finite groups in which the nonnormal subgroups have nontrivial intersection*, J. Algebra **3** (1966), 30–37.

[3] R.H.Bruck, *Contributions to the theory of loops*, Trans. Amer. Math. Soc. **60** (1946), 245–354.

[4] J.H.Conway, *A simple construction for the Fischer-Griess monster group*, Invent. Math. **79** (1985), 513–540.

[5] P.Csörgö and M.Niemenmaa, *Solvability conditions for loops and groups*, J. Algebra **232** (2000), 336–342.

[6] A.Drapal and T.Kepka, *Alternating groups and latin squares*, Europ. J. Combinatorics **10(2)** (1989), 175–180.

[7] A.Drapal, *Multiplication groups of finite loops that fix at most two points*, J. Algebra **235** (2001), 154–175.

[8] A.Drapal, *Orbits of inner mapping groups*, submitted.

[9] W.Feit and J.G.Thompson, *Solvability of groups of odd order*, Pacific J. Math. **13** (1963), 775–1029.

[10] T.Foguel and A.A.Ungar, *Involutary decomposition of groups into twisted subgroups and subgroups*, J. Group Theory **3(1)** (2000), 27–46.

[11] T.Foguel and A.A.Ungar, *Gyrogroups and the decomposition of groups into twisted subgroups and subgroups*, Pacific J. Math. **197** (2001), 1–11.

[12] G.Glauberman, *On loops of odd order II*, J. Algebra **8** (1968), 393–414.

[13] T.Ihringer, *Quasigroups, loops and centralizer rings*, Proceedings of the Vienna conference (1984) in: Contributions to general algebra (1985), 211–224.

[14] M.W.Liebeck, *The classification of finite simple Moufang loops*, Math. Proc. Camb. Phil. Soc. **102** (1987), 33–47.

[15] K.Myllylä, *On connected transversals to dihedral subgroups*, Acta Univ. Ouluensis **A 350** (2000), 1–23.

[16] M.Niemenmaa and T.Kepka, *On multiplication groups of loops*, J. Algebra **135** (1990), 112–122.

[17] M.Niemenmaa and T.Kepka, *On connected transversals to abelian subgroups*, Bull. Australian Math. Soc. **49** (1994), 121–128.

[18] M.Niemenmaa, *On loops which have dihedral 2-groups as inner mapping groups*, Bull. Australian Math. Soc. **52** (1995), 153–160.

[19] M.Niemenmaa, *On the structure of the inner mapping groups of loops*, Comm. Alg. **24(1)** (1996), 135–142.

[20] M.Niemenmaa, *On abelian inner mapping groups of finite loops*, Comment. Math. Univ. Carolinae **41(4)** (2000), 687–691.

[21] J.P.Paige, *A class of simple Moufang loops*, Proc. Amer. Math. Soc. **7** (1956), 471–482.

[22] H.O.Pflugfelder, *Historical notes on loop theory*, Comment. Math. Univ. Carolinae **41(2)** (2000), 359–370.

[23] A.Vesanen, *On connected transversals in PSL(2,q)*, Ann. Acad. Sci. Fenn., Ser. A, I. Mathematica, Dissertationes, **84** (1992).

[24] A.Vesanen, *On p-groups as loop groups*, Arch. Math. **61** (1993), 1–6.

[25] A.Vesanen, *The group PSL(2,q) is not the multiplication group of a loop*, Comm. Alg. **22(4)** (1994), 1177–1195.

[26] A.Vesanen, *Finite classical groups and multiplication groups of loops*, Math. Proc. Camb. Phil. Soc. **117** (1995), 425–429.

[27] A.Vesanen, *Solvable groups and loops*, J. Algebra **180**, (1996), 862–876.

GROUPS AND LATTICES

PÉTER P. PÁLFY

Department of Algebra and Number Theory, Eötvös University
Budapest, P.O.Box 120, H–1518 Hungary

Dedicated to the memory of my father, József Pálfy (1922–2001)

Abstract

In this survey paper we discuss some topics from the theory of subgroup lattices. After giving a general overview, we investigate the local structure of subgroup lattices. A major open problem asks if every finite lattice occurs as an interval in the subgroup lattice of a finite group. Next we investigate laws that are valid in normal subgroup lattices. Then we sketch the proof that every finite distributive lattice is the normal subgroup lattice of a suitable finite solvable group. Finally, we discuss how far the subgroup lattice of a direct power of a finite group can determine the group.

1 Introduction

This survey paper is the written version of my four talks given at the Groups – St Andrews 2001 in Oxford conference. I selected some topics on subgroup lattices and normal subgroup lattices according to my personal taste and interest. These topics, of course, cannot cover all interesting and important parts of the theory. For a more complete overview the reader should consult the small book of Michio Suzuki [60] from 1956 and the more recent monograph by Roland Schmidt [54]. The latter one is a thick volume of 541 pages including 384 references. So it is clearly impossible to give a comprehensive survey here. My choice of topics was partly guided by the review of Schmidt's book by Ralph Freese [13].

The study of subgroup lattices has quite a long history, starting with Richard Dedekind's work [10] in 1877, including Ada Rottlaender's paper [47] from 1928 and later numerous important contributions by Reinhold Baer, Øystein Ore, Kenkichi Iwasawa, Leonid Efimovich Sadovskii, Michio Suzuki, Giovanni Zacher, Mario Curzio, Federico Menegazzo, Roland Schmidt, Stewart Stonehewer, Giorgio Busetto, and many-many others.

In Section 2 we will list some of the most remarkable results on subgroup lattices. Hints to the contents of Sections 3–6 will also be given there. These later sections are surveys of some particular topics, therefore proofs are very rarely given, and even then, they will be quite sketchy.

The lattice formed by all subgroups of a group will be denoted by $\mathrm{Sub}(G)$ and will be called the *subgroup lattice* of the group G. It is a *complete lattice*: any number of subgroups H_i have a *meet* (greatest lower bound) $\bigwedge H_i$, namely their intersection

$\bigcap H_i$, and a *join* (least upper bound) $\bigvee H_i$, namely the subgroup generated by all of them together. Notice that we denote the lattice operations by $\wedge$ and $\vee$.

An element $c \in \mathcal{L}$ in a complete lattice is called *compact* if

$$c \leq \bigvee_{i \in I} a_i \quad \text{implies} \quad c \leq \bigvee_{i \in J} a_i \text{ for a finite subset } J \subseteq I.$$

It is easy to see that $H \in \mathrm{Sub}(G)$ is compact if and only if H is a finitely generated subgroup of G. A complete lattice is called *algebraic* if every element is a join of compact elements. We see that subgroup lattices are always algebraic.

If the group is finite, it is a convenient way to visualize the lattice using its *Hasse diagram*, where the bottom element represents the identity subgroup 1, the top element the group itself, and between two elements of the lattice a line segment is drawn whenever the lower subgroup is a maximal subgroup in the upper one. An example is shown in Figure 1.

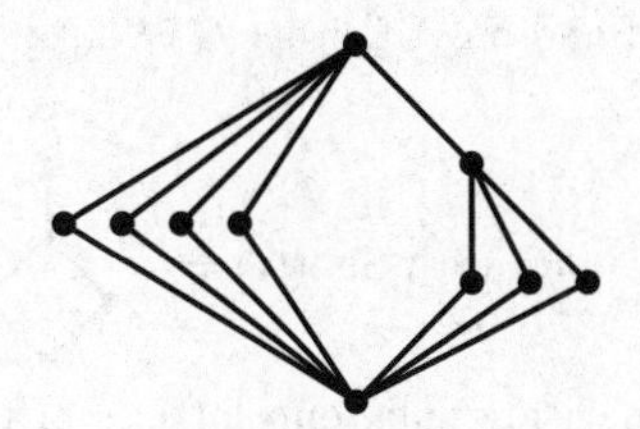

Figure 1. Hasse diagram of $\mathrm{Sub}(A_4)$

Our notation is mostly standard. For $H \leq G$ we denote by $\mathbf{N}_G(H)$, $\mathbf{C}_G(H)$ the normalizer and the centralizer of H in G, respectively. The center and the commutator subgroup of G is denoted by $\mathbf{Z}(G)$ and G'. The automorphism group and the inner automorphism group are written as $\mathrm{Aut}\, G$ and $\mathrm{Inn}\, G$. For normal subgroups we use the notation $N \lhd G$. The set of normal subgroups is a sublattice in $\mathrm{Sub}(G)$, it will be denoted by $\mathrm{Norm}(G)$. Intervals in lattices will be defined in Section 2, and in subgroup lattices they will be denoted as $\mathrm{Int}[H; G]$. The cyclic group of order n will be written as C_n, the dihedral group of degree n (and order $2n$) as D_n, the alternating and symmetric groups as A_n and S_n. Furthermore, $\mathrm{GF}(q)$ will denote the q-element field, and $F^\times$ the multiplicative group of a field F.

2 Overview

We start with some simple observations concerning subgroup lattices. Since we will mainly deal with finite groups, let us remark that the subgroup lattice $\mathrm{Sub}(G)$ is finite if and only if the group G is finite. For some small lattices it is easy to determine all groups that have the given lattice as subgroup lattice. For example,

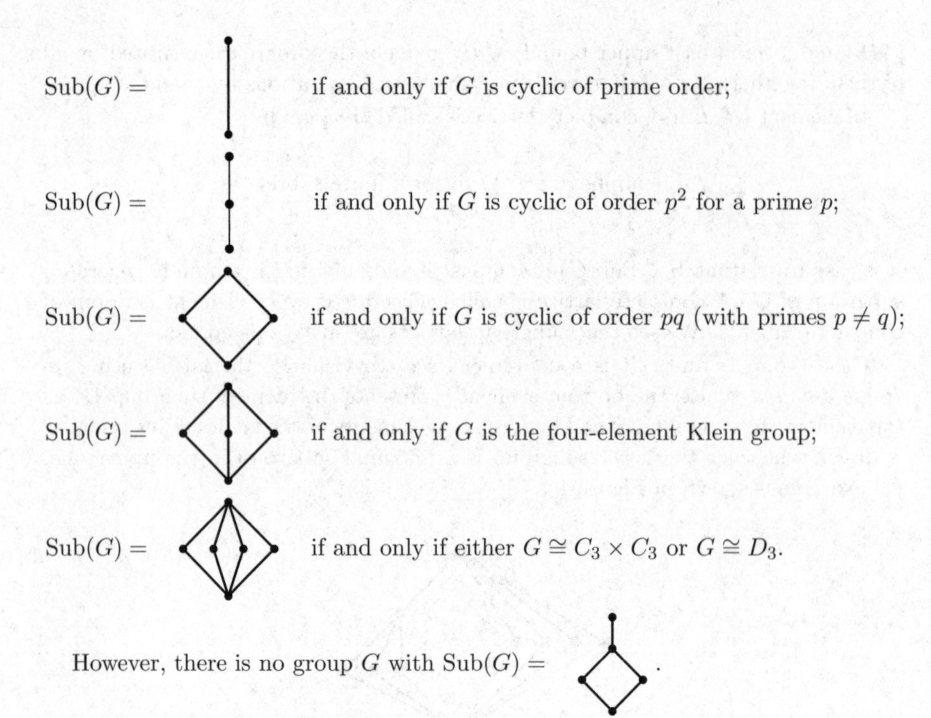

$\mathrm{Sub}(G) =$ if and only if G is cyclic of prime order;

$\mathrm{Sub}(G) =$ if and only if G is cyclic of order p^2 for a prime p;

$\mathrm{Sub}(G) =$ if and only if G is cyclic of order pq (with primes $p \neq q$);

$\mathrm{Sub}(G) =$ if and only if G is the four-element Klein group;

$\mathrm{Sub}(G) =$ if and only if either $G \cong C_3 \times C_3$ or $G \cong D_3$.

However, there is no group G with $\mathrm{Sub}(G) =$.

So there are lattices which are subgroup lattices of infinitely many, of finitely many, of a unique, or of no group. That is, the correspondence $G \mapsto \mathrm{Sub}(G)$ is neither injective, nor surjective. This fact gives rise to two questions:

1. Which groups are uniquely determined by their subgroup lattices?
2. Which lattices are subgroup lattices?

The answer to the second question is very complicated, as given by B. V. Yakovlev [64] in 1974, based on his description of the subgroup lattices of free groups. We would be interested rather in the local structure of subgroup lattices, that is we would like to know what are the possible intervals in subgroup lattices. If $a < b$ are elements of a lattice $\mathcal{L}$, by the *interval* $\mathrm{Int}[a;b]$ we mean the sublattice formed by the intermediate elements:

$$\mathrm{Int}[a;b] = \{x \in \mathcal{L} \mid a \leq x \leq b\}.$$

In Section 3 we will see that every algebraic lattice can occur as an interval in the subgroup lattice of an infinite group. For finite groups, however, it is not known whether every finite lattice can be found as an interval in the subgroup lattice of a suitable finite group. The main subject of Section 3 will be a survey of results concerning this open problem.

An important line of investigations deals with groups whose subgroup lattices satisfy certain laws. As it follows from the following basic result, there is no non-trivial law that holds in the subgroup lattice of every group.

Theorem 2.1 (Whitman [63], 1946) *Every lattice is isomorphic to a sublattice of the subgroup lattice of some group.*

There is also a remarkable finite version of this embedding theorem.

Theorem 2.2 (Pudlák and Tůma [46], 1980) *Every finite lattice is isomorphic to a sublattice of the subgroup lattice of some finite group.*

The most familiar lattice law is the distributivity. Recall that a lattice $\mathcal{L}$ is called *distributive* if the following equivalent conditions hold for every $x, y, z \in \mathcal{L}$:

- $x \vee (y \wedge z) = (x \vee y) \wedge (x \vee z)$;
- $x \wedge (y \vee z) = (x \wedge y) \vee (x \wedge z)$;
- $(x \vee y) \wedge (x \vee z) \wedge (y \vee z) = (x \wedge y) \vee (x \wedge z) \vee (y \wedge z)$.

One of the nicest results in the theory of subgroup lattices characterizes those groups which have distributive subgroup lattices.

Theorem 2.3 (Ore [37], 1937–38) *The subgroup lattice* $\mathrm{Sub}(G)$ *is distributive if and only if the group G is locally cyclic.*

Recall that a group G is said to be *locally cyclic*, if every finitely generated subgroup of G is cyclic. There are not too many such groups: a group G is locally cyclic if and only if it is isomorphic to a subgroup of either the additive group of the rationals $\mathbb{Q}$ or of its quotient group $\mathbb{Q}/\mathbb{Z}$.

Cyclic groups can be characterized by the properties that $\mathrm{Sub}(G)$ is distributive and satisfies the *ascending chain condition* (i.e., it contains no infinite chain of subgroups $H_1 < H_2 < H_3 < \ldots$). If $n = p_1^{k_1} \cdots p_r^{k_r}$, then the subgroup lattice of the cyclic group of order n is the direct product of chains of lengths $k_1, \ldots, k_r$, independently of the primes p_i.

The description of groups with modular subgroup lattices is quite complicated. As it is well known, a lattice $\mathcal{L}$ is called *modular* if for all $x, y, z \in \mathcal{L}$

$$x \geq z \Rightarrow x \wedge (y \vee z) = (x \wedge y) \vee z.$$

(In Section 4 we will give some equivalent conditions as well.) The subgroup lattices of abelian groups are modular, as it was discovered by Richard Dedekind [10] in 1877 for the case of the subgroup lattice of the additive group of the complex numbers. So we make the assumption that

$$G \text{ is nonabelian and } \mathrm{Sub}(G) \text{ is a modular lattice.}$$

The characterization consists of several pieces.

Theorem 2.4 (Iwasawa [23], 1943) *If G has elements of infinite order, then the torsion subgroup $T(G)$ of G is abelian, $G/T(G)$ is a torsion-free abelian group of rank one, etc.*

For the omitted details see [54, 2.4.11 Theorem].

Theorem 2.5 (Schmidt [53], 1986) *If G is a torsion group, then G is a direct product of Tarski groups, extended Tarski groups and a locally finite group, such that elements from different direct factors have coprime orders.*

A *Tarski group* is an infinite group in which every proper nontrivial subgroup has prime order. Tarski groups were first constructed by Olshanskii [35] in 1979. An *extended Tarski group* is such that $G/\mathbf{Z}(G)$ is a Tarski group of exponent p for some prime p, $\mathbf{Z}(G)$ is cyclic of order $p^r > 1$, and for every subgroup $H \leq G$, either $H \leq \mathbf{Z}(G)$ or $H \geq \mathbf{Z}(G)$ holds. The subgroup lattice of an extended Tarski group is shown in Figure 2. Note that extended Tarski groups do exist if p is sufficiently large (see Olshanskii [36]).

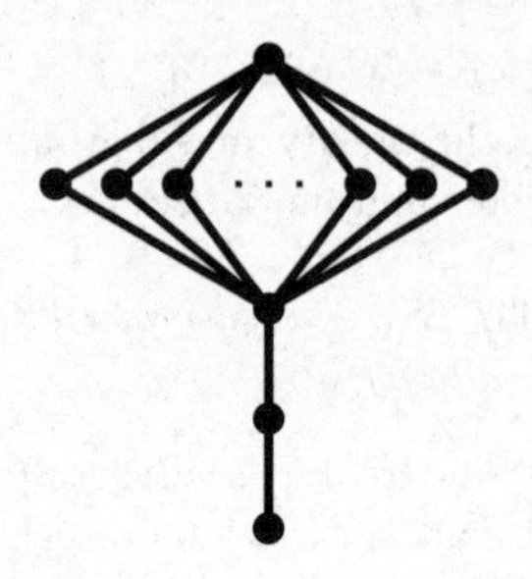

Figure 2. The subgroup lattice of an extended Tarski group

Theorem 2.6 (Iwasawa [23], 1943) *If G is a locally finite group, then G is a direct product of P^*-groups and locally finite p-groups, such that elements from different direct factors have coprime orders.*

By definition, a *P^*-group* is a semidirect product of an elementary abelian normal subgroup A with a cyclic group $\langle t \rangle$ of prime power order such that t induces a power automorphism ($tat^{-1} = a^r$ with a fixed r for all $a \in A$) of prime order on A. In fact, all these groups have modular subgroup lattices.

Theorem 2.7 (Iwasawa [23], 1943) *If G is a locally finite p-group, then either G is a direct product of the quaternion group with an elementary abelian 2-group, or G contains an abelian normal subgroup A of exponent p^k with cyclic quotient group G/A of order p^m and there exist an element $b \in G$ with $G = A\langle b \rangle$ and an integer s (which is at least 2 if $p = 2$) such that $s < k \leq s + m$ and $bab^{-1} = a^{1+p^s}$ for all $a \in A$.*

Again, all these groups have modular subgroup lattices.

Theorems 2.4, 2.5, 2.6, 2.7 together yield a complete characterization of non-abelian groups with modular subgroup lattices.

As we have seen, some lattice theoretic properties may correspond to some simple group theoretic ones, but sometimes the description of groups with subgroup lattices of certain type (such as modular lattices) are awkward. Unfortunately, nice characterisations are rather rare. We list some (maybe all) of them now.

The *Jordan–Dedekind chain condition* means that all maximal chains in the lattice have the same length. Finite groups with such subgroup lattices have a beautiful description.

Theorem 2.8 (Iwasawa [22], 1941) *For a finite group G, the subgroup lattice* $\mathrm{Sub}(G)$ *satisfies the Jordan–Dedekind chain condition if and only if G is supersolvable.*

A lattice $\mathcal{L}$ is called *sectionally complemented* if for every $b < c \in \mathcal{L}$ there exists a $d \in \mathcal{L}$ such that $b \wedge d = 0$ (the smallest element of the lattice) and $b \vee d = c$. This lattice theoretic property also has a neat group theoretic counterpart.

Theorem 2.9 (Bechtell [6], 1965) *For a finite group G, the subgroup lattice* $\mathrm{Sub}(G)$ *is sectionally complemented if and only if every Sylow subgroup of G is elementary abelian.*

Sectionally complemented lattices are more general than *relatively complemented lattices*, in which for every $a < b < c \in \mathcal{L}$ the existence of a $d \in \mathcal{L}$ with $b \wedge d = a$ and $b \vee d = c$ is required (see Figure 3). So here we need an additional condition. By definition, G is a *T^*-group* if being a normal subgroup is a transitive relation among the subgroups of G, that is $A \lhd B \lhd C \leq G$ implies $A \lhd C$.

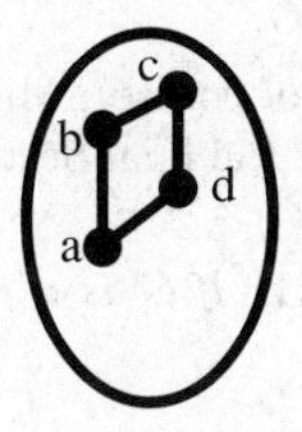

Figure 3. Relatively complemented lattice

Theorem 2.10 (Zacher [65], 1952) *For a finite group G, the subgroup lattice* $\mathrm{Sub}(G)$ *is relatively complemented if and only if every Sylow subgroup of G is elementary abelian and G is a T^*-group.*

We shall also investigate laws in normal subgroup lattices. One basic fact is that $\mathrm{Norm}(G)$ is always modular. There are even stronger laws that hold in normal subgroup lattices, as — for example — the arguesian law. We will deal with this subject in Section 4. Our main concern will be whether there exist laws that distinguish subgroup lattices of abelian groups from normal subgroup lattices in general.

In Section 5 we are going to show that every finite distributive lattice is the normal subgroup lattice of a finite solvable group. It does not seem feasible to describe which lattices can be normal subgroup lattices.

In general $\mathrm{Sub}(G)$ does not determine G uniquely. A lattice isomorphism between subgroup lattices $\mathrm{Sub}(G)$ and $\mathrm{Sub}(H)$ is called a *projectivity*. We restrict our attention — unless stated otherwise — to finite groups. Although there are cases when infinitely many groups share the same subgroup lattice, these lattices can be singled out easily.

Theorem 2.11 (Suzuki [59], 1951) *If* $\mathrm{Sub}(G)$ *has no chain as a direct factor, then there are only finitely many nonisomorphic groups* H *with* $\mathrm{Sub}(H) \cong \mathrm{Sub}(G)$.

Also certain — but by far not all — properties of groups are preserved by projectivities. Assume that $\mathrm{Sub}(G) \cong \mathrm{Sub}(H)$ for finite groups G, H.

- If G is cyclic, then H is also cyclic.
- If G is abelian, then H need not be abelian (even nilpotent).
- If G is a p-group which is neither cyclic, nor elementary abelian, then H is also a p-group (Suzuki [59], 1951).
- If G is solvable, then H is also solvable (Suzuki [59], 1951; Zappa [67], 1951; Schmidt [50], 1968).
- If G is simple, then H is also simple (Suzuki [59], 1951; extension to infinite groups: Zacher [66], 1982). Moreover, using the classification of finite simple groups, it follows that $H \cong G$.

Without using the classification (of course), Michio Suzuki proved a result which will motivate our investigations in the final Section 6.

Theorem 2.12 (Suzuki [59], 1951) *If* G *is a finite simple group and* $\mathrm{Sub}(H) \cong \mathrm{Sub}(G \times G)$, *then* $H \cong G \times G$.

We will look at the question, whether the lattice $\mathrm{Sub}(G \times \cdots \times G)$ can be used to characterize the group G.

3 Local stucture

In this section we are going to study the local structure of subgroup lattices, that is, the possible intervals $\mathrm{Int}[H;K] = \{X \mid H \le X \le K\}$ in subgroup lattices $\mathrm{Sub}(G)$ (where $H < K \le G$). Clearly, we can restrict our attention to *top intervals* (what lattice theorists call *principal filters*), where $K = G$ (see Figure 4). Also, if $N \lhd G$ with $N \le H$, then obviously $\mathrm{Int}[H;G] \cong \mathrm{Int}[H/N;G/N]$, hence we may — and will — always assume that H is *core-free*, i.e., $\bigcap_{g\in G} gHg^{-1} = 1$.

It is easy to see, that intervals in algebraic lattices are algebraic lattices themselves, hence every interval in a subgroup lattice is an algebraic lattice. Namely, a subgroup $X \in \mathrm{Int}[H;G]$ is a compact element of the interval if and only if it

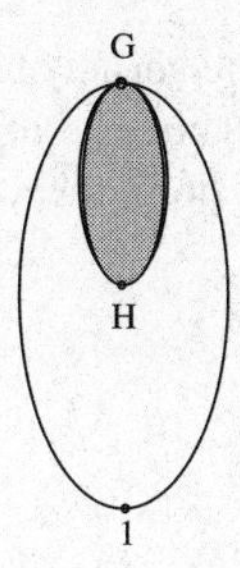

Figure 4. Top interval in Sub(G)

is finitely generated over H, i.e., $X = \langle H, g_1, \ldots, g_k \rangle$ for a suitable finite set of elements $g_1, \ldots, g_k \in G$.

Indeed, there is nothing more one can say about the local structure of subgroup lattices as the following deep result of Jiří Tůma shows.

Theorem 3.1 (Tůma [62], 1989) *For every algebraic lattice $\mathcal{L}$ there exist groups $H < G$ such that* $\text{Int}[H; G] \cong \mathcal{L}$.

It should be noted that Tůma's ingenious construction always yields infinite groups G, even for finite lattices $\mathcal{L}$. Hence we have the following open problem.

Problem 3.2 Is it true that for every finite lattice $\mathcal{L}$ there exist finite groups $H < G$ such that $\text{Int}[H; G] \cong \mathcal{L}$?

The problem actually originates from universal algebra, so let us make a short detour to this area. The reader should be reminded of Graham Higman's witty remarks about Cohn's *Universal Algebra* [19][1]:

"Universal algebra is something everyone ought to know about, though nobody should specialize in it (from which it might appear to follow that though everyone ought to read this book, nobody should have written it). From the point of view of the working algebraist, its main function is to remind him that there are several levels of generality at which work can profitably be done, and that, to get the best out of a method, it is necessary to set it at the right level."

In what follows I try to present a problem in universal algebra which had made me think that it is worthwhile specializing in universal algebra, before I realized that it is in fact a problem in group theory.

By an *algebra* $\mathbf{A} = (A; F)$ we mean a nonempty set A equipped with a set of operations F, that is, each $f \in F$ is a map $f : A^{n(f)} \to A$ for a suitable $n(f)$, called the *arity* of f. For example, in a group we have three operations: multiplication (binary), inverse (unary), and the identity element (considered a nullary operation $G^0 \to G$). In a lattice we have two binary operations: join and meet. The obvious definitions for subalgebras, homomorphisms, direct products make sense in this general setting as well.

---

[1] I thank Πeter Neumann for calling my attention to Higman's review.

However, for arbitrary algebras the kernel of a homomorphism $\varphi : \mathbf{A} \to \mathbf{B}$ cannot be defined in the way it is done in the case of groups, namely, as a preimage of a specific element of $\mathbf{B}$; not even for a homomorphism between lattices (cf. Figure 5).

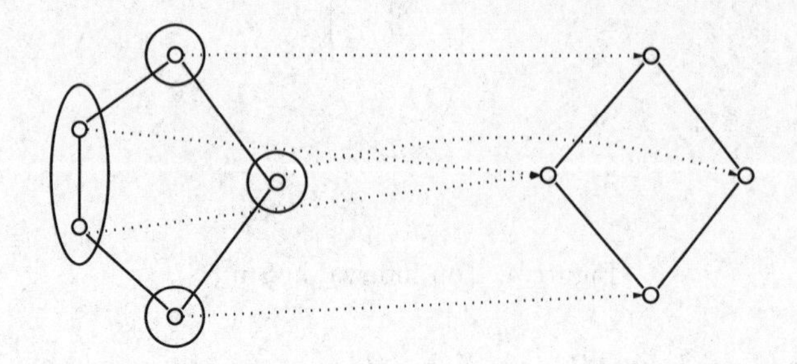

Figure 5. Kernel of a lattice homomorphism

Instead, the appropriate definition of the *kernel* gives a binary relation

$$\ker\varphi = \{(a, a') \in A^2 \mid \varphi(a) = \varphi(a')\}.$$

In fact, $\alpha = \ker\varphi$ is a *congruence relation*, that is, an equivalence relation compatible with all operations $f \in F$, i.e., if f is n-ary, and $a_1, \ldots, a_n, a_1', \ldots, a_n' \in A$ are such that $(a_i, a_i') \in \alpha$ for all $i = 1, \ldots, n$ then

$$\Big(f(a_1, \ldots, a_n), f(a_1', \ldots, a_n')\Big) \in \alpha.$$

All congruence relations of an algebra $\mathbf{A}$ form an algebraic lattice $\operatorname{Con}\mathbf{A}$, the *congruence lattice* of $\mathbf{A}$. For a group G the congruence lattice is essentially the same as the normal subgroup lattice $\operatorname{Norm}(G)$. Apart from being algebraic there is no other general property of congruence lattices as the following classical result of universal algebra tells us.

Theorem 3.3 (Grätzer and Schmidt [15], 1963) *For every algebraic lattice $\mathcal{L}$ there exists an algebra $\mathbf{A}$ such that $\operatorname{Con}\mathbf{A} \cong \mathcal{L}$.*

If we have a group G and a subgroup $H < G$, then we can consider the permutation representation of G on the left cosets by H as a multi-unary algebra $\mathbf{A} = (G/H; G)$, where $g \in G$ as a unary operation sends the coset $xH \in G/H$ to gxH. Now it is easy to see that the congruence lattice of this multi-unary algebra, $\operatorname{Con}\mathbf{A} \cong \operatorname{Int}[H; G]$. So Tůma's Theorem 3.1 yields a new proof for the Grätzer–Schmidt Theorem.

Both of these proofs are inherently infinite. Namely, the basic idea — without going into technical details — can be summarized in the following steps, constructing recursively an infinite sequence of algebras (or groups) $\mathbf{A}_1$, $\mathbf{A}_2$, $\mathbf{A}_3$, ... (see Figure 6):

1. take $\mathbf{A}_i$ and list all the "troubles" occurring in $\mathbf{A}_i$;

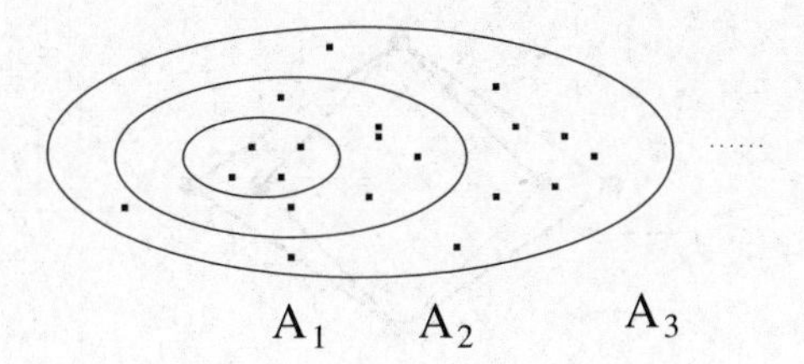

Figure 6. Infinite procedure

2. add new elements and extend the operations so that these "troubles" are eliminated and call the extended algebra $\mathbf{A}_{i+1}$;
3. repeat.

Finally, in $\mathbf{A} = \bigcup_{i=1}^{\infty} \mathbf{A}_i$ all "troubles" disappear, and it will have the required congruence lattice.

Clearly, every finite lattice is algebraic. Therefore, the finite version of the congruence lattice representation problem arises naturally.

Problem 3.4 Is it true that for every finite lattice $\mathcal{L}$ there exists a finite algebra $\mathbf{A}$ such that $\mathrm{Con}\,\mathbf{A} \cong \mathcal{L}$?

Although arbitrary algebras are allowed here, the core of the problem is group theoretic, as the following result shows.

Theorem 3.5 (Pálfy and Pudlák [42], 1980) *The following are equivalent:*
(i) Every finite lattice occurs as the congruence lattice of some finite algebra.
(ii) Every finite lattice occurs as an interval in the subgroup lattice of some finite group.

Using the multi-unary algebra arising from the permutation representation of G on the left cosets by H, it is obvious that (ii) implies (i). The point here is that the converse is also true. Note that, in the case when both statements were false, we do not claim that every finite lattice which is a congruence lattice of a finite algebra is in fact an interval in the subgroup lattice of a finite group. Quite possibly, there can be more lattices that are not intervals in subgroup lattices of finite groups, than lattices that are not congruence lattices of finite algebras.

The "tame congruence theory" developed by David Hobby and Ralph McKenzie [21] shows that in every finite algebra with a certain type of congruence lattice there are some subsets (the so-called minimal sets) on which the induced algebras are actually permutation groups with the same congruence lattice as the one of the original algebra. Therefore, Problem 3.4 in fact belongs to group theory, not to universal algebra.

Special attention has been given to finite lattices with the simplest structure. These lattices $\mathcal{M}_n$ consist of a smallest, a greatest, and n pairwise incomparable elements (see Figure 7).

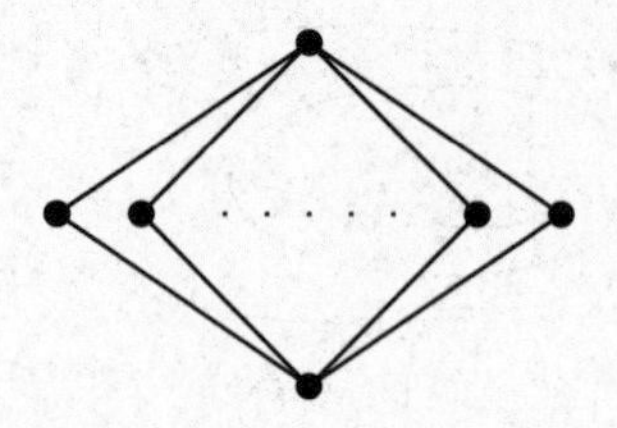

Figure 7. The lattice $\mathcal{M}_n$

Some of these lattices are easy to represent as intervals, even as subgroup lattices. Namely, $\mathcal{M}_1 = \mathrm{Sub}(C_{p^2})$ for any prime p; $\mathcal{M}_2 = \mathrm{Sub}(C_{pq})$ for any pair of distinct primes p, q; $\mathcal{M}_{p+1} = \mathrm{Sub}(C_p \times C_p)$ for each prime p, but also this is the subgroup lattice of any nonabelian group of order pq, where q is a prime divisor of $p-1$.

If $n = p^k + 1$ for some prime p and exponent $k \geq 1$, then it is also possible to find suitable intervals $\mathrm{Int}[H; G] = \mathcal{M}_n$. Namely, let V be the 2-dimensional vector space over the Galois field $\mathrm{GF}(p^k) = F$ and take

$$G = \{x \mapsto \lambda x + v \mid \lambda \in F^\times, v \in V\}, \quad H = \{x \mapsto \lambda x \mid \lambda \in F^\times\}.$$

Then every intermediate subgroup $H \leq K \leq G$ has the form

$$K = \{x \mapsto \lambda x + v \mid \lambda \in F^\times, v \in U\}$$

for a suitable subspace $U \leq V$, so $\mathrm{Int}[H; G]$ is just the lattice of subspaces of V.

For quite a while it had been conjectured that these are the only $\mathcal{M}_n$'s occurring as intervals in subgroup lattices of finite groups. However, this is not the case, as it was first pointed out by Walter Feit. In formulating his observation we shall use the following notation. If $p > 2$ is a prime and $d \mid (p-1)/2$ then up to conjugacy there is a unique subgroup of order pd in the alternating group A_p, which we will denote simply by $p \cdot d$.

Example 3.6 (Feit [11], 1983)

(1) $\mathcal{M}_7 = \mathrm{Int}[31 \cdot 5, A_{31}]$, the intermediate subgroups are the normalizer $31 \cdot 15$ of $31 \cdot 5$ and six subgroups isomorphic to $\mathrm{GL}_5(2)$.

(2) $\mathcal{M}_{11} = \mathrm{Int}[31 \cdot 3, A_{31}]$, the intermediate subgroups are the normalizer $31 \cdot 15$ of $31 \cdot 3$ and ten subgroups isomorphic to $\mathrm{PSL}_3(5)$.

These examples cannot be generalized (see also [40]).

Theorem 3.7 (Basile [5], 2001) *If* $\mathrm{Int}[H; G] \cong \mathcal{M}_n$ *with* $G = S_d$ *or* A_d, *then either* $n \leq 3$ *or one of the following holds:* $n = 5$, $d = 13$; $n = 7$, $d = 31$; $n = 11$, $d = 31$.

Much later a new series of examples was found by Andrea Lucchini.

Theorem 3.8 (Lucchini [31], 1994) *There exist intervals $\mathcal{M}_n$ in subgroup lattices of finite groups with*

$$n = q + 2 \quad \text{or} \quad n = \frac{q^t + 1}{q + 1} + 1,$$

where q is a prime power and t is an odd prime.

At present the smallest cases for which no occurrence of $\mathcal{M}_n$ is known are $n = 16$, 23, 35, ... In a seminal paper Baddeley and Lucchini [2] analyse the structure of a hypothetical group providing an example of an interval $\mathcal{M}_n$ with n not belonging to the set of known values. More precisely, they make the following assumptions.

Assumptions. Let $n > 50$,

$$n \notin \left\{ q+1,\ q+2,\ \frac{q^t+1}{q+1} + 1 \;\middle|\; q \text{ a prime power, } t \text{ an odd prime} \right\},$$

and assume that there exist finite groups $H < G$ such that $\mathrm{Int}[H; G] \cong \mathcal{M}_n$. Furthermore, let G be the smallest one among all groups with this property.

Then, by a result of Peter Köhler [27], G has a unique minimal normal subgroup M, and M is nonabelian (see [42]). So M is a direct product of isomorphic nonabelian simple groups. Let F denote one of the simple factors.

The case when $M \cap H \neq 1$ was dealt with by Lucchini [32] in 1994. He proved that in this case M itself is simple, so G is an *almost simple* group. So we suppose that $M \cap H = 1$. We distinguish two cases, namely, whether $MH = G$ or $MH < G$.

It can be shown that M is complemented in the second case as well, that is, $G = MK$ and $M \cap K = 1$ with a suitable subgroup $K > H$. Now G has the structure of a *twisted wreath product* of F and K.

Baddeley and Lucchini derive the following properties of the ingredients of this twisted wreath product:

- K is an almost simple group,
- H is a core-free maximal subgroup of K,
- $Q = \mathbf{N}_K(F)$ is a core-free subgroup of K,
- $K = QH$,
- the homomorphism $\varphi : Q \to \mathrm{Aut}\, F$ satisfies $\varphi(Q \cap H) \geq \mathrm{Inn}\, F$,
- $\varphi|_{Q \cap H}$ has no extension to any subgroup of H properly containing $Q \cap H$.

Furthermore, $n - 1$ is the number of those homomorphisms $\psi : Q \to \mathrm{Aut}\, F$ for which $\psi|_{Q\cap H} = \varphi|_{Q\cap H}$ and $\tilde{\psi} = \tilde{\varphi}$ hold, where $\tilde{\varphi}$ denotes the composition of φ with the natural homomorphism onto the outer automorphism group $\mathrm{Out}\, F = \mathrm{Aut}\, F/\mathrm{Inn}\, F$.

It should be noted, however, that no such example is known with $n \geq 3$.

The case $MH = G$ leads to even more complex technical conditions, which we cannot reproduce here in full detail. We only mention that in this case H has a unique minimal normal subgroup N, which is a direct product of isomorphic copies of a nonabelian simple group E, and F is isomorphic to a section of E.

These reductions raise several problems about simple groups. We quote only two of them here.

Problem 3.9 (Baddeley and Lucchini [2], 1997) Describe the maximal nonabelian simple sections of the nonabelian simple groups.

Problem 3.10 (Baddeley and Lucchini [2], 1997) Describe all pairs (F, L) where F is a nonabelian simple group and L is a group of automorphisms of F such that there is exactly one proper nontrivial L-invariant subgroup of F.

Another important development concerning the local structure of subgroup lattices of finite groups is a recent result of Ferdinand Börner. He was able to reduce Problem 3.2 to two special cases.

Theorem 3.11 (Börner [8], 1999) *Every finite lattice is an interval in the subgroup lattice of some finite group if and only if at least one of the following statements is true:*

(C) For every finite lattice $\mathcal{L}$ there exist finite groups $H < G$ such that $\mathrm{Int}[H; G] \cong \mathcal{L}$*, with the following properties: G has a unique minimal normal subgroup M, $M \cap H = 1$, $MH = G$, M is nonabelian, and if F denotes one of the simple direct factors of M and $Q = \mathbf{N}_H(F)$, then Q induces all inner automorphisms of F and Q is core-free in H.*

(D) For every finite lattice $\mathcal{L}$ which is generated by its coatoms (maximal elements) there exist finite groups $H < G$ such that $\mathrm{Int}[H; G] \cong \mathcal{L}$*, where G is an almost simple group and H is core-free in G.*

The condition on the lattice in (D) is not very restrictive, as every finite lattice can be embedded as an interval into a finite lattice which is generated by its coatoms. The key of Börner's tricky construction is to embed the given lattice as an interval into a larger lattice as it is vaguely sketched in Figure 8. If this larger lattice occurs as an interval in the subgroup lattice of a finite group, then this group must have a very restricted structure.

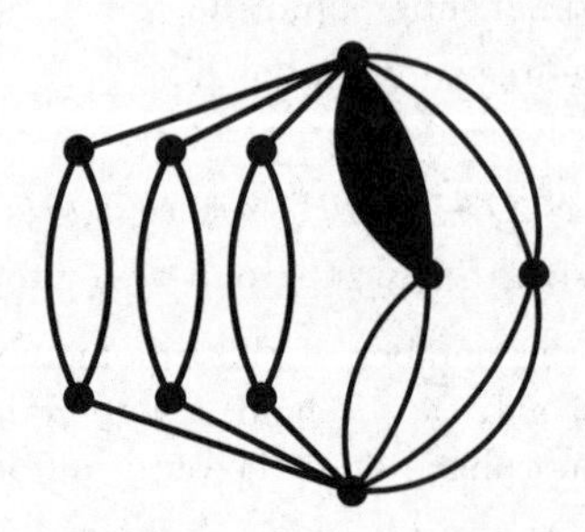

Figure 8. Börner's construction

Similar arguments can be found in a paper of Robert Baddeley [1]. The key words in these investigations are *quasiprimitive groups* and *twisted wreath products*.

Let me end this section with some speculation concerning statement (C). It is so restrictive that probably it can be proved to be false, and then the problem would be reduced to the case of almost simple groups. Then it would remain to do a case-by-case analysis, like it has been done for the alternating and symmetric groups by Alberto Basile (see Theorem 3.7).

Although it seems unlikely that (C) is true, but if it is, it may be proved "combinatorially", without relying much on the structure of the simple group F (similarly as in Lucchini's construction proving Theorem 3.8).

Here we could summarize only the most important developments concerning Problem 3.2. Some other aspects of it are discussed in more detail in [41].

4 Laws in normal subgroup lattices

It is well-known that normal subgroup lattices are modular. Modular lattices can be defined via a number of equivalent conditions. The most useful form is an implication (Horn-formula):

$$X \geq Z \quad \Rightarrow \quad (X \wedge Y) \vee Z = X \wedge (Y \vee Z).$$

Since the left hand side is always smaller than or equal to the right hand side, it can be formulated as an inequality as well:

$$X \geq Z \quad \Rightarrow \quad (X \wedge Y) \vee Z \geq X \wedge (Y \vee Z).$$

The assumption $X \geq Z$ can be eliminated by replacing Z with $X \wedge Z$, thus obtaining the *modular law:*

$$(X \wedge Y) \vee (X \wedge Z) = X \wedge [Y \vee (X \wedge Z)].$$

A characterization of modular lattices can be given by a forbidden sublattice as well:

A lattice is modular if and only if it contains no sublattice 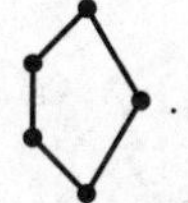 .

The modular law was discovered by Richard Dedekind [10] in 1877. He studied the subgroup lattice of the additive group of complex numbers, but the proof is clearly the same for the subgroup lattice of any abelian group. Dedekind called a subgroup "Modul" and denoted the join of two subgroups by $\mathfrak{a} + \mathfrak{b}$ and their meet by $\mathfrak{a} - \mathfrak{b}$. So the modular law in Dedekind's work appears in the form

$$(\mathfrak{a} - \mathfrak{b}) + (\mathfrak{a} - \mathfrak{c}) = \mathfrak{a} - (\mathfrak{b} + (\mathfrak{a} - \mathfrak{c}))$$

and its dual

$$(\mathfrak{a} + \mathfrak{b}) - (\mathfrak{a} + \mathfrak{c}) = \mathfrak{a} + (\mathfrak{b} - (\mathfrak{a} + \mathfrak{c}))$$

(see [10, p. 17]).

We will consider laws of normal subgroup lattices for various classes of groups. Let $\mathcal{V}$ be a class of groups, P, Q terms in the language of lattices (i.e., elements of the free lattice). $P \leq Q$ is a *law* in the normal subgroup lattices of $\mathcal{V}$ if for every

$G \in \mathcal{V}$ the inequality $P \leq Q$ holds in Norm(G) if we arbitrarily substitute normal subgroups of G for the variables in the terms P, Q and evaluate these terms in Norm(G). Of course, in Norm(G) the lattice operations are given by intersection and product:

$$X \wedge Y = X \cap Y, \quad X \vee Y = XY.$$

We will often use inequalities instead of equalities, but these are certainly equivalent to each other:

$$P \leq Q \iff P \vee Q = Q, \quad P = Q \iff P \vee Q \leq P \wedge Q.$$

There exist laws of normal subgroup lattices that are even stronger than modularity. The most important one is the *arguesian law* introduced by Bjarni Jónsson [24] in 1954. (The idea appeared earlier in a paper of Schützenberger [56] in 1945.) This is a translation of Desargues' Theorem from projective geometry into the language of lattices. Among the several equivalent formulations we prefer the following form:

$$X_1 \wedge \{Y_1 \vee [(X_2 \vee Y_2) \wedge (X_3 \vee Y_3)]\} \leq [(Q_{12} \vee Q_{23}) \wedge (Y_1 \vee Y_3)] \vee X_3,$$

where $Q_{ij} = (X_i \vee X_j) \wedge (Y_i \vee Y_j)$.

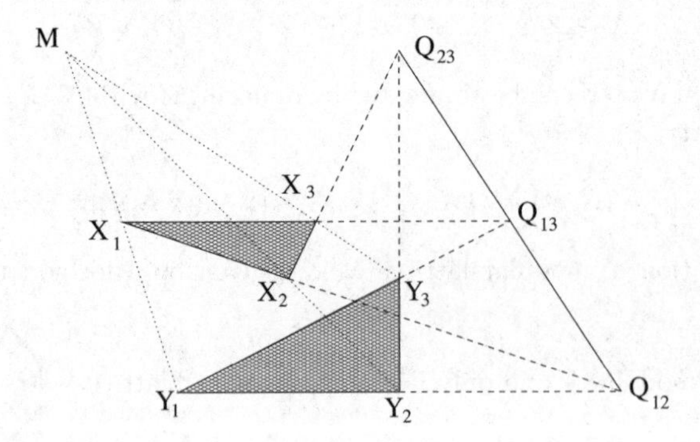

Figure 9. Desargues' Theorem

Consider the subspace lattice of a projective plane. There the join of two points is the line connecting them, the meet of two lines is their intersection point. Now if $X_1X_2X_3$ and $Y_1Y_2Y_3$ are triangles in a projective plane that are *perspective with respect to a point* (i.e., the lines X_1Y_1, X_2Y_2, X_3Y_3 go through a common point) then the left hand side of the arguesian law yields the point X_1, otherwise it yields the empty set. At the same time, the right hand side is $\geq X_1$ if and only if the two triangles are *perspective with respect to a line* (i.e., if the intersection of the corresponding sides X_iX_j and Y_iY_j are denoted by Q_{ij}, then the three points Q_{12}, Q_{13} and Q_{23} lie on one line). The reader is strongly advised to check this using Figure 9. It is a tedious but dull task to show that if the arguesian law holds for the points of the subspace lattice of a projective geometry then it holds for arbitrary substitution of elements of the lattice (see [14, p. 207]). So the arguesian law holds in the subspace lattice of a projective plane if and only if Desargues' Theorem is true in the geometry. Since there are nonarguesian planes, the arguesian law is stronger than the modular law, as the subspace lattice is always modular.

Theorem 4.1 (Jónsson [24], 1954) *The arguesian law holds in the normal subgroup lattice of every group.*

In fact the arguesian law holds in every lattice consisting of permuting equivalence relations. (Two equivalence relations α and β are said to *permute* if $\beta \circ \alpha = \alpha \circ \beta$, where the *relational product* is defined by

$$\alpha \circ \beta = \{(a, b) \mid \exists c : (a, c) \in \alpha, (c, b) \in \beta\}.$$

For every normal subgroup $N \lhd G$ there corresponds an equivalence relation $\alpha_N = \{(a, b) \mid a, b \in G, a^{-1}b \in N\}$, and $\alpha_K \circ \alpha_N = \alpha_{KN} = \alpha_N \circ \alpha_K$, since $KN = NK$ for normal subgroups $N, K \lhd G$.)

Proof Let $X_1, X_2, X_3, Y_1, Y_2, Y_3 \lhd G$ and $x_1 \in X_1 \wedge \{Y_1 \vee [(X_2 \vee Y_2) \wedge (X_3 \vee Y_3)]\} = X_1 \cap Y_1[X_2Y_2 \cap X_3Y_3]$. Then there exist elements $x_i \in X_i$, $y_i \in Y_i$, $m \in G$ such that $x_1 = my_1^{-1}$ (using $Y_1M = MY_1^{-1}$) and $m = x_2y_2 = x_3y_3$. So we have $x_1y_1 = x_2y_2 = x_3y_3$, $x_2^{-1}x_1 = y_2y_1^{-1} \in Q_{12} = (X_1 \vee X_2) \wedge (Y_1 \vee Y_2)$, similarly $x_3^{-1}x_2 = y_3y_2^{-1} \in Q_{23}$, and multiplying these equations we obtain $x_3^{-1}x_1 = y_3y_1^{-1} \in Q_{12}Q_{23} \cap Y_1Y_3$. Taking the product with X_3 we see that $x_1 \in X_3(Q_{12}Q_{23} \cap Y_1Y_3)$ indeed. □

Mark Haiman [17] in 1987 discovered a sequence of laws, the *higher arguesian identities*

$$X_1 \wedge \left[Y_1 \vee \bigwedge_{i=2}^{n}(X_i \vee Y_i)\right] \leq \left[\bigvee_{i=1}^{n-1} Q_{i,i+1} \wedge (Y_1 \vee Y_n)\right] \vee X_n,$$

where $Q_{ij} = (X_i \vee X_j) \wedge (Y_i \vee Y_j)$, each one being strictly stronger than the previous one, that all hold in every lattice consisting of permuting equivalence relations. Later it was proved by Ralph Freese [12] that there is no finite basis for the laws of the class of all normal subgroup lattices. Like the modular and the arguesian laws, the higher arguesian identities hold not only in subgroup lattices of abelian groups (as suggested by the underlying geometry), but also in normal subgroup lattices of arbitrary groups. Based on such experiences a positive solution of the following problem had been expected.

Problem 4.2 (Jónsson [24], 1954; Birkhoff [7, p. 179], 1967) Can one embed the normal subgroup lattice of an arbitrary group into the subgroup lattice of an abelian group? Do all the laws of subgroup lattices of abelian groups hold in normal subgroup lattices?

Another reason pointing towards a positive solution was the following observation:

Proposition 4.3 *If $\mathcal{M}_3$ is a sublattice of* Norm(G) *with top element N, bottom element M, then N/M is abelian (see Figure 10).*

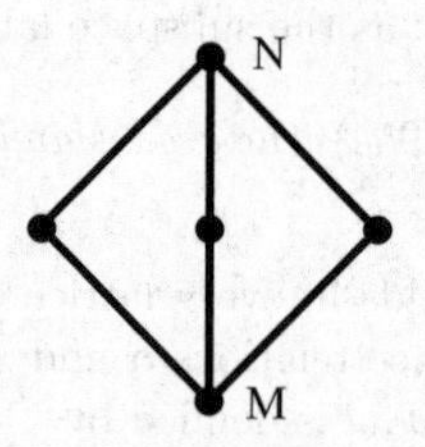

Figure 10. Abelian section in a normal subgroup lattice

However, together with my student Csaba Szabó we found another lattice identity with geometric background that shows that the normal subgroup lattice of some groups cannot be embedded into the subgroup lattice of any abelian group. A short proof for the nonembeddability of Norm(G) for a certain group G of order 2^9 is given in [25].

Theorem 4.4 (Pálfy and Szabó [44], 1995) *The six-cross law*

$$X_1 \wedge \{Y_1 \vee [(X_2 \vee Y_2) \wedge (X_3 \vee Y_3) \wedge (X_4 \vee Y_4)]\} \leq$$
$$(\{[(P_{12} \vee P_{34}) \wedge (P_{13} \vee P_{24})] \vee P_{23}\} \wedge \{X_4 \vee Y_1\}) \vee Y_4,$$

where $P_{ij} = (X_i \vee Y_j) \wedge (Y_i \vee X_j)$, holds in the subgroup lattice of every abelian group but fails in the normal subgroup lattice of the free group on five generators.

Another version of this law was given in [43].

Again, our six-cross law is a lattice theoretic translation of a geometric property (see Figure 11). Take four lines through a point, and two points X_i, Y_i on each of these lines ($i = 1, 2, 3, 4$). For each pair of lines define the cross point P_{ij} as the intersection of the lines X_iY_j and Y_iX_j. We say that the *six-cross theorem* holds in the projective plane, if the three lines $P_{12}P_{34}$, $P_{13}P_{24}$, and $P_{14}P_{23}$ go through a common point. (Actually, we need a more precise definition handling the degenerate cases as well, for example if some of the cross points coincide. An interesting case occurs when $P_{12} = P_{34}$, $P_{13} = P_{24}$, $P_{14} = P_{23}$. This is the famous Reye-configuration, see [20, §22]. For these details we refer to [44].) For a projective geometry the six-cross theorem is equivalent to Desargues' theorem, but their lattice theoretic counterparts differ. The six-cross law implies the arguesian law, but the converse does not hold.

Although in the formulation of Theorem 4.4 we used a free group, actually there exist finite quotients of the free group of rank 5 whose normal subgroup lattices do not satisfy the six-cross law. Moreover, it is enough to consider finite nilpotent groups, or p-groups, as the following lemma (see [34, p. 41]) shows.

Lemma 4.5 *Let $|G| = p_1^{k_1} \cdots p_n^{k_n}$, and P_i a Sylow p_i-subgroup of G for each $i = 1$, $\ldots$, n. Then* Norm(G) *is embedded into* Norm$(P_1 \times \cdots \times P_n)$ *via $N \mapsto (N \cap P_1) \times \cdots \times (N \cap P_n)$.*

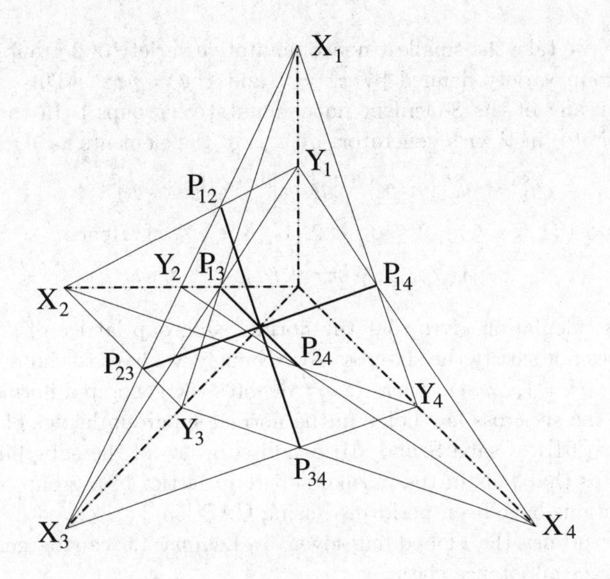

Figure 11. Six-cross theorem

So if the normal subgroup lattice of a finite group does not satisfy a certain law, then the same is true for at least one of its Sylow subgroups. If we want to work with a group of nilpotence class 2, then we better look for 2-groups, since for groups of odd order the following trick of Reinhold Baer yields an embedding of the normal subgroup lattice into the subgroup lattice of an abelian group defined on the same set of elements.

Lemma 4.6 (Baer [3], 1938) *If G of odd order has nilpotence class 2, then*

$$x + y = x^{1/2} y x^{1/2} \qquad (x^{1/2} = x^{(|G|+1)/2})$$

defines an abelian group operation.

Corollary 4.7 *If G of odd order has nilpotence class 2, then* $\mathrm{Norm}(G)$ *can be embedded into the subgroup lattice of an abelian group.*

Proof Denote by G^+ the abelian group defined in Lemma 4.6. Clearly every normal subgroup of G is a subgroup of G^+. We have to check that for every $X, Y \lhd G$ the lattice operations are the same in $\mathrm{Norm}(G)$ and in $\mathrm{Sub}(G^+)$. This is trivially true for the meet (intersection), and it is also true for the join, because the join in $\mathrm{Norm}(G)$ is a common upper bound for X and Y in $\mathrm{Sub}(G^+)$ as well, and the order formula yields that it is indeed the least upper bound:

$$|X \vee Y| = |XY| = \frac{|X| \cdot |Y|}{|X \cap Y|} = |X + Y|.$$

□

Therefore, we take the smallest noncommutative variety of 2-groups, namely let $\mathcal{V}$ be the group variety defined by $x^4 = 1$ and $x^2y = yx^2$. (This is the variety generated by any of the 8-element noncommutative groups.) In the (relatively) free groups $F_{\mathcal{V}}(r)$ in $\mathcal{V}$ with generators $g_1, \ldots, g_r$ the elements have a normal form

$$g_1^{\alpha_1} \cdots g_r^{\alpha_r} [g_1, g_2]^{\beta_{12}} [g_1, g_3]^{\beta_{13}} \cdots [g_{r-1}, g_r]^{\beta_{r-1,r}}$$

with $0 \le \alpha_i < 4$ $(1 \le i \le r)$, $0 \le \beta_{ij} < 2$ $(1 \le i < j \le r)$. Hence

$$|F_{\mathcal{V}}(r)| = 4^r 2^{r(r-1)/2} = 2^{(r^2+3r)/2}.$$

A tedious calculation gives that the normal subgroup lattice of $G = F_{\mathcal{V}}(5)$ of order 2^{20} does not satisfy the six-cross law. Namely, we have to choose $X_i = \langle g_i \rangle^G$, $Y_i = \langle g_i g_5 \rangle^G$ $(i = 1, \ldots, 4)$, where $\langle \ldots \rangle^G$ denotes the generated normal subgroup.

However, the six-cross law holds in the normal subgroup lattice of every group of odd order ([61]). Csaba Szabó [61] exhibited a law of the subgroup lattices of abelian groups that fails in the normal subgroup lattice of a group of order 3^{140}. (The calculations have been performed using GAP [55].)

For larger primes the embedding given by Lemma 4.6 can be generalized for groups of larger nilpotence class:

Theorem 4.8 (Groves [16], 1976) *If the nilpotence class of a finite p-group is less than p, then there exists a term defining an abelian group operation. Therefore, the normal subgroup lattice of such a group can be embedded into the subgroup lattice of an abelian group.*

In fact, the formula for the abelian group operation can be obtained from Lazard's inversion of the *Baker–Campbell–Hausdorff formula* (see [30])

$$x + y = xy[x, y]^{-1/2}[[x, y], x]^{1/12}[[x, y], y]^{-1/12} \cdots.$$

If the nilpotence class of the p-group is less than p, then this infinite product can be truncated at commutators of weight p or less, and the necessary roots in the group exist (as the denominators are not divisible by p). It is even enough to assume that every 3-generated subgroup has nilpotence class less than p. Is this the limit indeed?

Problem 4.9 For every prime p find a law of the subgroup lattices of abelian groups that fails in the normal subgroup lattice of a finite p-group of nilpotence class p.

We end this section by mentioning a nice result that the exponent of a group variety can be recovered from the laws satisfied by the normal subgroup lattices.

Theorem 4.10 (Herrmann and Huhn [18], 1975) *The law*

$$(X_1 \vee \ldots \vee X_n) \wedge (Y \vee Z) \le \bigvee_{i=1}^{n} [(X_1 \vee \ldots \vee X_{i-1} \vee Y \vee X_{i+1} \vee \ldots \vee X_n) \wedge (X_i \vee Z)]$$

holds in $\mathrm{Norm}(G)$ *for every group of exponent dividing n (i.e., $\forall g \in G : g^n = 1$), but if k does not divide n then it fails in* $\mathrm{Norm}(C_k^{n+1})$.

An excellent survey article related to the topics discussed in this section was written by Robert Burns and Sheila Oates-Williams [9].

5 Distributive normal subgroup lattices

It seems to be a difficult task to describe the possible normal subgroup lattices. In Section 4 we discussed certain laws that must hold in every normal subgroup lattice. However, these are not sufficient to characterize normal subgroup lattices, since this class of lattices is not closed for sublattices, as one can easily check that $\mathcal{M}_5$ is not isomorphic to the normal subgroup lattice of any group, although for each prime number p one has $\mathcal{M}_p \cong \mathrm{Norm}(C_p \times C_p)$.

In this section we have a modest goal to represent finite distributive lattices as normal subgroup lattices.

Theorem 5.1 *For every finite distributive lattice $\mathcal{D}$ there exists a finite group G such that* $\mathrm{Norm}(G) \cong \mathcal{D}$.

This result has an interesting history. It was first announced by Kuntzmann [28] in 1947. However, his proof was in error, and it could not be corrected, since he tried to construct a supersolvable group G for any finite distributive lattice $\mathcal{D}$. In fact, it is not difficult to show that not every finite distributive lattice can be represented as the normal subgroup lattice of a supersolvable group (see Figure 12). E. T. Schmidt [49, p. 101] listed Theorem 5.1 as an open problem, noting that Kuntzmann's proof was not correct.

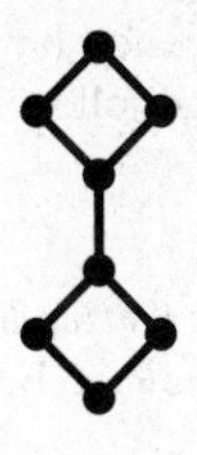

Figure 12. This is not the normal subgroup lattice of any supersolvable group

Finally, Howard Silcock [57] proved the theorem in 1977. He constructed a suitable group G as an iterated wreath product of nonabelian simple groups.

In 1986 I gave a natural construction which yields solvable groups (see [38], [39]). Now I am going to sketch this construction.

First recall a condition characterizing when $\mathrm{Norm}(G)$ is distributive. Remember that the *socle* of a group is the product of its minimal normal subgroups, and that two normal subgroups N_1 and N_2 are said to be *G-isomorphic*, if there is an isomorphism $\varphi : N_1 \to N_2$ such that $\varphi(gxg^{-1}) = g\varphi(x)g^{-1}$ for all $g \in G$ and $x \in N_1$.

Theorem 5.2 (Pazderski [45], 1987) *For a finite group G the normal subgroup lattice* $\mathrm{Norm}(G)$ *is distributive if and only if the socle of every quotient group G/N is a direct product of pairwise non-G/N-isomorphic minimal normal subgroups.*

We shall give a recursive proof of Theorem 5.1. In the groups we are going to construct every chief factor will be a Sylow subgroup, and so Pazderski's criterion will obviously be satisfied, hence the normal subgroup lattice will become distributive.

Let us start by choosing an atom $a \in \mathcal{D}$ and let $a' \in \mathcal{D}$ be a *pseudocomplement* of a, that is, a' is the largest element with $a' \wedge a = 0$. (By distributivity, if $b \wedge a = 0$ and $c \wedge a = 0$, then $(b \vee c) \wedge a = (b \wedge a) \vee (c \wedge a) = 0 \vee 0 = 0$, so the pseudocomplement exists.) Now by induction we can find a group H with $\mathrm{Norm}(H) \cong \mathrm{Int}[a; 1]$. Let us denote by $K \lhd H$ the normal subgroup corresponding to $a \vee a'$ at this isomorphism.

Next we choose a prime p not dividing $|H|$. Since in H/K no two minimal normal subgroups are isomorphic we can invoke a result of Kochendörffer [26] from 1948 guaranteeing the existence of a faithful irreducible representation of H/K over the p-element field. Let us denote the underlying module by V, and let us take the natural action of H on V with kernel K.

We claim that the semidirect product $G = VH$ will have the required normal subgroup lattice $\mathrm{Norm}(G) \cong \mathcal{D}$.

By the irreducibility of the action, V is a minimal normal subgroup of G. If $N \lhd G$, then either $N \geq V$ or $N \cap V = 1$. In the first case $N/V \lhd G/V \cong H$. In the second case $N \leq \mathbf{C}_G(V) = V \times K$, so $N \leq K$, since $(|N|, |V|) = 1$. Now it is clear that $\mathrm{Norm}(G) \cong \mathcal{D}$.

Since the subgroups of $G \times G$ containing the diagonal subgroup correspond to the normal subgroups of G, namely they have the form $H_N = \{(a, b) \in G \times G \mid aN = bN\}$ for $N \lhd G$, we obtain the following corollary related to the results of Section 3.

Corollary 5.3 *For every finite distributive lattice $\mathcal{D}$ there exists a finite group G such that for $H = \{(g, g) \mid g \in G\}$ we have* $\mathrm{Int}[H; G \times G] \cong \mathcal{D}$.

A similar idea, using higher powers, can be used to turn intervals upside-down (called the *dual lattice*).

Theorem 5.4 (Kurzweil [29], 1985) *The class of intervals in subgroup lattices of finite groups is closed under taking dual lattices.*

Proof Let $\mathcal{L} = \mathrm{Int}[H; G]$, with $|G : H| = n$ and take an arbitrary nonabelian simple group S. Consider the permutation representation of G on the cosets of H and let this permutation group act on S^n by permuting the coordinates. If $G^* = S^n G$ and $H^* = \{(s, \ldots, s) \mid s \in S\} G$, then it can be checked that $\mathrm{Int}[H^*; G^*]$ is isomorphic to the dual lattice of $\mathcal{L}$. □

6 Subgroup lattices of direct powers

The problem we are going to discuss in this section is motivated by the following classical result of Michio Suzuki.

Theorem 6.1 (Suzuki [59], 1951) *If G is a finite simple group and* $\mathrm{Sub}(H) \cong \mathrm{Sub}(G \times G)$*, then $H \cong G \times G$.*

This result was later generalized by Roland Schmidt.

Theorem 6.2 (Schmidt [51], 1981) *If G is a finite group with $G' = G$, $\mathbf{Z}(G) = 1$, and* $\mathrm{Sub}(H) \cong \mathrm{Sub}(G \times G)$*, then $H \cong G \times G$.*

However, we cannot always expect that $\mathrm{Sub}(H) \cong \mathrm{Sub}(G \times G)$ would imply $H \cong G \times G$. Namely, let p be a prime and q another prime, dividing $p - 1$. Then for any $d \geq 2$, $\mathrm{Sub}(C_p^d) \cong \mathrm{Sub}(G)$, where $G = A\langle b\rangle$ is a so-called *P-group* with $A = C_p^{d-1}$ and b of order q acting by a power automorphism on A ($bab^{-1} = a^r$, where $r^q \equiv 1, r \not\equiv 1 \pmod{p}$). Instead, we can ask the following question.

Problem 6.3 Does $\mathrm{Sub}(G \times G) \cong \mathrm{Sub}(H \times H)$ imply that G and H are isomorphic groups? In other words, does the subgroup lattice of the direct square uniquely determine the group?

Suzuki's Theorem 6.1 shows that for simple groups this is indeed the case. On the other end of the spectrum there are the abelian groups.

Theorem 6.4 (Lukács and Pálfy [33], 1986) *For a finite abelian group G, if* $\mathrm{Sub}(G \times G) \cong \mathrm{Sub}(H \times H)$*, then G and H are isomorphic.*

This result follows easily from the following observation.

Theorem 6.5 (Lukács and Pálfy [33], 1986) $\mathrm{Sub}(G \times G)$ *is modular if and only if G is abelian.*

Nevertheless, Problem 6.3 has a negative answer. Counterexamples are provided by the *Rottlaender groups* introduced by Ada Rottlaender [47] in 1928. Let $p, q \geq 5$ be primes with $q \mid p - 1$. Then there exists $r \not\equiv 1 \pmod{p}$ such that $r^q \equiv 1 \pmod{p}$. For each $\lambda \in \{2, \ldots, q-2\}$ we define the group

$$R_\lambda = \langle x, y, a \mid x^p = y^p = a^q = [x, y] = 1, axa^{-1} = x^r, aya^{-1} = y^{r^\lambda}\rangle.$$

Roland Schmidt observed the following.

Example 6.6 (Schmidt [51], 1981) $\mathrm{Sub}(R_\lambda \times R_\lambda) \cong \mathrm{Sub}(R_\mu \times R_\mu)$, but $R_\lambda \cong R_\mu$ only if $\lambda = \mu$ or $\lambda\mu \equiv 1 \pmod{q}$

However, it can be checked that for nonisomorphic Rottlaender groups R_λ, R_μ the third powers have nonisomorphic subgroup lattices:

$$\mathrm{Sub}(R_\lambda \times R_\lambda \times R_\lambda) \not\cong \mathrm{Sub}(R_\mu \times R_\mu \times R_\mu).$$

This indicated that perhaps the answer to the following question might be positive.

Problem 6.7 Does $\mathrm{Sub}(G \times G \times G)$ uniquely determine G? (That is, assuming $\mathrm{Sub}(G \times G \times G) \cong \mathrm{Sub}(H \times H \times H)$, does it follow that G and H are isomorphic?)

Another result where third powers play a crucial role is the following one.

Theorem 6.8 (Baer [4], 1939) *Let G be an abelian p-group, H an abelian group. Assume that for each prime power p^k, if G contains an element of order p^k, then it contains at least three independent elements of this order (i.e., a subgroup isomorphic to $C_{p^k} \times C_{p^k} \times C_{p^k}$). Then every projectivity (lattice isomorphism) between* $\mathrm{Sub}(G)$ *and* $\mathrm{Sub}(H)$ *is induced by an isomorphism between the groups G and H.*

It should be noted, however, that one cannot hope to prove that every isomorphism between $\mathrm{Sub}(G \times \cdots \times G)$ and $\mathrm{Sub}(H \times \cdots \times H)$ is induced by a group isomorphism, as the following example shows.

Example 6.9 (Schmidt [52], 1982) Let p be a prime, q and r prime divisors of $p-1$, and let G be the direct product of the nonabelian groups of order pq and pr. Then for every $d \geq 1$, $\mathrm{Sub}(G^d)$ has lattice automorphisms (autoprojectivities) that are not induced by group automorphisms.

Note that the order of elements can be recovered from $\mathrm{Sub}(G \times G)$, and then from the subgroup lattices of higher powers as well.

Lemma 6.10 *If* $\mathrm{Sub}(X) \cong \mathcal{M}_n$*, then either*

(1) *$X \cong C_{p^2}$ and $n = 1$; or*

(2) *$X \cong C_{pq}$ and $n = 2$; or*

(3) *$X \cong C_p \times C_p$ and $n = p+1$; or*

(4) *X is a nonabelian group of order pq, $q \mid p-1$ and $n = p+1$.*

Corollary 6.11 *If $P < G \times G$ has order p, then there exists $P < X_0 \leq G \times G$ such that* $\mathrm{Int}[1; X_0] \cong \mathcal{M}_{p+1}$ *(in fact, $X_0 \cong C_p \times C_p$), and if $P < X \leq G \times G$ is such that* $\mathrm{Int}[1; X] \cong \mathcal{M}_n$ *then either $n \leq 2$ or $n \geq p+1$. Hence*

$$p = \min\{n-1 \mid n > 2, \exists X > P : \mathrm{Int}[1; X] \cong \mathcal{M}_n\}.$$

So we can find the order of each minimal subgroup in $\mathrm{Sub}(G \times G)$. Furthermore, if $\mathrm{Int}[1; H]$ is a chain of length k, then $|H| = p^k$, where p is the order of the unique minimal subgroup contained in H. Thus p-subgroups can be identified, and the order of every subgroup can be determined.

Schmidt [54, 7.6.11 Problem] considers a closely related question. Let φ : $\mathrm{Sub}(G_1 \times \cdots \times G_n) \to \mathrm{Sub}(H)$ be a lattice isomorphism (projectivity), where all direct factors $G_1, \ldots, G_n$ are isomorphic to a given group G. Assume that $H = \varphi(G_1) \times \cdots \times \varphi(G_n)$. Does it follow that $\varphi(G_1) \cong G_1$? Concerning this problem there are some positive results, for example in the following cases:

- if G has a self-centralizing normal Hall subgroup (Schmidt [52], 1982);
- if G is a finite p-group and one of the the following holds:
  - (a) G has nilpotence class 2, $p \neq 2$, $n \geq 3$;
  - (b) G has class ≤ 4 and exponent p, $n \geq 2$;
  - (c) G is metabelian of exponent p, $n \geq p - 2$ (Schenke [48], 1987).

Only after the meeting in Oxford I heard about an apparently forgotten paper by Anne Penfold Street [58][2] that implicitly contains the negative solution of Problem 6.7.

Example 6.12 (Street [58], 1968) There exist nonisomorphic groups $(G, \circ)$ and $(G, *)$ on the same base set such that for each $n \geq 1$ their direct powers $(G, \circ)^n$ and $(G, *)^n$ have exactly the same subgroups.

Proof Let us choose prime numbers p and q subject to the following restrictions: $q \equiv 1 \pmod 3$, $p \equiv 1 \pmod{3q}$. Furthermore, let m have order 3 modulo p and let n have order 3 modulo q (i.e., $m^3 \equiv 1 \pmod p$ and $m \not\equiv 1 \pmod p$). Consider the group

$$G(m, n) = \langle s, t, u \mid s^p = t^q = u^3 = [s, t] = 1, usu^{-1} = s^m, utu^{-1} = t^n \rangle$$

and denote the operation also by $\circ$. If we define $x * y = xy[x, y]^{p-1}$ then it turns out that $*$ is also a group operation and the group defined this way is $G(m^2, n)$ which is not isomorphic to $G(m, n)$. Now $\circ$ can be expressed in the same way from $*$ (in the language of universal algebra $(G, \circ)$ and $(G, *)$ are *term equivalent*), therefore exactly the same subsets of the cartesian powers of the base set are closed for the operation $\circ$ that are closed for $*$, that is the powers of both groups have the same subgroup lattice. For the calculations the reader is referred to [58, Example V.(i)]. □

Thus even the subgroup lattices of all powers of G are not sufficient to determine the isomorphism type of G. However, it still may be true that if the subgroup lattices of some power distinguish two groups than already the third powers do.

Problem 6.13 If $\mathrm{Sub}(G \times G \times G) \cong \mathrm{Sub}(H \times H \times H)$, does it follow that G and H are term equivalent groups?

Acknowledgements. I would like to thank the organizers for inviting me to give a series of talks at the Groups – St Andrews 2001 in Oxford conference. The financial support from the organizers as well as from the Hungarian National Research Fund (OTKA) under grant no. T29132 is gratefully acknowledged.

[2]Thanks to Keith Kearnes for discovering this paper.

References

[1] R. Baddeley. *A new approach to the finite lattice representation problem.* Period. Math. Hungar. **36** (1998), 17–59.

[2] R. Baddeley and A. Lucchini. *On representing finite lattices as intervals in subgroup lattices of finite groups.* J. Algebra **196** (1997), 1–100.

[3] R. Baer. *Groups with abelian central quotient group.* Trans. Amer. Math. Soc. **44** (1938), 357–386.

[4] R. Baer. *The significance of the system of subgroups for the structure of the group.* Amer. J. Math. **61** (1939), 1–44.

[5] A. Basile. *Second maximal subgroups of the finite alternating and symmetric groups.* D.Phil. Thesis, Australian National University, Canberra, April 2001.

[6] H. Bechtell. *Elementary groups.* Trans. Amer. Math. Soc. **114** (1965), 355–362.

[7] G. Birkhoff. *Lattice Theory.* 3rd ed. American Mathematical Society Colloquium Publications, Vol. XXV, Amer. Math. Soc., Providence, RI, 1967.

[8] F. Börner. *A remark on the finite lattice representation problem.* Contributions to general algebra, 11 (Olomouc/Velké Karlovice, 1998), Verlag Johannes Heyn, Klagenfurt, 1999, 5–38.

[9] R. G. Burns and S. Oates-Williams. *Varieties of groups and normal-subgroup lattices — a survey.* Algebra Universalis **32** (1994), 145–152.

[10] R. Dedekind. *Über die Anzahl der Ideal-classen in den verschiedenen Ordnungen eines endlichen Körpers.* Festschrift zur Saecularfeier des Geburtstages von C. F. Gauss, Vieweg, Braunschweig, 1877, 1–55; see Ges. Werke, Band I, Vieweg, Braunschweig, 1930, 105–157.

[11] W. Feit. *An interval in the subgroup lattice of a finite group which is isomorphic to* M_7. Algebra Universalis **17** (1983), 220–221.

[12] R. Freese. *Finitely based modular congruence varieties are distributive.* Algebra Universalis **32** (1994), 104–114.

[13] R. Freese. *Subgroup lattices of groups by R. Schmidt.* Book review, Bull. Amer. Math. Soc. **33** (1996), 487–492.

[14] G. Grätzer. *General Lattice Theory.* Pure and Applied Mathematics, 75, Academic Press, New York–London; Lehrbücher und Monographien aus dem Gebiete der exakten Wissenschaften, Mathematische Reihe, Band 52, Birkhäuser-Verlag, Basel–Stuttgart, 1978.

[15] G. Grätzer and E. T. Schmidt. *Characterizations of congruence lattices of abstract algebras.* Acta Sci. Math. (Szeged) **24** (1963), 34–59.

[16] J. R. J. Groves. *Regular p-groups and words giving rise to commutative group operations.* Israel J. Math. **24** (1976), 73–77.

[17] M. D. Haiman. *Arguesian lattices which are not linear.* Bull. Amer. Math. Soc. (N.S.) **16** (1987), 121–123.

[18] C. Herrmann und A. Huhn. *Zum Begriff der Charakteristik modularer Verbände.* Math. Z. **144** (1975), 185–194.

[19] G. Higman. *Universal algebra by P. M. Cohn.* Book review, J. London Math. Soc. **41** (1966), 760.

[20] D. Hilbert und S. Cohn-Vossen. *Anschauliche Geometrie.* Die Grundlehren der mathematischen Wissenschaften in Einzeldarstellungen, Band 37, Julius Springer, Berlin, 1932.

[21] D. Hobby and R. McKenzie. *The Structure of Finite Algebras.* Contemporary Mathematics, 76, Amer. Math. Soc., Providence, RI, 1988.

[22] K. Iwasawa. *Über die endlichen Gruppen und die Verbände ihrer Untergruppen.* J. Fac. Sci. Imp. Univ. Tokyo Sect. I **4** (1941), 171–199.

[23] K. Iwasawa. *On the structure of infinite M-groups.* Jap. J. Math. **18** (1943), 709–728.

[24] B. Jónsson. *Modular lattices and Desargues' theorem.* Math. Scand. **2** (1954), 295–314.

[25] E. W. Kiss and P. P. Pálfy. *A lattice of normal subgroups that is not embeddable into the subgroup lattice of an abelian group.* Math. Scand. **83** (1998), 169–176.

[26] R. Kochendörffer. *Über treue irreduzible Darstellungen endlicher Gruppen.* Math. Nachr. **1** (1948), 25–39.

[27] P. Köhler. *M_7 as an interval in a subgroup lattice.* Algebra Universalis **17** (1983), 263–266.

[28] J. Kuntzmann. *Contribution à l'étude des chaînes principales d'un groupe fini.* Bull. Sci. Math. (2) **71** (1947), 155–164.

[29] H. Kurzweil. *Endliche Gruppen mit vielen Untergruppen.* J. Reine Angew. Math. **356** (1985), 140–160.

[30] M. Lazard. *Sur les groupes nilpotents et les anneaux de Lie.* Ann. Sci. École Norm. Sup. (3) **71** (1954), 101–190.

[31] A. Lucchini. *Representation of certain lattices as intervals in subgroup lattices.* J. Algebra **164** (1994), 85–90.

[32] A. Lucchini. *Intervals in subgroup lattices of finite groups.* Comm. Algebra **22** (1994), 529–549.

[33] E. Lukács and P. P. Pálfy. *Modularity of the subgroup lattice of a direct square.* Arch. Math. (Basel) **46** (1986), 18–19.

[34] R. McKenzie. *Some interactions between group theory and the general theory of algebras.* Groups–Canberra 1989, Lecture Notes Math., vol. 1456, Springer, Berlin, 1990, 32–48.

[35] A. Yu. Olshanskii. *Infinite groups with cyclic subgroups.* Dokl. Akad. Nauk SSSR **245** (1979), 785–787 (in Russian); Translation in Soviet Math. Dokl. **20** (1979), 343–346.

[36] A. Yu. Olshanskii. *Geometry of Defining Relations in Groups.* Mathematics and its Applications (Soviet Series), 70, Kluwer Academic Publishers Group, Dordrecht, 1991.

[37] Ø. Ore. *Structures and group theory, I–II.* Duke Math. J. **3** (1937), 149–174; **4** (1938), 247–269.

[38] P. P. Pálfy. *On partial ordering of chief factors in solvable groups.* Manuscripta Math. **55** (1986), 219–232.

[39] P. P. Pálfy. *Distributive congruence lattices of finite algebras.* Acta Sci. Math. (Szeged) **51** (1987), 153–162.

[40] P. P. Pálfy. *On Feit's examples of intervals in subgroup lattices.* J. Algebra **116** (1988), 471–479.

[41] P. P. Pálfy. *Intervals in subgroup lattices of finite groups.* Groups'93 Galway/St Andrews, vol. 2, London Math. Soc. Lecture Note Ser., vol. 212, Cambridge University Press, Cambridge, 1995, 482–494.

[42] P. P. Pálfy and P. Pudlák. *Congruence lattices of finite algebras and intervals in subgroup lattices of finite groups.* Algebra Universalis **11** (1980), 22–27.

[43] P. P. Pálfy and Cs. Szabó. *An identity for subgroup lattices of abelian groups.* Algebra Universalis **33** (1995), 191–195.

[44] P. P. Pálfy and Cs. Szabó. *Congruence varieties of groups and abelian groups.* Lattice Theory and Its Applications (Darmstadt, 1991), Res. Exp. Math., 23, Heldermann Verlag, Lemgo, 1995, 163–183.

[45] G. Pazderski. *On groups for which the lattice of normal subgroups is distributive.* Beiträge Algebra Geom. **24** (1987), 185–200.

[46] P. Pudlák and J. Tůma. *Every finite lattice can be embedded in a finite partition lattice.* Algebra Universalis **10** (1980), 74–95.

[47] A. Rottlaender. *Nachweis der Existenz nicht-isomorpher Gruppen von gleicher Situ-*

ation der Untergruppen. Math. Z. **28** (1928), 641–653.

[48] M. Schenke. *Analoga des Fundamentalsatzes der projektiven Geometrie in der Gruppentheorie, I–II.* Rend. Sem. Mat. Univ. Padova **77** (1987), 255–303; **78** (1987), 175–225.

[49] E. T. Schmidt. *Kongruenzrelationen algebraischer Strukturen.* Mathematische Forschungsberichte, vol. XXV, VEB Deutscher Verlag der Wissenschaften, Berlin, 1969.

[50] R. Schmidt. *Eine verbandstheoretische Charakterisierung der auflösbaren und der überauflösbaren endlichen Gruppen.* Arch. Math. (Basel) **19** (1968), 449–452.

[51] R. Schmidt. *Der Untergruppenverband des direkten Produktes zweier isomorpher Gruppen.* J. Algebra **73** (1981), 264–272.

[52] R. Schmidt. *Untergruppenverbände endlicher Gruppen mit elementarabelschen Hallschen Normalteilern.* J. Reine Angew. Math. **334** (1982), 116–140.

[53] R. Schmidt. *Gruppen mit modularem Untergruppenverband.* Arch. Math. (Basel) **46** (1986), 118–124.

[54] R. Schmidt. *Subgroup Lattices of Groups.* de Gruyter Expositions in Mathematics, 14, Walter de Gruyter and Co., Berlin, 1994.

[55] M. Schönert et al. *Groups, Algorithms and Programming.* Lehrstuhl D für Mathematik, RWTH Aachen, 1992.

[56] M. Schützenberger. *Sur certains axiomes de la théorie des structures.* C. R. Acad. Sci. Paris **221** (1945), 218–220.

[57] H. L. Silcock. *Generalized wreath products and the lattice of normal subgroups of a group.* Algebra Universalis **7** (1977), 361–372.

[58] A. P. Street. *Subgroup-determining functions on groups.* Illinois J. Math. **12** (1968), 99–120.

[59] M. Suzuki. *On the lattice of subgroups of finite groups.* Trans. Amer. Math. Soc. **70** (1951), 345–371.

[60] M. Suzuki. *Structure of a Group and the Structure of its Lattice of Subgroups.* Ergebnisse der Mathematik und ihrer Grenzgebiete, Neue Folge, Heft 10, Springer-Verlag, Berlin–Göttingen–Heidelberg, 1956.

[61] Cs. Szabó. *Congruence varieties of abelian groups and groups* (in Hungarian). Thesis, Hungarian Academy of Sciences, Budapest, 1992.

[62] J. Tůma. *Intervals in subgroup lattices of infinite groups.* J. Algebra **125** (1989), 367–399.

[63] P. M. Whitman. *Lattices, equivalence relations, and subgroups.* Bull. Amer. Math. Soc. **52** (1946), 507–522.

[64] B. V. Yakovlev. *Conditions under which a lattice is isomorphic to the lattice of subgroups of a group.* Algebra i Logika **13** (1974), 694–712 (in Russian); Translation in Algebra and Logic **13** (1975), 400–412.

[65] G. Zacher. *Determinazione dei gruppi d'ordine finito relativemente complementati.* Rend. Accad. Sci. Fis. Mat. Napoli (4) **19** (1952), 200–206.

[66] G. Zacher. *Sulle immagini dei sottogruppi normali nelle proiettività.* Rend. Sem. Mat. Univ. Padova **67** (1982), 39–74.

[67] G. Zappa. *Sulla risolubilità dei gruppi finiti in isomorfismo reticolare con un gruppo risolubile.* Giorn. Mat. Battaglini (4) **4(80)** (1951), 213–225.

FINITE GENERALIZED TETRAHEDRON GROUPS WITH A CUBIC RELATOR

GERHARD ROSENBERGER*, MARTIN SCHEER*
and RICHARD M. THOMAS‡

* Fachbereich Mathematik, University of Dortmund, 44227 Dortmund, Germany
‡ Department of Mathematics and Computer Science, University of Leicester,
Leicester LE1 7RH, England

Abstract

An *ordinary tetrahedron group* is a group with a presentation of the form

$$\langle x, y, z \mid x^{e_1} = y^{e_2} = z^{e_3} = (xy^{-1})^{f_1} = (yz^{-1})^{f_2} = (zx^{-1})^{f_3} = 1\rangle$$

where $e_i \geq 2$ and $f_i \geq 2$ for all i. Following Vinberg, we call groups defined by a presentation of the form

$$\langle x, y, z \mid x^{e_1} = y^{e_2} = z^{e_3} = R_1(x, y)^{f_1} = R_2(y, z)^{f_2} = R_3(z, x)^{f_3} = 1\rangle,$$

where each $R_i(a, b)$ is a cyclically reduced word involving both a and b, *generalized tetrahedron groups.* These groups appear in many contexts, not least as subgroups of generalized triangle groups. In this paper, we build on previous work of Coxeter, Edjvet, Fine, Howie, Levin, Metaftsis, Roehl, Rosenberger, Stille, Thomas, Tsaranov and Vinberg (amongst others) to give a classification of the finite generalized tetrahedron groups with a cubic relator.

1 Introduction and preliminary results

Consider a tetrahedron in either Euclidean, spherical or hyperbolic 3-space. The reflections at the faces generate a group of isometries and we consider the subgroup consisting of the orientation-preserving isometries. This subgroup is called an ordinary tetrahedron group and is the obvious analogy in 3-dimensional geometry of an ordinary triangle group.

Coxeter showed (see [2], [3], [4]) that such a group has a presentation of the form

$$\langle x, y, z \mid x^{e_1} = y^{e_2} = z^{e_3} = (xy^{-1})^{f_1} = (yz^{-1})^{f_2} = (zx^{-1})^{f_3} = 1\rangle.$$

So we have:

Definition 1.1 An *ordinary tetrahedron group* is a group G with a presentation of the form

$$\langle x, y, z \mid x^{e_1} = y^{e_2} = z^{e_3} = (xy^{-1})^{f_1} = (yz^{-1})^{f_2} = (zx^{-1})^{f_3} = 1\rangle$$

where $e_i \geq 2$ and $f_i \geq 2$ for all i. □

Coxeter showed that an ordinary tetrahedron group is finite if and only if the matrix

$$\begin{pmatrix} 1 & -\cos(\frac{\pi}{e_1}) & -\cos(\frac{\pi}{e_2}) & -\cos(\frac{\pi}{e_3}) \\ -\cos(\frac{\pi}{e_1}) & 1 & -\cos(\frac{\pi}{f_1}) & -\cos(\frac{\pi}{f_3}) \\ -\cos(\frac{\pi}{e_2}) & -\cos(\frac{\pi}{f_1}) & 1 & -\cos(\frac{\pi}{f_2}) \\ -\cos(\frac{\pi}{e_3}) & -\cos(\frac{\pi}{f_3}) & -\cos(\frac{\pi}{f_2}) & 1 \end{pmatrix}$$

has positive determinant (see [2], [3], [4]). We refer to this as the *Coxeter criterion.*

Definition 1.2 Following Vinberg, a *generalized tetrahedron group* is a group G with a presentation of the form

$$\langle x, y, z \mid x^{e_1} = y^{e_2} = z^{e_3} = R_1(x,y)^{f_1} = R_2(y,z)^{f_2} = R_3(z,x)^{f_3} = 1 \rangle,$$

with $e_1, e_2, e_3, f_1, f_2, f_3 \geq 2$ and $R_1(x,y)$ a cyclically reduced word in the free product on x, y, which involves both x and y, $R_2(y,z)$ a cyclically reduced word in the free product on y, z, which involves both y and z, and $R_3(z,x)$ a cyclically reduced word in the free product on z, x, which involves both z and x. Further each R_i, $i = 1, 2, 3$, is not a proper power in the free product on the generators it involves. □

We call two such presentations of generalized tetrahedron groups *equivalent* if one can be obtained from the other by a sequence of operations of the following type:

1. replace a generator u of order e_i by a new generator $d = u^k$, where k is coprime to e_i, and then amend the relations accordingly;
2. apply a permutation to the generators x, y and z;
3. if $S(u,v)$ is a cyclically reduced conjugate of $R(u,v)$ in the free group on u and v, then replace the relator $R(u,v)^{f_i}$ by $S(u,v)^{f_i}$;
4. replace the relator $R(u,v)^{f_i}$ by $S(u,v)^{f_i}$, where $S(u,v)$ is the inverse of $R(u,v)$ in the free group on u and v;
5. if u is a generator of order 2, if v is a generator of order e, if k and l are coprime to e, and if we have a relator of the form $R = (uv^k)^2$, then replace R by $(uv^l)^2$.

It is clear that, if two presentations are equivalent, then they define isomorphic groups.

The most general result for large exponents f_i was given by Edjvet, Howie, Rosenberger and Thomas in [5]:

Theorem 1.3 *Let G be a finite generalized tetrahedron group with a presentation of the form*

$$\wp = \langle x, y, z \mid x^{e_1} = y^{e_2} = z^{e_3} = R_1(x,y)^{f_1} = R_2(y,z)^{f_2} = R_3(z,x)^{f_3} = 1 \rangle,$$

where at least one of the f_i is greater than 3. Then $\wp$ is equivalent to a presentation of an ordinary tetrahedron group.

In this paper we will show that, if $2 \leq f_1, f_2, f_3 \leq 3$, at least one of the f_i equals 3 and G is finite, then the presentation for G is either equivalent to that for an ordinary tetrahedron group or to one of the six presentations (1)-(6) listed below.

For the benefit of the reader we will state some more results (see, for instance, [5]). We will use the following concepts:

a) a representation $\rho : G \to L$, where L is a linear group, is called *essential* if $\rho(x)$ has order e_1, $\rho(y)$ has order e_2, $\rho(z)$ has order e_3, and $\rho(R_j)$ has order f_j for $j = 1, 2, 3$;

b) a subgroup of $\mathrm{PSL}(2, \mathbb{C})$ is called *non-elementary* if it contains a free subgroup of rank 2.

Theorem 1.4 *Every generalized tetrahedron group admits an essential representation in $PSL(2, \mathbb{C})$.*

Theorem 1.5 *Let G be the generalized tetrahedron group defined by the presentation*

$$\langle x, y, z \mid x^{e_1} = y^{e_2} = z^{e_3} = R_1(x, y)^{f_1} = R_2(y, z)^{f_2} = R_3(z, x)^{f_3} = 1 \rangle,$$

and let G_1 be the generalized triangle group defined by the presentation

$$\langle x, y \mid x^{e_1} = y^{e_2} = R_1(x, y)^{f_1} = 1 \rangle.$$

Suppose that ρ is an essential representation of G_1 in $PSL(2, \mathbb{C})$, that $X = \rho(x)$ and $Y = \rho(y)$, and that one of the following two possibilities occurs:

1. *$tr[X, Y] \neq 2$;*
2. *$(e_3, f_2, f_3) \neq (2, 2, 2)$ and $\langle X, Y \rangle$ is an infinite metabelian subgroup of $PSL(2, \mathbb{C})$.*

Then there is an essential representation $\tilde{\rho} : G \to PSL(2, \mathbb{C})$ such that $\tilde{\rho}(x) = X$ and $\tilde{\rho}(y) = Y$.

In particular, if $\langle X, Y \rangle$ is infinite, then G is infinite.

Corollary 1.6 *Every finite generalized tetrahedron group admits an essential representation in $PSL(2, \mathbb{C})$ with image $\mathcal{D}_{2n}$ $(n \geq 2)$, $\mathcal{A}_4$, $\mathcal{S}_4$ or $\mathcal{A}_5$.*

Here $\mathcal{D}_{2n}$ is the dihedral group of order $2n$, $\mathcal{A}_n$ the alternating group of degree n, and $\mathcal{S}_n$ the symmetric group of degree n.

Theorem 1.7 *Let G be the generalized tetrahedron group defined by the presentation*

$$\langle x, y, z \mid x^{e_1} = y^{e_2} = z^{e_3} = R_1(x, y)^{f_1} = R_2(y, z)^{f_2} = R_3(z, x)^{f_3} = 1 \rangle,$$

and suppose that $e_1 \leq e_2$ and that

$$R_1(x, y) = x^{\alpha_1} y^{\beta_1} \ldots x^{\alpha_{k_1}} y^{\beta_{k_1}},$$

where $k_1 > 1$ and $R_1(x, y)$ is not a proper power in the free product of the cyclic groups generated by x and y. Assume further that one of the following holds:

1. $e_2 \geq 4$ *and* $f_1 \geq 3$*;*

2. $f_1 \geq 4$*;*

3. $e_1 \geq 3$ *and* $f_1 \geq 3$*.*

Then G *is infinite.*

Theorem 1.8 *A generalized tetrahedron group* $G(e_1, e_2, e_3, f_1, f_2, f_3)$ *is infinite if* $\frac{1}{f_1} + \frac{1}{f_2} + \frac{1}{f_3} \leq 1$.

Theorem 1.9 *A generalized tetrahedron group* $G(e_1, e_2, e_3, f_1, f_2, f_3)$ *is infinite if* $\frac{1}{e_1} + \frac{1}{e_2} + \frac{1}{f_1} < 1$.

Theorem 1.10 *A generalized tetrahedron group* $G(e_1, e_2, e_3, 2, 2, 2)$ *is infinite if* $\frac{1}{e_1} + \frac{1}{e_2} \leq \frac{1}{2}$.

Theorem 1.11 (Spelling Theorem)
Let $H = \langle x, y \mid x^p = y^q = W^r = 1\rangle$ *be a generalized triangle group with*

$$W = x^{\alpha_1} y^{\beta_1} \cdots x^{\alpha_k} y^{\beta_k} \ (0 < \alpha_i < p,\, 0 < \beta_i < q),$$

and let

$$V(x, y) = x^{\gamma_1} y^{\delta_1} \cdots x^{\gamma_\ell} y^{\delta_\ell}, \ (0 < \gamma_i < p,\, 0 < \delta_i < q)$$

be a word that is equal to 1 in H*. Then* $\ell \geq k \cdot (r - 1) + 1$.

In the following we use the methods of Gersten and Stallings (see [16]).

If we have two subgroups A, B of a group H, then the inclusions $A \hookrightarrow H$ and $B \hookrightarrow H$ determine a homomorphism

$$\phi : \quad A * B \longrightarrow H.$$

The *angle* $(H; A, B)$ between A and B is defined as follows:

- if ϕ injective, let $(H; A, B) = 0$;
- otherwise let $(H; A, B) = \pi/n$, where $2n$ is the minimal length of an element of $Ker(\phi)$.

Remark 1.12 A generalized tetrahedron group

$$\langle x, y, z \mid x^{e_1} = y^{e_2} = z^{e_3} = R_1(x, y)^{f_1} = R_2(y, z)^{f_2} = R_3(z, x)^{f_3} = 1\rangle,$$

can be realized as a *triangle of groups*, that is as the colimit of the diagram of groups and injective homomorphisms (see figure)

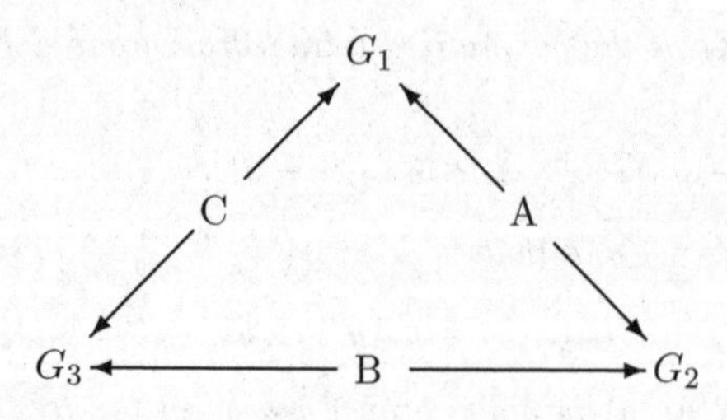

in which

$$G_1 = \langle x_1, y_1 \mid x_1^{e_1} = y_1^{e_2} = R_1(x_1, y_1)^{f_1} = 1\rangle,$$

$$G_2 = \langle y_2, z_1 \mid y_2^{e_2} = z_1^{e_3} = R_2(y_2, z_1)^{f_2} = 1\rangle,$$
$$G_3 = \langle z_2, x_2 \mid z_2^{e_3} = x_2^{e_1} = R_3(z_2, x_2)^{f_3} = 1\rangle,$$
$$A = \langle y_1\rangle \cong \langle y_2\rangle, \qquad B = \langle z_1\rangle \cong \langle z_2\rangle, \qquad C = \langle x_1\rangle \cong \langle x_2\rangle.$$

We refer to the groups G_1, G_2 and G_3 as *vertex groups.*

Definition 1.13 Such a triangle of groups is said to be *spherical* if

$$(G_1; C, A) + (G_2; A, B) + (G_3; B, C) > \pi,$$

and *non-spherical* otherwise. □

Proposition 1.14 *If H is the colimit of a non-spherical triangle of groups then H is infinite.*

Remark 1.15 For any representation $\rho : \langle a, b \mid a^{2p}, b^{2q}\rangle \to \mathrm{SL}(2,\mathbb{C})$ we have that the trace of $\rho(W)$ is given by a polynomial in the traces of $\rho(a)$, $\rho(b)$ and $\rho(ab)$:

$$f_W\left(tr(\rho(a)), tr(\rho(b)), tr(\rho(ab))\right).$$

The degree of the polynomial f in the third variable is at most k, that is, half the (free product) length of $W = a^{\alpha_1}b^{\beta_1}\cdots a^{\alpha_k}b^{\beta_k}$ (see Lemma 1 in [8]). If the first two variables of f are fixed as $2\cos(\pi/p)$ and $2\cos(\pi/q)$ respectively, then an easy argument shows that f becomes a polynomial of degree precisely k in the single variable $\mathrm{tr}(\rho(ab))$. We refer to this as the *trace polynomial* of W.

Theorem 1.16 (A slight generalization of Theorem 2.26 of [13]) *Let G be a generalized tetrahedron group with a presentation of the form*

$$G = \langle x, y, z \mid x^{e_1} = y^{e_2} = z^{e_3} = R_1^2(x, y) = R_2^{f_2}(y, z) = R_3^{f_3}(z, x) = 1\rangle,$$

where $(e_1, e_2) \neq (2, 2)$, and let

$$G_1 = \langle x, y \mid x^{e_1} = y^{e_2} = R_1^2(x, y) = 1\rangle.$$

Assume that the trace polynomial of $R_1(x, y)$ has a multiple root β. Then G is infinite.

Proof If there is an essential representation $\rho : G \to \mathrm{PSL}(2,\mathbb{C})$ with $\rho(G)$ cyclic or infinite metabelian then G is infinite.

Assume now that there is no such representation. If, for some $1 \leq i \leq 3$, there is an essential representation $\rho_i : G_i \to \mathrm{PSL}(2,\mathbb{C})$ with $\rho_i(G_i)$ non-elementary then G is infinite (see Theorem 1.5). Assume now that, for $1 \leq i \leq 3$, there is no essential representation $\rho_i : G_i \to \mathrm{PSL}(2,\mathbb{C})$ with $\rho_i(G_i)$ non-elementary.

Without loss of generality we may assume that $e_1 \leq e_2$. G is infinite if

$$\frac{1}{e_1} + \frac{1}{e_2} < \frac{1}{2} \text{ and } (f_2, f_3) \neq (2, 2) \text{ or if } \frac{1}{e_1} + \frac{1}{e_2} \leq \frac{1}{2} \text{ and } (f_2, f_3) = (2, 2).$$

If $(e_3, f_2, f_3) \neq (2,2,2)$ then G is infinite if there exists an essential representation $\rho_1 : G_1 \to \mathrm{PSL}(2,\mathbb{C})$ with $\rho_1(G_1)$ cyclic or infinite metabelian (see Theorem 1.5).

Assume first that

$$\frac{1}{e_1} + \frac{1}{e_2} = \frac{1}{2},\ (f_2, f_3) \neq (2,2)$$

and there is no essential representation $\rho_1 : G_1 \to \mathrm{PSL}(2,\mathbb{C})$ with $\rho_1(G_1)$ cyclic or infinite metabelian. If $\rho_1 : G_1 \to \mathrm{PSL}(2,\mathbb{C})$ is an essential representation then certainly $\rho_1(G_1)$ is not a dihedral group. Now, if

$$\frac{1}{e_1} + \frac{1}{e_2} = \frac{1}{2} \text{ and } e_1 \leq e_2,$$

then $(e_1, e_2) = (4,4)$ or $(3,6)$. G is infinite if $(e_1, e_2) = (3,6)$ by Corollary 1.6 because here, if $\rho : G \to \mathrm{PSL}(2,\mathbb{C})$ is an essential representation, then $\rho(G)$ is not dihedral. For $(e_1, e_2) = (4,4)$ it is possible that $\rho_1(G_1) = S_4$ for an essential representation $\rho_1 : G_1 \to \mathrm{PSL}(2,\mathbb{C})$.

Now assume that $(e_1, e_2) = (4,4)$. Without loss of generality we may assume that $f_2 \geq 3$ since $(f_2, f_3) \neq (2,2)$. If $e_3 \geq 3$ then G is infinite because

$$\frac{1}{e_1} + \frac{1}{e_3} + \frac{1}{f_2} < 1$$

if $e_3 \geq 3$. Hence let $e_3 = 2$. Now we use Theorem 1.3, Theorem 1.11, Lemma 3.1 of [13] and the assumption that, for no $1 \leq i \leq 3$, there is an essential representation $\rho_i : G_i \to \mathrm{PSL}(2,\mathbb{C})$ with $\rho_i(G_i)$ non-elementary. Therefore G is, as a triangle of groups, non-spherical and hence infinite by Proposition 1.14, unless, up to equivalence,

$$G = \langle x, y, z \mid x^4 = y^4 = z^2 = R_1^2(x,y) = (x^\delta z)^3 = (y^\gamma z)^2 = 1\rangle$$

with $1 \leq \delta, \gamma < 4$. Recall that the trace polynomial of $R_1(x,y)$ has (at least) the multiple root β.

If $\delta = 2$ then G_2 has an essential non-elementary image in $\mathrm{PSL}(2,\mathbb{C})$ and therefore G is infinite. If $\gamma = 2$ then we introduce the relations $x^2 = y^2 = 1$ to get an infinite factor group. Hence, without loss of generality, we may assume that $\delta = \gamma = 1$.

Here G is infinite unless the only essential representation $\rho_1 : G_1 \to \mathrm{PSL}(2,\mathbb{C})$ with $\mathrm{tr}\rho_1(x) = \mathrm{tr}\rho_1(y) = \sqrt{2}$ is given by $\mathrm{tr}\rho_1(xy) = 1$; that is, $\langle \rho_1(x), \rho_1(y)\rangle = S_4$, and, if

$$R_1(x,y) = x^{\alpha_1} y^{\beta_1} \cdots x^{\alpha_{k_1}} y^{\beta_{k_1}},\ 1 \leq \alpha_i, \beta_i < 4,\ k_1 \geq 2$$

(up to equivalence), then the trace polynomial of $R_1(x,y)$ is just $\tau(t) = a \cdot (t-1)^{k_1}$ with $a \neq 0$.

Recall that $k_1 \geq 2$ because the trace polynomial has a multiple root. G is infinite if $k_1 \geq 5$ by Theorem 1.11 and Proposition 1.14. Let $\bar{R}_1(x,y)$ be the cyclically reduced word which we get from $R_1(x,y)$ if we introduce in G_1 the relations $x^2 = y^2 = 1$ and make free reductions. If $\bar{R}_1(x,y)$ is equal to 1 or x or y then we get an infinite factor group of G. If this does not occur and if

$2 \leq k_1 \leq 4$ then, by [11] and [12], the trace polynomial of $R_1(x, y)$ is not of the form $\tau(t) = a \cdot (t-1)^{k_1}$. Hence, altogether, G is infinite if $1/e_1 + 1/e_2 \leq 1/2$.

Hence we may assume that $1/e_1 + 1/e_2 > 1/2$.

Since $(e_1, e_2) \neq (2, 2)$ we have

$$e_1 = 2,\ e_2 \geq 3 \quad \text{or}\ e_1 = 3,\ 3 \leq e_2 \leq 5.$$

Certainly G is infinite unless $\beta = \alpha + \alpha^{-1}$ for some $2m^{\text{th}}$ root of unity $\alpha \neq \pm 1$. For the following let β always be of this form. G is infinite if

$$e_1 = 3,\ 3 \leq e_2 \leq 5 \ \text{ and } \ m \geq 6$$

by Corollary 2.4 of [5] because, if $\rho_1 : G_1 \to \mathrm{PSL}(2, \mathbb{C})$ is an essential representation, then $\rho_1(G_1)$ cannot be dihedral: if G were finite then G would have a factor group

$$\bar{G} = \langle x, y, z \mid x^3 = y^{e_2} = z^2 = R_1^2(x, y) = (xz)^2 = (yz)^2 = 1\rangle$$

which is infinite, a contradiction.

If $e_1 = 3$, $3 \leq e_2 \leq 5$ and $2 \leq m \leq 5$ then there is no essential representation $\rho_1 : G_1 \to \mathrm{PSL}(2, \mathbb{C})$ with $\mathrm{tr}\rho_1(xy) = \beta$ and $\rho_1(G_1)$ cyclic or infinite metabelian because $f_1 = 2$.

Assume now that there is an essential representation $\rho_1 : G_1 \to \mathrm{PSL}(2, \mathbb{C})$ with $\mathrm{tr}\rho_1(xy) = \beta$ and $\rho_1(G_1)$ cyclic or infinite metabelian, and, without loss of generality, let $\rho_1(G)$ be infinite metabelian. As pointed out above we must then have that $e_1 = 2$. G is infinite if $(e_3, f_2, f_3) \neq (2, 2, 2)$ by Theorem 1.5.

Let $(e_3, f_2, f_3) = (2, 2, 2)$. Assume that G has a factor group

$$\bar{G} = \langle x, y, z \mid x^2 = y^{e_2} = z^2 = R_1^2(x, y) = (xz)^2 = (y^\gamma z)^2 = 1\rangle$$

with $1 \leq \gamma \leq e_2$. Let $d = \gcd(\gamma, e_2)$. If $d \geq 2$ then we introduce the relation $y^d = 1$ and get an infinite factor group of $\bar{G}$. If $d = 1$ then we may assume that $\gamma = 1$.

We consider the subgroup $H = \langle x, y\rangle$ of $\bar{G}$. H has presentation

$$H = \langle x, y \mid x^2 = y^{e_2} = R_1^2(x, y) = R_1^2(x^{-1}, y^{-1}) = 1\rangle$$

and G is infinite since ρ_1 factorizes via H.

Assume now that G does not have such a factor group $\bar{G}$. Here we use the construction from [8] and [9].

For the construction of the extension $\rho : G \to \mathrm{PSL}(2, \mathbb{C})$ of

$$\rho_1 : G_1 \to \mathrm{PSL}(2, \mathbb{C})$$

we just may replace $\mathrm{tr}\rho(yz) \neq 0$ by $-\mathrm{tr}\rho(yz)$, if necessary, to get an infinite image of G in $\mathrm{PSL}(2, \mathbb{C})$. So, from now on, we may assume that there is no essential representation $\rho_1 : G_1 \to \mathrm{PSL}(2, \mathbb{C})$ with $\mathrm{tr}\rho_1(xy) = \beta$ and $\rho_1(G_1)$ cyclic or infinite metabelian. Then G is infinite if $m \geq 6$ by Theorem 2.8 and Corollary 2.4 of [5].

Now let $2 \leq m \leq 5$. Then we may argue as in the proof of Theorem 2.26 of [13] (see also [10]), and we write $p = e_1$ and $q = e_2$. Let

$$\lambda = 2\cos\frac{k\pi}{p}, \quad \mu = 2\cos\frac{l\pi}{q},$$

where k and l are integers coprime to p and q respectively. We may choose k and l so that both the inequalities

$$\left[\cos\left(\frac{k\pi}{p}\right) + \cos\left(\frac{l\pi}{q}\right)\right]^2 \neq \beta + 2 \quad \text{and} \quad \cos\left(\frac{k\pi}{p}\right) \neq \cos\left(\frac{l\pi}{q}\right).$$

hold.

Let $\gamma \in \mathbb{C}$ with

$$(-\gamma + \lambda)(\gamma + \mu) - 2 = \beta.$$

Define matrices $X, Y \in \mathrm{SL}(2, \mathbb{C}[t])$ by

$$X = \begin{pmatrix} 0 & 1 \\ -1 & \lambda \end{pmatrix} \quad \text{and} \quad Y = \begin{pmatrix} 1 & t-\gamma \\ 0 & 1 \end{pmatrix} \begin{pmatrix} 0 & 1 \\ -1 & \mu \end{pmatrix} \begin{pmatrix} 1 & \gamma - t \\ 0 & 1 \end{pmatrix}.$$

Then

$$f(t) = \ \mathrm{tr}XY = (-t + \gamma + \lambda)(t - \gamma + \mu) - 2 \ \text{ and } \ f(2\gamma) = \beta.$$

Hence we get a representation

$$\rho_1 : G_1 \to \mathrm{PSL}(2, \Lambda), \quad \Lambda = \frac{\mathbb{C}[t]}{((t-\gamma)^2)}$$

with $\rho_1(G_1)$ infinite (see [10]). By Theorem 2.25 of [13] there exists an extension $\rho : G \to \mathrm{PSL}(2, \Lambda)$, and in particular G is infinite. □

In the following we classify the finite generalized tetrahedron groups with a cubic relator. At a few places we show the infiniteness via the abelian invariants of a subgroup which we found by using the computer algebra system GAP [15].

2 The long relator case

Let

$$G = \langle x, y, z \mid x^{e_1} = y^{e_2} = z^{e_3} = R_1^{f_1}(x,y) = R_2^{f_2}(x,z) = R_3^{f_3}(y,z) = 1 \rangle,$$

with

$$2 \leq e_1, e_2, e_3, f_1, f_2, f_3; \quad 2 \leq f_1, f_2, f_3 \leq 3;$$
$$R_1(x,y) = x^{\alpha_1} y^{\beta_1} \cdots x^{\alpha_{k_1}} y^{\beta_{k_1}}, \ k_1 \geq 1, \ R_1 \text{ not a proper power},$$
$$R_2(x,z) = x^{\gamma_1} z^{\delta_1} \cdots x^{\gamma_{k_2}} z^{\delta_{k_2}}, \ k_2 \geq 1, \ R_2 \text{ not a proper power},$$
$$R_3(y,z) = y^{\varepsilon_1} z^{\eta_1} \cdots y^{\varepsilon_{k_3}} z^{\eta_{k_3}}, \ k_3 \geq 1, \ R_3 \text{ not a proper power},$$
$$1 \leq \alpha_i, \gamma_i < e_1, \ 1 \leq \beta_i, \varepsilon_i < e_2, \ 1 \leq \delta_i, \eta_i < e_3.$$

Let at least one f_i be equal to 3. Let

$$\begin{aligned} G_1 &= \langle x, y \mid x^{e_1} = y^{e_2} = R_1^{f_1}(x,y) = 1 \rangle, \\ G_2 &= \langle x, z \mid x^{e_1} = z^{e_3} = R_2^{f_2}(x,z) = 1 \rangle, \\ G_3 &= \langle y, z \mid y^{e_2} = z^{e_3} = R_3^{f_3}(y,z) = 1 \rangle \end{aligned}$$

be the corresponding vertex groups.

Proposition 2.1 *Suppose that $k_i \geq 2$ for at least one value of i. If at least two of the f_i are equal to 3, then G is infinite.*

Proof If $f_1 = f_2 = f_3 = 3$, then G is infinite by Theorem 1.8. So assume now that exactly two f_i are equal to 3.

Without loss of generality we may assume that $k_1 \geq 2$ and that $e_1 \leq e_2$. If $f_2 = f_3 = 3$, then G is infinite by Proposition 3.1 of [6] and Lemma 3.1 of [13] because G_1 has a non-elementary image in $\mathrm{PSL}(2, \mathbb{C})$ or the angle at G_1 is at most $\pi/4$.

Assume now that $f_1 = 3$. By Theorem 1.5, G is infinite if $e_1 \geq 3$ or $e_2 \geq 4$ since $k_1 \geq 2$. So let $e_1 = 2$ and $e_3 = 3$. If $k_1 \geq 3$ then

$$\frac{1}{2k_1+1} + \frac{1}{k_2(f_2-1)+1} + \frac{1}{k_3(f_3-1)+1} \leq \frac{1}{7} + \frac{1}{3} + \frac{1}{2} < 1$$

because $f_2 = 3$ or $f_3 = 3$, and G is infinite by Proposition 3.1 of [6]. So we may assume that $k_1 = 2$.

Since $R_1(x, y)$ is not a proper power we have, up to equivalence, that

$$R_1(x, y) = xyxy^2.$$

We choose an essential representation $\rho_1 : G_1 \to \mathrm{PSL}(2, \mathbb{C})$ with

$$\mathrm{tr}\rho_1(x) = 0, \ \ \mathrm{tr}\rho_1(y) = 1, \ \ \text{and} \ \ \mathrm{tr}\rho_1(xyxy^2) = -1.$$

If $s = \mathrm{tr}\rho_1(xy)$ it follows that $s^2 = 0$. So 0 is a double root of $\tau(s) = \mathrm{tr}\rho_1(xyxy^2)+1$.

We now use, for this special case, a slight extension of Theorem 1.16. We choose $\gamma \in \mathbb{C}$ such that

$$\gamma^2 + \gamma + 2 = 0.$$

Let I be the ideal in $\mathbb{C}[t]$ generated by $(t-\gamma)^2$ and then let $\Lambda = \mathbb{C}[t]/I$. As in [10] we get an essential representation $\rho_1 : G_1 \to \mathrm{PSL}(2, \mathbb{C})$ with $\rho_1(G_1)$ infinite. By Theorem 2.25 of [13] there exists an extension $\rho : G \to \mathrm{PSL}(2, \Lambda)$, and, in particular, G is infinite. □

Proposition 2.2 *Let $k_i \geq 2$ for at least two values of i and precisely one f_i be equal to 3. Then G is infinite.*

Proof Assume, without loss of generality, that $k_1 \geq 2$ and $k_2 \geq 2$.

If $f_3 = 3$, then G is infinite by Proposition 3.1 of [6] and Lemma 3.1 of [13] because $f_1 = f_2 = 2$, and the angle at G_1 is at most $\pi/4$; therefore we either have $1/4 + 1/3 + 1/3 < 1$ or else that G_1 has a non-elementary image in $\mathrm{PSL}(2, \mathbb{C})$.

Assume now that $f_1 = 3$ or $f_2 = 3$; without loss of generality, let $f_1 = 3$. Then $f_2 = f_3 = 2$. If G_2 has a non-elementary image in $\mathrm{PSL}(2, \mathbb{C})$ then G is infinite. Otherwise the angle at G_2 is at most $\pi/4$, and G is infinite because

$$\frac{1}{k_1(f_1-1)+1} + \frac{1}{4} + \frac{1}{2} \leq \frac{1}{5} + \frac{1}{4} + \frac{1}{2} < 1.$$

This completes the proof. □

Proposition 2.3 *Let $k_1 \geq 2$ and $f_1 = 3$, and suppose that G is finite. Then G is one of the following (up to equivalence):*

(1) $\langle x, y, z \mid x^2 = y^3 = z^2 = (xyxyxy^2)^3 = (xz)^2 = (yz)^2 = 1 \rangle$ *of order* 2880,
(2) $\langle x, y, z \mid x^2 = y^3 = z^3 = (xyxyxy^2)^3 = (xz)^2 = (yz)^2 = 1 \rangle$ *of order* 864000.

Proof Assume, without loss of generality, that $e_1 \leq e_2$.

If $e_1 \geq 3$ or $e_2 \geq 4$, then G is infinite by Theorem 2.5 of [5]. So assume now that $e_1 = 2$ and $e_2 = 3$. G is infinite if $k_2 \geq 2$ or $k_3 \geq 2$ (see Propositions 2.1 and 2.2). So let $k_2 = k_3 = 1$.

Since there exists no essential representation $\rho : G \to \mathrm{PSL}(2, \mathbb{C})$ with cyclic image $\rho(G)$, we know, from [9] Theorem 7.3.2.4, that we only have to consider, up to equivalence, the following possibilities for $R_1(x, y)$:

1. $R_1(x, y) = xyxy^2$ and
2. $R_1(x, y) = xyxyxy^2$.

If $R_1(x, y) = xyxy^2$ then, for any essential representation

$$\rho_1 : G_1 \to \mathrm{PSL}(2, \mathbb{C}),\ x \mapsto X,\ y \mapsto Y \text{ with } \mathrm{tr}X = 0,\ \mathrm{tr}Y = 1,\ \mathrm{tr}XY = 0,$$

the polynomial $f(t) = \mathrm{tr}(XYXY^2) + 1 = t^2$, where $t = \mathrm{tr}XY$, has 0 as a double root, i.e. G is infinite by straightforward arguments as in the proof of Proposition 2.1.

So let $k_2 = k_3 = 1$ and $R_1(x, y) = xyxyxy^2$. Introducing the relation $(xy)^5 = 1$, we have the factor group

$$\bar{G} = \langle x, y, z \mid x^2 = y^3 = z^{e_3} = (xy)^5 = (xz^\delta)^2 = (yz^\eta)^2 = 1 \rangle$$

of G. $\bar{G}$ is infinite for $e_3 = 4$ and $e_3 \geq 6$, since there exists an essential representation $\rho : \bar{G} \to \mathrm{PSL}(2, \mathbb{C})$ with $\rho(\bar{G})$ infinite: $\rho(\bar{G})$ cannot be dihedral and, if $\langle x, y \rangle$ admits an essential representation ρ_1 with $\rho_1(\langle x, y \rangle)$ infinite metabelian, we can use Theorem 1.5. Moreover, a finite subgroup H of $\mathrm{PSL}(2, \mathbb{C})$ with H non-cyclic or dihedral cannot contain, at the same time, an element of order 4 and an element of order 5.

Now let $e_3 = 5$. Then we can assume that $\eta = 1$ and $\delta = 1$ in $\bar{G}$ since $\langle x, z \rangle$ is dihedral. Hence $\bar{G}$ is an infinite ordinary tetrahedron group by the Coxeter criterion and therefore G is infinite. For $2 \leq e_3 \leq 3$ we have, up to equivalence, the given finite groups. □

So we may assume, throughout the following, that $k_1 \geq 2$, $f_1 = 2$ and that $(f_2, f_3) = (2, 3)$ or $(f_2, f_3) = (3, 2)$. Further let $k_2 = k_3 = 1$. So

$$G = \langle x, y, z \mid x^{e_1} = y^{e_2} = z^{e_3} = R_1^2(x, y) = (x^\alpha z^\gamma)^{f_2} = (y^\beta z^\delta)^{f_3} = 1 \rangle,$$

with

$$e_1, e_2, e_3 \geq 2$$

$$R_1(x, y) = x^{\alpha_1} y^{\beta_1} \cdots x^{\alpha_{k_1}} y^{\beta_{k_1}},\ k_1 \geq 2,\ 1 \leq \alpha_i, \alpha < e_1,\ 1 \leq \beta_i, \beta < e_2,$$

$$1 \leq \gamma, \delta < e_3 \text{ and } (f_2, f_3) = (2,3) \text{ or } (f_2, f_3) = (3,2).$$

Assume, without loss of generality, that $e_1 \leq e_2$. If $k_1 \geq 5$, then G is infinite by Theorem 1.11 and Proposition 1.14 since

$$\frac{1}{k_1+1} + \frac{1}{3} + \frac{1}{2} \leq 1$$

for $k_1 \geq 5$. So let $2 \leq k_1 \leq 4$. Since $k_1 \geq 2$ it follows that $e_2 \geq 3$.

Proposition 2.4 *Let $e_1 = 2$, $e_2 = 3$, $f_1 = 2$, $k_1 \geq 2$ and $(f_2, f_3) = (2,3)$ or $(3,2)$. If G is finite, then G is equivalent to*

$$\langle x, y, z \mid x^2 = y^3 = z^2 = (xyxy^2)^2 = (xz)^3 = (yz)^2 = 1 \rangle \text{ of order } 360.$$

Proof Assume that G is finite. For $R_1(x,y)$ we have to consider, up to equivalence, the following four possibilities (see [1] and [11]):

1. $R_1(x,y) = xyxy^2$,
2. $R_1(x,y) = xyxyxy^2$,
3. $R_1(x,y) = xyxyxyxy^2$,
4. $R_1(x,y) = xyxyxy^2xy^2$.

In cases 3 and 4 the angle at G_1 is at most $\pi/5$ by Theorem 1.11 and Proposition 1.14. Actually it is at most $\pi/6$, since a relation of the form

$$xy^{s_1}xy^{s_2}\ldots xy^{s_5} = 1, \ 1 \leq s_i \leq 2$$

cannot hold in G_1 (if we set $y = 1$, then G_1 has the factor group $\langle x \mid x^2 = 1 \rangle$). Hence G is infinite in cases 3 and 4.

So, assume now that case 1 or 2 holds. In case 2, the angle at G_1 is again even at most $\pi/6$.

In either case, we have, that the angle is G_1 is at most $\pi/4$. As above there is no relation of the form

$$xy^{s_1}xy^{s_2}\ldots xy^{s_5} = 1, \ 1 \leq s_i \leq 2$$

in G_1. Assume now that, in G_1, a relation of the form

$$xy^{s_1}xy^{s_2}xy^{s_3}xy^{s_4} = 1, \ 1 \leq s_i \leq 2$$

holds. Without loss of generality we can assume that we have one of

$$\begin{array}{llll} \text{a)} & (xy)^4 = 1, & \text{b)} & xyxyxyxy^2 = 1, \\ \text{c)} & (xyxy^2)^2 = 1, & \text{d)} & xy^2xy^2xyxy = 1. \end{array}$$

We remark that G_1 is of order 48 (see [1]). Moreover, if $\rho_1 : G_1 \to \mathrm{PSL}(2,\mathbb{C})$ is an essential representation with

$$x \mapsto X, \ y \mapsto Y, \ \mathrm{tr}X = 0, \ \mathrm{tr}Y = 1,$$

then $\mathrm{tr}XY = 0$ or $\mathrm{tr}XY = \pm\sqrt{2}$. If $\mathrm{tr}XY = 0$, then $\langle X, Y\rangle$ is isomorphic to $\mathcal{S}_3$, and, if $\mathrm{tr}XY = \pm\sqrt{2}$, then $\langle X, Y\rangle$ is isomorphic to $\mathcal{S}_4$.

Case a) is not possible, since $24 = |\mathcal{S}_4| < 48 = |G_1|$.

For case b) we consider X, Y with $\mathrm{tr}XY = \sqrt{2}$; then $Y = 1$, which does not hold.

In case c) we consider X, Y with $\mathrm{tr}XY = 0$; then $Y^2 = Y^3 = 1$, so that $Y = 1$, which does not hold.

Case d) is not possible, since a group $\langle x, y\rangle$ with the relations

$$x^2 = y^3 = xy^2xy^2xyxy = 1$$

has order at most 18 (see [17]).

Hence, in total, the angle at G_1 for case 2 is at most $\pi/6$ and G is infinite.

There remains case 1) with $R_1(x, y) = xyxy^2$. We have, up to equivalence,

$$G = \langle x, y, z \mid x^2 = y^3 = z^{e_3} = (xyxy^2)^2 = (xz^\gamma)^{f_2} = (yz^\delta)^{f_3} = 1\rangle$$

with $1 \leq \gamma, \delta < e_3$ and $(f_2, f_3) = (3, 2)$ or $(f_2, f_3) = (2, 3)$.

I) Let $f_2 = 3$ and $e_3 \geq 3$. Then we consider the factor group

$$\bar{G} = \langle x, y, z \mid x^2 = y^3 = z^{e_3} = (xy)^3 = (xz^\gamma)^3 = (yz^\delta)^2 = 1\rangle.$$

If $e_3 = 3$, then $\bar{G}$ is an infinite ordinary tetrahedron group by the Coxeter criterion. If $e_3 = 4$ and $\gamma = 2$ or $\delta = 2$, then $\bar{G}$ is infinite, since G_2 and G_3 respectively have essential non-elementary images in $\mathrm{PSL}(2, \mathbb{C})$. If $e_3 = 4$ and $\gamma \neq 2 \neq \delta$, then $\bar{G}$ is an infinite ordinary tetrahedron group by the Coxeter criterion.

Assume now that $e_3 = 5$. We can assume that $\delta = 1$ and that $a \leq \gamma \leq 2$. For $\gamma = 1$, $\bar{G}$ is an infinite ordinary tetrahedron group by the Coxeter criterion; for $\gamma = 2$, we showed that $\bar{G}$ is infinite by GAP.

Assume now that $e_3 = 6$. If $\gamma = 2, 3$ or 4, then G_1 has an essential non-elementary image in $\mathrm{PSL}(2, \mathbb{C})$. So let (without loss of generality) $\gamma = 1$. If $\delta = 2, 3$ or 4, then G_2 has an essential non-elementary image in $\mathrm{PSL}(2, \mathbb{C})$. If $\delta = 1$ or 5, then G_1 is an infinite ordinary tetrahedron group by the Coxeter criterion. If $e_3 \geq 7$, then $\bar{G}$ is infinite, since

$$\frac{1}{2} + \frac{1}{3} + \frac{1}{7} < 1.$$

II) Let $f_2 = 3$ and $e_3 = 2$. Then, without loss of generality,

$$G = \langle x, y, z \mid x^2 = y^3 = z^2 = (xyxy^2)^2 = (xz)^3 = (yz)^2 = 1\rangle,$$

and G is finite of order 360.

III) Let $f_3 = 3$. Hence $e_3 = 2$, since, for $e_3 \geq 4$, we have

$$\frac{1}{3} + \frac{1}{3} + \frac{1}{4} < 1,$$

and, for $e_3 = 3$, we have the factor group

$$\bar{G} = \langle x, y, z \mid x^2 = y^3 = z^3 = (xy)^3 = (xz)^2 = (yz)^3 = 1 \rangle$$

which is an infinite ordinary tetrahedron group by the Coxeter criterion. So, without loss of generality,

$$G = \langle x, y, z \mid x^2 = y^3 = z^2 = (xyxy^2)^2 = (xz)^2 = (yz)^3 = 1 \rangle.$$

Here G is infinite by GAP.

This concludes the proof. □

Proposition 2.5 *Let $e_1 = e_2 = 3$, $k_1 \geq 2$, $f_1 = 2$ and $(f_2, f_3) = (2,3)$ or $(3,2)$. If G is finite, then G is equivalent to*

$$\langle x, y, z \mid x^3 = y^3 = z^2 = (xyxy^2)^2 = (xz)^3 = (yz)^2 = 1 \rangle$$

of order 7200.

Proof G is infinite for $e_3 \geq 3$ by Theorem 2.2 of [5], since $f_2 = 3$ or $f_3 = 3$ and $e_1 = e_2 = 3$. So let $e_3 = 2$. Then, up to equivalence,

$$G = \langle x, y, z \mid x^3 = y^3 = z^2 = R_1(x,y)^2 = (xz)^{f_3} = (yz)^{f_2} = 1 \rangle.$$

If G is finite, we have, by [12] and [11], the following possibilities for $R_1(x,y)$ (up to equivalence):

1. $R_1(x,y) = xyxy^2$,
2. $R_1(x,y) = xyx^2y^2$,
3. $R_1(x,y) = xyxyx^2y^2$,
4. $R_1(x,y) = x^2y^2xyxyx^2y$,
5. $R_1(x,y) = xy^2x^2y^2xyx^2y$.

If $R_1(x,y) = xyxy^2$, then G is finite of order 7200 for $f_2 = 3$ and infinite for $f_3 = 3$ by GAP.

If $R_1(x,y) = xyx^2y^2$, then G is infinite by GAP (note that, by symmetry, we do not have to distinguish between $f_2 = 3$ and $f_2 = 2$).

If $R_1(x,y) = x^2y^2xyxyx^2y$, the trace polynomial of $R_1(x,y)$ has $\lambda = 2\cos\pi/5$ as a double root and therefore G is infinite by Theorem 1.16.

Suppose that $R_1(x,y) = xy^2x^2y^2xyx^2y$. The trace polynomial of $R_1(x,y)$ has roots 0, 1, $\lambda = 2\cos\pi/5$, and $1 - \lambda$. Additionally we have a non-essential, non-trivial representation $\rho : G_1 \to \mathrm{PSL}(2,\mathbb{C})$ with

$$x \mapsto X = \begin{pmatrix} e^{\frac{\pi i}{3}} & 0 \\ t & e^{-\frac{\pi i}{3}} \end{pmatrix}, \quad y \mapsto Y = \begin{pmatrix} e^{-\frac{\pi i}{3}} & 0 \\ -t & e^{\frac{\pi i}{3}} \end{pmatrix}, \quad t \in \mathbb{C},$$

where $\langle X, Y \rangle$ is isomorphic to $\mathbb{Z}_3$.

If we consider this representation, we can improve the estimate of the Spelling Theorem to $\ell \geq 6$, i.e. the angle at G_1 is at most $\pi/6$. Since

$$\frac{1}{6} + \frac{1}{3} + \frac{1}{2} = 1,$$

we have that G is infinite in this case.

We now consider case 3 where

$$G_1 = \langle x, y \mid x^3 = y^3 = (xyxyx^2y^2)^2 = 1 \rangle$$

and we use the Gersten-Stallings method of angles. If $\rho_1 : G_1 \to \mathrm{PSL}(2, \mathbb{C})$ is an essential representation with (as usual) $x \mapsto X$, $y \mapsto Y$, $\mathrm{tr}X = 1$, $\mathrm{tr}Y = 1$, then the trace polynomial has roots

$$\mathrm{tr}XY = 0, \quad \mathrm{tr}XY = \lambda = 2\cos\frac{\pi}{5}, \quad \text{and} \quad \mathrm{tr}XY = 1 - \lambda.$$

For $\mathrm{tr}XY = 0$ we have that $\langle X, Y \rangle$ is isomorphic to $\mathcal{A}_4$ and, for $\mathrm{tr}XY = \lambda$ or $1 - \lambda$, we have that $\langle X, Y \rangle$ is isomorphic to $\mathcal{A}_5$. G_1 admits a non-essential, non-trivial representation

$$\varphi : G_1 \to \mathrm{PSL}(2, \mathbb{C}), \quad x \mapsto X = \begin{pmatrix} e^{\frac{\pi i}{3}} & 0 \\ t & e^{-\frac{\pi i}{3}} \end{pmatrix},$$

$$y \mapsto Y = \begin{pmatrix} e^{-\frac{\pi i}{3}} & 0 \\ -t & e^{\frac{\pi i}{3}} \end{pmatrix}, \quad t \in \mathbb{C}.$$

If we consider this representation, we can improve the estimate of Theorem 1.11 to $\ell \geq 5$, i.e. the angle at G_1 is at most $\pi/5$. This is insufficient for our purpose, and so we consider equations of the form

$$(\star) \quad x^{r_1}y^{s_1}x^{r_2}y^{s_2}x^{r_3}y^{s_3}x^{r_4}y^{s_4}x^{r_5}y^{s_5} = 1, \quad 1 \leq r_i, s_i \leq 2.$$

In equation $(\star)$ x has to appear as often as y and x^2 as often as y^2, since G_1 admits the above representation to $\mathbb{Z}_3$. Using symmetry, as well as cyclic conjugations and inversion of words of the form in $(\star)$, we can restrict ourselves to the following equations:

a) $(xy)^5 = 1$, b) $(xy)^4x^2y^2 = 1$,
c) $(xy)^3x^2yxy^2 = 1$, d) $(xy)^2x^2(yx)^2y^2 = 1$,
e) $(xy)^3(x^2y^2)^2 = 1$, f) $(xy)^2x^2yxy^2x^2y^2 = 1$,
g) $xyx^2(yx)^2y^2x^2y^2 = 1$, h) $xyx^2yxy^2xyx^2y^2 = 1$,
i) $xyx^2y^2(xy)^2x^2y^2 = 1$, j) $xy^2x^2(yx)^2yx^2y^2 = 1$,
k) $xy^2xyx^2yxyx^2y^2 = 1$, l) $xy^2(xy)^2x^2yx^2y^2 = 1$,
m) $xy^2x^2yxyx^2yxy^2 = 1$, n) $xy^2x^2yx^2(yx)^2y^2 = 1$,
o) $xy^2xyx^2yx^2yxy^2 = 1$, p) $xyxy^2xyx^2yx^2y^2 = 1$,
q) $(xy)^2xy^2x^2yx^2y^2 = 1$, r) $xyxy^2x^2yxyx^2y^2 = 1$.

In cases a), b), d), e), f), g), h), i), k), m), o) and p) we consider X, Y with $\mathrm{tr}XY = 0$ and get $XY = 1$ in $\mathcal{A}_4$, a contradiction.

In case c) $(xy)^3x^2yxy^2 = 1$, we consider X and Y with $\mathrm{tr}XY = \lambda$ and, after inserting $(XY)^5 = 1$, we get $XYXYX^2Y^2 = 1$, which is impossible in $\mathcal{A}_5$, since we have an essential representation.

In case j) $xy^2x^2(yx)^2yx^2y^2 = 1$, we replace $xyxyx^2y^2$ by $yxy^2x^2y^2x^2$ and get an equation $x^2y^2xy^2x^2y = 1$ in G_1, which is a contradiction because the angle at G_1 is at most $\pi/5$. Analogously we get the same contradiction in case q).

In case l) $xy^2(xy)^2x^2yx^2y^2 = 1$, we replace y^2xyxyx^2 by $xy^2x^2y^2x^2y$ to get $(xy)^4 = 1$ in G_1, which does not hold. Analogously we consider case n).

In case r) $xyxy^2x^2yxyx^2y^2 = 1$, we again replace $xyxyx^2y^2$ by $yxy^2x^2y^2x^2$ and get $(xy)^2 = 1$ in G_1, which does not hold. So, indeed, the angle at G_1 is at most $\pi/6$. Hence G is infinite, since $1/6 + 1/3 + 1/2 = 1$. □

Proposition 2.6 *If $k_1 \geq 2$ and $e_2 = 4$, then G is infinite.*

Proof

I) Assume first that $e_1 \geq 3$. Let $f_2 = 3$. If $e_3 \geq 3$, then

$$\frac{1}{e_1} + \frac{1}{e_3} + \frac{1}{3} \leq 1,$$

and so G is infinite, since $k_2 = 1$, $e_2 = 4$ and

$$\langle x, z \mid x^3 = z^3 = (xz)^3 = 1\rangle$$

has an essential cyclic, and therefore essential infinite metabelian, image in $\mathrm{PSL}(2, \mathbb{C})$.

So let $e_3 = 2$. If the trace polynomial of $R_1(x, y)$ has a multiple root, then G is infinite by Theorem 1.16. Hence G is infinite if $e_1 \geq 4$ (G is infinite, if G admits an essential infinite metabelian image in $\mathrm{PSL}(2, \mathbb{C})$ because $f_2 = 3$). So let $e_1 = 3$. We have then

$$G = \langle x, y, z \mid x^3 = y^4 = z^2 = R_1^2(x, y) = (xz)^3 = (y^\beta z)^2 = 1\rangle$$

with $1 \leq \beta \leq 2$. If G_1 admits an essential cyclic or essential infinite metabelian image in $\mathrm{PSL}(2, \mathbb{C})$, then G is infinite since $f_2 = 3$. Introducing the relation $y^2 = 1$, we see that G has image

$$\bar{G} = \langle \bar{x}, \bar{y}, \bar{z} \mid \bar{x}^3 = \bar{y}^2 = \bar{z}^2 = \bar{R}_1^2(\bar{x}, \bar{y}) = (\bar{x}\bar{z})^3 = (\bar{y}^\beta \bar{z})^2 = 1\rangle,$$

where $\bar{R}_1(\bar{x}, \bar{y})$ is the cyclically reduced word, that we derive from $R_1(x, y)$ after using the relation $y^2 = 1$ and free reduction.

If $\bar{R}_1(\bar{x}, \bar{y}) = 1$ or $\bar{y}$, then $\bar{G}$, and hence G, is infinite. If the trace polynomial of $R_1(x, y)$ admits a multiple root then G is infinite by Theorem 1.16. If G_1 admits an essential non-elementary image in $\mathrm{PSL}(2, \mathbb{C})$, then G is infinite by Theorem 2.2 of [5]. By [12] and [11] we only have, up to equivalence, to consider the case $R_1(x, y) = xyxy^{-1}$. So we have

$$G = \langle x, y, z \mid x^3 = y^4 = z^2 = (xyxy^{-1})^2 = (xz)^3 = (y^\beta z)^2 = 1\rangle$$

with $1 \leq \beta \leq 2$. If $\beta = 2$, we introduce the relation $y^2 = 1$ and get an infinite factor group of G. If $\beta = 1$, then G is infinite by GAP.
Assume now that $f_2 = 2$ and $f_3 = 3$. If $e_3 \geq 3$, then

$$\frac{1}{e_2} + \frac{1}{e_3} + \frac{1}{3} < 1,$$

and so G is infinite. Assume now that $e_3 = 2$.
If $e_1 \geq 4$ and $f_3 = 3$, then G is infinite by [12] and [11], Theorem 1.5 and Theorem 1.16 (last, if the trace polynomial of $R_1(x, y)$ has a multiple root). Hence let $e_1 = 3$. Then, up to equivalence,

$$G = \langle x, y, z \mid x^3 = y^4 = z^2 = R_1^2(x, y) = (xz)^2 = (y^\beta z)^3 = 1 \rangle$$

with $1 \leq \beta \leq 2$. If $\beta = 2$ then

$$G_3 = \langle y, z \mid y^4 = z^2 = (y^2 z)^3 = 1 \rangle$$

admits an essential non-elementary image $\langle Z, Y \rangle$ in $\mathrm{PSL}(2, \mathbb{C})$ with

$$\mathrm{tr}Z = 0, \ \mathrm{tr}Y = \sqrt{2}, \ \text{and } \mathrm{tr}YZ = -\frac{1}{\sqrt{2}}.$$

Hence G is infinite by Theorem 1.5.
Assume now that $\beta = 1$. If G_1 admits an essential cyclic or essential infinite metabelian image in $\mathrm{PSL}(2, \mathbb{C})$, then G is infinite since $f_3 = 3$. If we introduce the relation $y^2 = 1$ then G has image

$$\bar{G} = \langle \bar{x}, \bar{y}, \bar{z} \mid \bar{x}^3 = \bar{y}^2 = \bar{z}^2 = \bar{R}_1^2(\bar{x}, \bar{y}) = (\bar{x}\bar{z})^2 = (\bar{y}\bar{z})^3 = 1 \rangle,$$

where $\bar{R}_1(\bar{x}, \bar{y})$ is the cyclically reduced word that we derive from $R_1(x, y)$ after using the relation $y^2 = 1$ and free reduction.
If $\bar{R}_1(\bar{x}, \bar{y}) = 1$ or $\bar{y}$ then $\bar{G}$, and hence G, is infinite. If the trace polynomial of $R_1(x, y)$ admits a multiple root, then G is infinite by Theorem 1.16. If G_1 admits an essential non-elementary image in $\mathrm{PSL}(2, \mathbb{C})$, then G is infinite by Theorem 1.5. By [12] and [11] we only have, up to equivalence, to consider the case $R_1(x, y) = xyxy^{-1}$. Here G is infinite by GAP.
So, altogether, G is infinite for $e_1 \geq 3$.

II) Assume now that $e_1 = 2$ and that $f_3 = 3$. Analogously to I), G is infinite for $e_3 \geq 3$. So let $e_3 = 2$, that is

$$G = \langle x, y, z \mid x^3 = y^4 = z^2 = R_1^2(x, y) = (xz)^2 = (y^\beta z)^3 = 1 \rangle$$

with $1 \leq \beta \leq 2$. If $\beta = 2$, then G is infinite as in I). So let $\beta = 1$. Introducing the relation $y^2 = 1$, we derive, as in I), the factor group

$$\bar{G} = \langle \bar{x}, \bar{y}, \bar{z} \mid \bar{x}^2 = \bar{y}^2 = \bar{z}^2 = \bar{R}_1^2(\bar{x}, \bar{y}) = (\bar{x}\bar{z})^2 = (\bar{y}\bar{z})^3 = 1 \rangle.$$

If $\bar{R}_1(\bar{x}, \bar{y}) = \bar{x}$, $\bar{y}$ or 1 then $\bar{G}$, and hence G, is infinite. If G has the factor group

$$\bar{\bar{G}} = \langle x, y, z \mid x^2 = y^4 = z^2 = (xy)^4 = (xz)^2 = (yz)^3 = 1 \rangle,$$

then $\bar{\bar{\bar{G}}}$, and therefore G, is infinite by the Coxeter criterion. If the trace polynomial of $R_1(x,y)$ admits a multiple root, then G is infinite by Theorem 1.16. By [12] and [11] we only have to consider, up to equivalence, the case $R_1(x,y) = xyxyxy^3$.
So let $R_1(x,y) = xyxyxy^3$. The subgroup H of index 2 in G, generated by $a = y$, $b = xyx$, $c = z$ and $d = xzx$, has presentation

$$\langle a, b, c \mid a^4 = b^4 = c^2 = (ac)^3 = (bc)^3 = bab^3aba^3 = 1 \rangle$$

(note that $cd = 1$). Introducing the relations $b^2 = a^2 = 1$ to H, we get the infinite factor group

$$\bar{H} = \langle a, b, c \mid a^2 = b^2 = c^2 = (ac)^3 = (bc)^3 = (ab)^3 = 1 \rangle.$$

Hence G is infinite.

III) Assume now that $e_1 = 2$ and $f_2 = 3$; then $f_3 = 2$. Assume first that $e_3 = 2$. Then

$$G = \langle x, y, z \mid x^3 = y^4 = z^2 = R_1^2(x,y) = (xz)^2 = (y^\beta z)^2 = 1 \rangle$$

with $1 \leq \beta \leq 2$. If $\beta = 2$, we introduce the relation $y^2 = 1$ and get an infinite factor group. So let $\beta = 1$. With the same argument as in II) we have to consider, up to equivalence, only the case $R_1(x,y) = xyxyxy^3$.
So let $R_1(x,y) = xyxyxy^3$. We consider the subgroup H of index 2 of G, generated by $a = x$, $b = yxy^{-1}$, $c = y^2$, $d = z$ and $e = yzy^{-1}$. Then H has presentation

$$\langle a, b, c, d, e \mid a^2 = b^2 = c^2 = d^2 = e^2 = abcacbcab = (ad)^3 = (be)^3 = ecd = 1 \rangle.$$

If we introduce the relation $c = 1$ then we get the infinite factor group

$$\bar{H} = \langle a, b, d \mid a^2 = b^2 = d^2 = (ab)^3 = (ad)^3 = (bd)^3 = 1 \rangle.$$

Hence G is infinite.
So assume now that $e_3 \geq 3$. We then have

$$G = \langle x, y, z \mid x^2 = y^4 = z^{e_3} = R_1^2(x,y) = (xz^\gamma)^3 = (y^\beta z^\delta)^2 = 1 \rangle$$

with $1 \leq \beta \leq 2$ and $1 \leq \gamma, \beta < e_3$.
If $e_3 \geq 5$, then G is infinite because

$$\frac{1}{4} + \frac{1}{e_3} + \frac{1}{2} < 1.$$

Hence let $e_3 = 4$. If $\gamma = 2$, we consider

$$G_2 = \langle x, z \mid x^2 = z^4 = (xz^2)^3 = 1 \rangle.$$

G_2 has an essential non-elementary image in $\mathrm{PSL}(2, \mathbb{C})$; hence G is infinite for $\gamma = 2$. If $\gamma = 1$ or 3, we can assume (via inversion of xz^γ), that $\gamma = 1$.

So let $\gamma = 1$. If $\beta = 2$ or $\delta = 2$, we introduce the relations $y^2 = z^2 = 1$ to get an infinite factor group. Hence let $\beta \neq 2 \neq \delta$. We can now assume that $\beta = 1 = \delta$. G_3 has an essential infinite metabelian image in $\mathrm{PSL}(2,\mathbb{C})$; hence G is infinite by Theorem 1.5 because $f_2 = 3 > 2$.
Assume now that $e_3 = 3$. Up to equivalence, we have

$$G = \langle x, y, z \mid x^2 = y^4 = z^3 = R_1^2(x,y) = (xz)^3 = (y^\beta z)^2 = 1\rangle$$

with $1 \leq \beta \leq 2$. If $\beta = 2$, then

$$G_3 = \langle y, z \mid y^4 = z^3 = (y^2 z)^2 = 1\rangle$$

has an essential non-elementary image $\langle Y, Z\rangle$ in $\mathrm{PSL}(2,\mathbb{C})$ with

$$\mathrm{tr}Y = \sqrt{2},\ \ \mathrm{tr}Z = 1 \ \text{ and } \ \mathrm{tr}YZ = \frac{1}{\sqrt{2}};$$

therefore G is infinite for $\beta = 2$.
Hence let $\beta = 1$. Analogously, as in II), we have to consider, up to equivalence, only the case $R_1(x,y) = xyxyxy^3$. We have then

$$G = \langle x, y, z \mid x^2 = y^4 = z^3 = (xyxyxy^3)^2 = (xz)^3 = (yz)^2 = 1\rangle.$$

We consider the subgroup H of index 2, generated by

$$a = x,\ b = yxy^{-1},\ c = y^2,\ d = z \text{ and } e = yzy^{-1}.$$

Then H has presentation

$$\langle a, b, c, d, e \mid a^2 = b^2 = c^2 = d^3 = e^3 = abcacbcab = (ad)^3 = (be)^3 = ecd = 1\rangle.$$

If we introduce the relation $c = 1$ then we get the infinite factor group

$$\bar{H} = \langle a, b, d \mid a^2 = b^2 = d^3 = (ab)^3 = (ad)^3 = (bd)^3 = 1\rangle.$$

Hence G is infinite.
This completes the proof. □

Proposition 2.7 *If $e_2 = 5$, $k_1 \geq 2$ and exactly one f_i is equal to 3 then G is finite if and only if, up to equivalence,*

$$G = \langle x, y, z \mid x^2 = y^5 = z^2 = (xyxy^2)^2 = (xz)^3 = (yz)^2 = 1\rangle$$

of order 43200.

Proof

I) Assume that $e_1 = 2$ and $k_1 \geq 3$. We have, up to equivalence,

$$G = \langle x, y, z \mid x^2 = y^5 = z^{e_3} = R_1(x, y)^2 = (xz^\gamma)^{f_2} = (y^\beta z^\delta)^{f_3} = 1 \rangle,$$

with $1 \leq \beta < 5$, $1 \leq \gamma, \delta < e_3$ and $(f_2, f_3) = (2, 3)$ or $(3, 2)$. G is infinite for $e_3 \geq 4$ because, then,

$$\frac{1}{5} + \frac{1}{e_3} + \frac{1}{2} < 1.$$

So we can assume that $\gamma = \delta = 1$ and $1 \leq \beta \leq 2$. If $k_1 \geq 5$ then G is infinite because

$$\frac{1}{k_1 + 1} + \frac{1}{f_2} + \frac{1}{f_3} \leq 1.$$

If $k_1 = 4$ then G admits an essential non-elementary image in $\mathrm{PSL}(2, \mathbb{C})$ and therefore G is infinite (see [11]). Therefore let $k_1 = 3$.
Then we have to consider (see [11]), up to equivalence:

1. $R_1(x, y) = xyxy^2xy^4$,
2. $R_1(x, y) = xyxyxy^4$ and
3. $R_1(x, y) = xyxyxy^3$ (otherwise G is infinite).

In case 3, $R_1(x, y) = xyxyxy^3$, the vertex group G_1 admits an essential, cyclic or infinite metabelian, image in $\mathrm{PSL}(2, \mathbb{C})$ and hence G is infinite, because $(f_2, f_3) \neq (2, 2)$.
We can handle cases 1 and 2 at the same time. In both cases the trace polynomial for $R_1(x, y)$ has roots 0, $\lambda - 1$ and $1 - \lambda$, where $\lambda = 2\cos\pi/5$. We consider now essential representations

$$\rho_1 : G_1 \to \mathrm{PSL}(2, \mathbb{C}), \quad x \mapsto X, \quad y \mapsto Y,$$

with

$$\mathrm{tr}X = 0, \ \mathrm{tr}Y = \lambda = 2\cos\frac{\pi}{5} \ \text{ and } \mathrm{tr}XY = 0, \lambda - 1 \text{ or } 1 - \lambda.$$

If $\mathrm{tr}XY = 0$ then $\langle X, Y \rangle$ is isomorphic to $\mathcal{D}_5$, the dihedral group, and, if $\mathrm{tr}XY = \pm(\lambda - 1)$, then $\langle X, Y \rangle$ is isomorphic to $\mathcal{A}_5$. If $\mathrm{tr}XY = 0$ then $\mathrm{tr}XY^\beta = 0$ for $1 \leq \beta \leq 4$. If $\mathrm{tr}XY = \lambda - 1$ then $\mathrm{tr}XY^2 = 1$, $\mathrm{tr}XY^3 = 1$ and $\mathrm{tr}XY^4 = \lambda - 1$. Anyway, the angle at G_1 is at most $\pi/4$. We now use the trace relations to show that the angle at G_1 is, in fact, equal to $\pi/6$. We cannot have an equation of the form

$$(2) \quad xy^{q_1}xy^{q_2}xy^{q_3}xy^{q_4}xy^{q_5} = 1, \ 1 \leq q_i \leq 4,$$

in G_1 because, if we introduce the relation $y = 1$ in G_1, we get the factor group $\langle x \mid x^2 = 1 \rangle = \mathbb{Z}_2$ which contradicts equation 1.
Now we have to consider equations of the form

$$(1) \quad xy^{s_1}xy^{s_2}xy^{s_3}xy^{s_4} = 1, \ 1 \leq s_i \leq 4,$$

in G_1. If we choose X, Y with tr$XY = 0$ we see that we must have

$$s_1 + s_3 \equiv s_2 + s_4 \mod 5.$$

Assume first that $s_i = 2$ or 3 for some i with $1 \leq i \leq 4$. Without loss of generality let $s_1 = 2$. We choose X, Y with tr$XY = \lambda - 1$. Then tr$XY^2 = 1$, that is, $(XY^2)^3 = 1$, and we get, in $\mathcal{A}_5$, the equation

$$XY^{s_2-2}XY^{s_3}XY^{s_4-2} = 1.$$

This automatically gives that $s_2 \neq 2$ and $s_4 \neq 2$ if $s_1 = 2$. If $s_2 = 4$ then we may argue as before to get $XY^{s_3-2}XY^{s_4-4} = 1$, that is,

$$XY^{s_3-2}X = Y^{4-s_4}$$

in $\mathcal{A}_5$. This holds only if $s_3 = 2$ and $s_4 = 4$, but then

$$s_1 + s_3 = 4 \not\equiv s_2 + s_4 \equiv 3 \mod 5,$$

a contradiction. Hence $s_2 \neq 4$ if $s_1 = 2$.
Analogously we get $s_4 \neq 4$ if $s_1 = 2$. Assume that $s_3 = 2$ or 3 if $s_1 = 2$, and, without loss of generality, let $s_3 = 2$. Then we get, in $\mathcal{A}_5$, the equation

$$XY^{s_2-4}XY^{s_4-4} = 1,$$

which gives a contradiction. Hence $s_3 \neq 2$ and $s_3 \neq 3$ if $s_1 = 2$, that is, $s_3 = 1$ or $s_3 = 4$ if $s_1 = 2$.
If $s_3 = 1$ then $s_1 + s_3 = 3$ but $s_2 + s_4 \equiv 1$, 2 or 4 (mod 5) which gives a contradiction. If $s_3 = 4$ then $s_1 + s_3 \equiv 0$ (mod 5) which also gives a contradiction. Therefore we have $s_i \neq 2$ and $s_i \neq 3$ for all $1 \leq i \leq 4$ in equation (1).
Assume now that $s_1 = s_2$ or $s_2 = s_3$ or $s_3 = s_4$ or $s_4 = s_1$. Without loss of generality let $s_1 = s_2 = 1$. Then $s_3 = s_4 = 1$ or $s_3 = s_4 = 4$ because

$$s_1 + s_3 \equiv s_2 + s_4 \mod 5.$$

We cannot have $s_3 = s_4 = 1$ because then $(xy)^4 = 1$ which gives a contradiction if we choose X, Y with tr$XY = \lambda - 1$, that is $(XY)^5 = 1$.
Let $s_3 = s_4 = 4$. Then we have, in G_1, the equation

$$(xy)^2(x^{-1}y^{-1})^2 = 1.$$

We choose X, Y with tr$XY = \lambda - 1$. But then $(XY)^2(X^{-1}Y^{-1})^2$ has order 5 in $\mathcal{A}_5$ which gives a contradiction. Hence, we finally also have $s_1 \neq s_2$, $s_2 \neq s_3$, $s_3 \neq s_4$ and $s_4 \neq s_1$. We know that $s_i = 1$ or 4 for all $1 \leq i \leq 4$. Therefore, without loss of generality, let $s_1 = 1$. Then we get $s_2 = 4$, $s_3 = 1$ and $s_4 = 4$. But then

$$2 \equiv s_1 + s_3 \not\equiv s_2 + s_4 \equiv 3 \mod 5.$$

Altogether we have that the angle at G_1 is $\pi/6$. Therefore G is infinite by Proposition 3.1 of [6] because

$$\frac{1}{6}+\frac{1}{3}+\frac{1}{3}=1.$$

II) Here we assume that $e_1 = 2$ and $k_1 = 2$. We have, up to equivalence,

$$G = \langle x, y, z \mid x^2 = y^5 = z^{e_3} = R_1(x,y)^2 = (xz^\gamma)^{f_2} = (y^\beta z^\delta)^{f_3} = 1\rangle,$$

with $1 \le \gamma, \delta < e_3, 1 \le \beta \le 4$, and $(f_2, f_3) = (2,3)$ or $(3,2)$. Up to equivalence we have to consider (see [11]) only $R_1(x,y) = xyxy^2$ because otherwise G is infinite.

A) Let $f_2 = 3$. G is infinite for $e_3 \ge 4$, since, in that case,

$$\frac{1}{5}+\frac{1}{e_3}+\frac{1}{2}<1.$$

So let $2 \le e_3 \le 3$. We may assume that $\gamma = \delta = 1$ and $1 \le \beta \le 2$. If $\beta = 1$ and $e_3 = 3$ then we introduce the relation $(xy)^3 = 1$ and get an infinite factor group by the Coxeter criterion, and hence G is infinite. If $\beta = 2$ and $e_3 = 3$ then G is infinite by GAP. If $e_2 = 2$ then we may assume that $\beta = 1$ because $\langle b, c\rangle$ is dihedral. Then

$$G = \langle x, y, z \mid x^2 = y^5 = z^2 = (xyxy^2)^2 = (xz)^3 = (yz)^2 = 1\rangle$$

is finite of order 43200.

B) Let $f_3 = 3$. Now G is infinite for $e_3 \ge 3$ because, in that case,

$$\frac{1}{5}+\frac{1}{e_3}+\frac{1}{3}<1.$$

So let $e_3 = 2$. We can, up to equivalence, assume that $\gamma = \beta = \delta = 1$. Then

$$G = \langle x, y, z \mid x^2 = y^5 = z^2 = (xyxy^2)^2 = (xz)^2 = (yz)^3 = 1\rangle.$$

We can show that G is infinite by GAP.

III) Assume that $e_1 \ge 3$. G is infinite if $e_1 \ge 4$ because

$$\frac{1}{5}+\frac{1}{4}+\frac{1}{2}<1.$$

Hence, let $e_1 = 3$. Analogously, as in the case $e_2 = 4$, G is infinite if $e_3 \ge 3$. Hence let $e_3 = 2$.

If G admits an essential representation $\rho_1 : G_1 \to \mathrm{PSL}(2,\mathbb{C})$ with $\rho_1(G_1)$ non-elementary cyclic or infinite metabelian, then G is infinite by Theorem 1.5 since $f_2 = 3$ or $f_3 = 3$. So we have, up to equivalence, to consider the following possibilities for $R_1(x,y)$ (see [12], [11]):

1. $R_1(x,y) = xyx^2y^2$

2. $R_1(x,y) = x^2yxy^{-1}xy^{-1}$
3. $R_1(x,y) = x^2yxy^3xy^{-1}$
4. $R_1(x,y) = x^{-1}yx^{-1}y^2xyxy^{-1}$
5. $R_1(x,y) = x^{-1}yxyx^{-1}y^2xy^{-1}$
6. $R_1(x,y) = x^{-1}y^2xyx^{-1}yxy^{-1}$
7. $R_1(x,y) = x^{-1}yx^{-1}y^3xyxy^{-1}$.

In case 1 $R_1(x,y) = xyx^2y^2$, we have that G is infinite by GAP.
In case 7, the trace polynomial has 1 as a double root, and so G is infinite by Theorem 1.16. In addition, G has an essential non-elementary image in $\mathrm{PSL}(2,\mathbb{C})$, and so G is infinite anyway.
In cases 4, 5 and 6, the trace polynomial has roots 0, 1, λ and $\lambda - 1$, where

$$\lambda = 2\cos\frac{\pi}{5}.$$

G_1 admits, in addition, a non-essential, non-trivial representation

$$\varphi : G_1 \to \mathrm{PSL}(2,\mathbb{C}),$$

given by

$$x \mapsto X = \begin{pmatrix} e^{\frac{\pi i}{3}} & 0 \\ t & e^{-\frac{\pi i}{3}} \end{pmatrix}, \quad y \mapsto Y = \begin{pmatrix} 1 & 0 \\ 0 & 1 \end{pmatrix}, \quad t \in \mathbb{C}.$$

If we consider this representation, we can improve the estimate of Theorem 1.11 to $\ell \geq 6$, i.e. the angle at G_1 is at most $\pi/6$. Since

$$\frac{1}{6} + \frac{1}{3} + \frac{1}{2} = 1,$$

G is infinite in cases 4, 5 and 6 by Proposition 3.1 of [6].
In case 3, $R_1(x,y) = x^2yxy^3xy^{-1}$, the trace polynomial of $R_1(x,y)$ has 1 as a double root; hence G is infinite by Theorem 1.16.
We now consider case 2, $R_1(x,y) = x^2yxy^{-1}xy^{-1}$. We can transform

$$x^2yxy^{-1}xy^{-1}$$

equivalently to $xyxyx^{-1}y^{-1}$. We now use the notation $R_1(x,y)$ to represent $xyxyx^{-1}y^{-1}$ as well. The trace polynomial for $xyxyx^{-1}y^{-1}$ has roots 0, 1 and $\lambda - 1$. We consider now essential representations

$$\rho_1 : G_1 \to \mathrm{PSL}(2,\mathbb{C}), \quad x \mapsto X, \quad y \mapsto Y$$

with

$$\mathrm{tr}X = 1, \ \mathrm{tr}Y = \lambda = 2\cos\frac{\pi}{5} \ \text{ and } \ \mathrm{tr}XY = 0, 1 \text{ or } \lambda - 1.$$

In any case $\langle X, Y\rangle$ is isomorphic to $\mathcal{A}_5$.

(i) $\mathrm{tr}XY = 0$. Then
$\mathrm{tr}XY^2 = -1$, $\mathrm{tr}XY^3 = -\lambda$, $\mathrm{tr}XY^4 = -\lambda$, $\mathrm{tr}X^2Y = -\lambda$,
$\mathrm{tr}X^2y^2 = -\lambda$, $\mathrm{tr}X^2Y^3 = -1$ and $\mathrm{tr}X^2Y^4 = 0$.

(ii) $\mathrm{tr}XY = 1$. Then
$\mathrm{tr}XY^2 = \lambda - 1$, $\mathrm{tr}XY^3 = 0$, $\mathrm{tr}XY^4 = 1 - \lambda$, $\mathrm{tr}X^2Y = 1 - \lambda$,
$\mathrm{tr}X^2y^2 = 0$, $\mathrm{tr}X^2Y^3 = \lambda - 1$ and $\mathrm{tr}X^2Y^4 = 1$.

(iii) $\mathrm{tr}XY = \lambda - 1$. Then
$\mathrm{tr}XY^2 = 0$, $\mathrm{tr}XY^3 = 1 - \lambda$, $\mathrm{tr}XY^4 = -1$, $\mathrm{tr}X^2Y = -1$,
$\mathrm{tr}X^2y^2 = 1 - \lambda$, $\mathrm{tr}X^2Y^3 = 0$ and $\mathrm{tr}X^2Y^4 = \lambda - 1$.

Anyway, the angle at G_1 is at most $\pi/4$. We now use the trace relations to show that the angle at G_1 is, in fact, equal to $\pi/6$. Therefore we have to consider equations of the form

$$(1) \quad W(x,y) := x^{r_1}y^{s_1}\dots x^{r_4}y^{s_4} = 1, \quad 1 \le r_i \le 2,\ 1 \le s_i \le 4, \text{ and}$$
$$(2) \quad V(x,y) := x^{p_1}y^{q_1}\dots x^{p_5}y^{q_5} = 1, \quad 1 \le p_j \le 2,\ 1 \le q_j \le 4,$$

and we have to show that such equations cannot hold in G_1.
If we can derive an equation of the form (1) or (2) from another one of the same form by inversion of $W(x,y)$ or $V(x,y)$, respectively, or by cyclic conjugation of $W^{\pm 1}(x,y)$ or $V^{\pm 1}(x,y)$, respectively, then we have to consider only one of the two equations. Nevertheless, the number of equations to be considered is very large. We call a term $x^{r_i}y^{s_i}$ or $x^{p_j}y^{q_j}$, respectively, a *block* of such an equation.
If, in $W(x,y)$ or $V(x,y)$, a cyclic conjugate S of

$$R_1^{\pm 1}(x,y) = (xyxyx^{-1}y^{-1})^{\pm 1}$$

appears as a subword, we replace S by S^{-1} and get, after conjugation, an equation of the form

$$x^{n_1}y^{m_1}\dots x^{n_k}y^{m_k} = 1, \quad 1 \le n_l \le 2,\ 1 \le m_l \le 4,$$

with $k \le 3$ or $k \le 4$, respectively, i.e. an equation with fewer blocks. So we have to examine equations of the form (1) or (2), in which each cyclic conjugate of $W^{\pm 1}(x,y)$ or $V^{\pm 1}(x,y)$ contains no cyclic conjugate of $R_1^{\pm 1}(x,y)$ as a subword and we have to show that such an equation cannot hold in G_1. Therefore we consider only such equations of the form (1) and (2).
If, in such an equation once $r_i = r_{i+1}$ or $p_j = p_{j+1}$, respectively, or $s_i = s_{i+1}$ or $q_j = q_{j+1}$, respectively, appears in a cyclic conjugate of $W(x,y)$ or $V(x,y)$, then we get, by a suitable choice of $X, Y \in \mathrm{PSL}(2,\mathbb{C})$, an equation, which does not hold in $\mathcal{A}_5$ and, hence, the original equation cannot hold in G_1. Here we consider, because of the large number of possible equations, in the following only three particular equations.

(α) $xy^2xy^3x^2y^2x^2y^4 = 1$.
Choose X, Y with $\mathrm{tr}XY = 0$, i.e. $(XY)^2 = 1$ in $\mathcal{A}_5$.

Then $(XY^2)^3 = (Y^2X^2)^5 = 1$ in $\mathcal{A}_5$. Hence we get

$$XY^2XY^3X^2Y^2X^2Y^4 = Y^{-2}X^{-1}YX^2Y^2X^2Y^4 = 1,$$

i.e. $X^2YX^2Y^2X^2Y^2 = 1$ and, further,

$$Y^{-1}Y^2X^2Y^2X^2Y^2 = 1$$

in $\mathcal{A}_5$. Hence $Y^{-1}XY^{-2}XY^{-2} = 1$, i.e.

$$XY^2XY^3 = Y^{-2}X^{-1}Y = 1,$$

i.e. $XY = 1$ in $\mathcal{A}_5$, a contradiction.

(β) $x^2y^4xy^4x^2yx^2yxy^3 = 1$.
Choose X, Y with $\text{tr}XY = \lambda - 1$, i.e. $(XY)^5 = 1$.
Then $(XY^4)^3 = (YX^2)^2 = 1$. Hence

$$X^2Y^4XY^4X^2YX^2YXY^3 = XY^{-4}X^{-1}X^2YX^2YXY^3$$

$$= XYXYX^2YXY^3 = 1,$$

and hence $XYXX^{-1}Y^3 = XY^4 = 1$ in $\mathcal{A}_5$, a contradiction.

(γ) $xyxy^2x^2y^3xy^4x^2y^3 = 1$.
Choose X, Y with $\text{tr}XY = 1$; then $(XY)^3 = (XY^3)^2 = 1$. Hence

$$XYXY^3X^2Y^3XY^4X^2Y^3 = Y^{-1}X^{-1}Y^2X^2Y^3XY^4X^2Y^3 = 1.$$

Hence

$$X^2Y^2X^2Y^3XY^4X^2Y^2 = 1,$$

so that $YXY^4X^2Y^2 = 1$, i.e.

$$XY^4X^2Y^3 = 1.$$

Hence $XY^4X^{-1} = Y^2$ in $\mathcal{A}_5$, a contradiction.

The other equations with $r_i = r_{i+1}$, $p_j = p_{j+1}$, $s_i = s_{i+1}$ or $q_j = q_{j+1}$, respectively, are handled analogously. The consideration of these cases is left to the reader

Assume, now, that $r_i = r_{i+1}$, $p_j = p_{j+1}$, $s_i = s_{i+1}$ or $q_j = q_{j+1}$ never appears. This shows that an equation of the form (2) cannot hold in G_1: since $x^3 = 1$ we have necessarily that $p_j = p_{j+1}$ at least once (maybe after a cyclic conjugation of $V(x, y)$).

We consider now an equation of the form (1) with $r_i \neq r_{i+1}$ and $s_i \neq s_{i+1}$ (modulo cyclic conjugation). Then, without loss of generality, we have that

$$xy^{s_1}x^2y^{s_2}xy^{s_3}x^2y^{s_4} = 1$$

in G_1 with $1 \leq s_i \leq 4$, $s_1 \neq s_2 \neq s_3 \neq s_4 \neq s_1$.

Case 1) $s_1 = s_3$ or $s_2 = s_4$. Without loss of generality we may assume that $s_1 = s_3$.
Assume first that $s_1 = s_3 = 1$, so

$$xyx^2y^{s_2}xyx^2y^{s_4} = 1$$

in G_1. We choose X, Y with $\mathrm{tr}XY = 0$. Then

$$XYX^2Y^{s_2}XYX^2Y^{s_4} = Y^{-1}XY^{s_2}Y^{-1}XY^{s_4} = 1,$$

and hence $XY^{s_2-1}XY^{s_4-1} = 1$. Then, necessarily, $s_2 = s_4 = 2$. Hence, in G_1, we have that $(xyx^2y^2)^2 = 1$. We choose now X, Y with $\mathrm{tr}XY = 1$. Then $\mathrm{tr}XYX^2Y^2 = -1$, i.e. XYX^2Y^2 is of order 3 in $\mathcal{A}_5$, a contradiction.
The cases $s_1 = s_3 = 2$, $s_1 = s_2 = 3$ and $s_1 = s_3 = 4$ are considered analogously by examination of xy^2, xy^3 and y^4x^2, respectively.

Case 2) $s_1 \neq s_3$ and $s_2 \neq s_4$. Without loss of generality we may assume that $s_1 = 1$ or $s_2 = 1$.
Assume first that $s_1 = 1$. Then we have

$$xyx^2y^{s_2}xy^{s_3}x^2y^{s_4} = 1$$

in G_1. We choose X, Y with $\mathrm{tr}XY = 0$. So

$$XY^{s_2}XY^{s_3}X^2Y^{s_4-1} = 1.$$

If $s_3 = 4$ then $XY^{s_2}X^2Y^{s_4-s_3-1} = 1$, i.e. $XY^{s_2}X^2 = Y^{s_3+1-s_4}$ in $\mathcal{A}_5$, a contradiction.
Hence $s_3 \neq 4$. Then $s_3 = 2$ or $s_3 = 3$. If $s_2 = 2$ then $s_3 = 3$ and

$$XY^2X = Y^{-2}X^{-1}Y^{-2},$$

and hence $X^{-1}YX^2Y^2 = 1$ in $\mathcal{A}_5$, a contradiction.
Hence $s_2 \neq 2$. Then $s_2 = 3$ or $s_2 = 4$. If $s_4 = 4$ then $s_2 = 3$ and $s_3 = 2$, so that

$$XY^3XY^2X^2Y^3 = 1$$

in $\mathcal{A}_5$. Since $(XY^3)^5 = 1$ it follows that $(XY^3)^2 = Y^{-1}X$ in $\mathcal{A}_5$. We have $\mathrm{tr}Y^{-1}X = \lambda$, but $(\mathrm{tr}(XY^3)^2 = \lambda - 1$ in $\mathcal{A}_5$, a contradiction.
Hence $s_4 \neq 4$. Then, necessarily, $s_2 = 4$. If $s_3 = 3$, then $s_4 = 2$ and $YXY^4 = XY^2X^2$ in $\mathcal{A}_5$, a contradiction.
Hence $s_3 \neq 3$. Then $s_3 = 2$ and $s_4 = 3$. We have the equation

$$XY^4XY^2X^2Y^2 = 1,$$

i.e. $XY^2X^2 = YX^{-1}Y^3$ in $\mathcal{A}_5$. We have $\mathrm{tr}XY^2X^2 = 1 - \lambda$, but $\mathrm{tr}YX^{-1}Y^3 = \mathrm{tr}X^{-1}Y^4 = 0$ in $\mathcal{A}_5$, a contradiction.
So $s_1 = 1$ is not possible. The case $s_2 = 1$ is considered analogously by examination of the equation $yxy^{s_3}x^2y^{s_4}xy^{s_1}x^2 = 1$. Therefore, in total, the angle at G_1 is equal to $\pi/6$, and G is infinite since

$$\frac{1}{6} + \frac{1}{3} + \frac{1}{2} = 1.$$

This completes the proof. □

For this last case (where $R_1(x,y) = xyxyx^{-1}y^{-1}$) we also found the infiniteness of G by using GAP.

Proposition 2.8 *Let $e_2 \geq 6$, $k_1 \geq 2$, with $2 \leq f_i \leq 3$ and exactly one f_i equal to 3. Then G is infinite.*

Proof If $f_1 = 3$, then G is infinite, since $k_1 \geq 2$ (see [5]). Hence, let $f_1 = 2$.

If $e_1 \geq 3$ then G is infinite if

$$\frac{1}{e_1} + \frac{1}{e_2} + \frac{1}{2} < 1.$$

Now let $1/e_1 + 1/e_2 + 1/2 = 1$. Then $e_1 = 3$ and $e_2 = 6$. If G_1 admits an essential infinite metabelian image in $\mathrm{PSL}(2,\mathbb{C})$ then G is infinite since $f_2 = 3$ or $f_3 = 3$ (see Theorem 1.5). If the trace polynomial of $R_1(x,y)$ has a multiple root then G is infinite by Theorem 1.16. Hence G is infinite by Corollary 1.6 and Theorem 1.5.

Assume now that

$$\frac{1}{e_1} + \frac{1}{e_2} + \frac{1}{2} > 1.$$

Then $e_1 = 2$.

1. Let $e_3 \geq 3$.
   G is infinite, if $e_3 \geq 4$ and if

$$\frac{1}{e_2} + \frac{1}{e_3} + \frac{1}{f_3} < 1.$$

   Assume now that $e_3 = 3$ and

$$\frac{1}{e_2} + \frac{1}{e_3} + \frac{1}{f_3} \geq 1.$$

   Then $f_3 = 2$ and $f_2 = 3$ as well as

$$\frac{1}{e_2} + \frac{1}{e_3} + \frac{1}{f_3} = 1.$$

   G is now infinite analogously to the case $e_1 = 3$.
2. Let $e_3 = 2$. Then

$$G = \langle x, y, z \mid x^2 = y^{e_2} = z^2 = R_1(x,y)^2 = (xz)^{f_2} = R_3(y,z)^{f_3} = 1 \rangle$$

   with $e_2 \geq 6$ and $(f_2, f_3) = (3,2)$ or $(f_2, f_3) = (3,2)$. As above, if G_1 admits an essential infinite metabelian image in $\mathrm{PSL}(2,\mathbb{C})$ or if the trace polynomial of $R_1(x,y)$ has a multiple root, then G is infinite. Hence G is infinite, since $k_1 \geq 2$ by Corollary 1.6 and Theorem 1.5.

This completes the proof. □

Theorem 2.9 (Summary) *Let*

$$G = \langle x, y, z \mid x^{e_1} = y^{e_2} = z^{e_3} = R_1^{f_1}(x,y) = R_2^{f_2}(x,z) = R_3^{f_3}(y,z) = 1 \rangle,$$

with

$$2 \leq e_1, e_2, e_3, f_1, f_2, f_3; \quad 2 \leq f_1, f_2, f_3 \leq 3;$$

$$R_1(x,y) = x^{\alpha_1} y^{\beta_1} \cdots x^{\alpha_{k_1}} y^{\beta_{k_1}},\ k_1 \geq 2,\ R_1 \text{ not a proper power},$$

$$R_2(x,z) = x^{\gamma_1} z^{\delta_1} \cdots x^{\gamma_{k_2}} z^{\delta_{k_2}},\ k_2 \geq 1,\ R_2 \text{ not a proper power},$$

$$R_3(y,z) = y^{\varepsilon_1} z^{\eta_1} \cdots y^{\varepsilon_{k_3}} z^{\eta_{k_3}},\ k_3 \geq 1,\ R_3 \text{ not a proper power},$$

$$1 \leq \alpha_i, \gamma_i < e_1,\ 1 \leq \beta_i, \varepsilon_i < e_2,\ 1 \leq \delta_i, \eta_i < e_3.$$

Let at least one f_i be equal to 3. If G is finite then G is equivalent to one of the following:

(1) $\langle x, y, z \mid x^2 = y^3 = z^2 = (xyxyxy^2)^3 = (xz)^2 = (yz)^2 = 1 \rangle$ *of order* 2880,
(2) $\langle x, y, z \mid x^2 = y^3 = z^3 = (xyxyxy^2)^3 = (xz)^2 = (yz)^2 = 1 \rangle$ *of order* 864000,
(3) $\langle x, y, z \mid x^2 = y^3 = z^2 = (xyxy^2)^2 = (xz)^3 = (yz)^2 = 1 \rangle$ *of order* 360,
(4) $\langle x, y, z \mid x^3 = y^3 = z^2 = (xyxy^2)^2 = (xz)^3 = (yz)^2 = 1 \rangle$ *of order* 7200,
(5) $\langle x, y, z \mid x^2 = y^5 = z^2 = (xyxy^2)^2 = (xz)^3 = (yz)^2 = 1 \rangle$ *of order* 43200.

Proof Let G be finite. Then $f_i = 3$ for precisely one $i \in \{1,2,3\}$ by Proposition 2.1 and $k_j \geq 2$ for precisely one $j \in \{1,2,3\}$ by Proposition 2.2. Hence $k_1 \geq 2$ and $k_2 = k_3 = 1$. If $k_1 \geq 2$ and $f_1 = 3$ then G is equivalent to (1) or (2) by Proposition 2.3.

Now, let $f_1 = 2$ and, without any loss of generality, assume that $f_2 = 3$. The remaining part is a case-by-case consideration which is given in Propositions 2.4, 2.5, 2.6, 2.7 and 2.8. □

3 The Tsaranov cases

Let

$$G = \langle x, y, z \mid x^{e_1} = y^{e_2} = z^{e_3} = (x^{\alpha} y^{\beta})^{f_1} = (x^{\gamma} z^{\delta})^{f_2} = (y^{\varepsilon} z^{\eta})^{f_3} = 1 \rangle$$

with

$$2 \leq e_1, e_2, e_3, f_1, f_2, f_3 \text{ and } 1 \leq \alpha, \gamma < r_1,\ 1 \leq \beta, \varepsilon < e_2,\ 1 \leq \delta, \eta < e_3.$$

We call such a presentation a *Tsaranov case.* We are only interested in finite Tsaranov cases which are not equivalent to presentations of finite ordinary tetrahedron groups, since those groups can be considered by the Coxeter criterion. So, in the following, let G be a Tsaranov case which is not equivalent to a presentation of a finite ordinary tetrahedron group.

Theorem 3.1 *Let G be a finite Tsaranov case, not equivalent to a presentation of an ordinary tetrahedron group. Let $2 \le f_1, f_2, f_3 \le 3$ and at least one f_i be equal to 3. Then G is equivalent to the following presentation:*

(6) $\langle x, y, z \mid x^2 = y^5 = z^3 = (xy)^3 = (xz)^2 = (y^2z)^2 = 1\rangle$ *of order* 7200.

Proof G is infinite for $f_1 = f_2 = f_3 = 3$, since

$$\frac{1}{3} + \frac{1}{3} + \frac{1}{3} = 1.$$

So, without loss of generality, let $f_1 = 3$ and $f_2 = 2$ or $f_3 = 2$. We can assume that $e_1 \le e_2$.

If $1/e_1 + 1/e_2 + 1/3 < 1$, then G is infinite; hence let

$$\frac{1}{e_1} + \frac{1}{e_2} + \frac{1}{3} \ge 1.$$

We have that $e_1 \le 3$ (since $e_1 \le e_2$) and we have $e_1 = 2$ if $e_2 \ge 4$. So we can assume that $\alpha = 1$.

If $\beta = 2$ and $e_2 = 4$, or $\beta = 2$, 3 or 4 and $e_2 = 6$, then G_1 admits an essential non-elementary image in $\mathrm{PSL}(2, \mathbb{C})$, and therefore G is infinite by Theorem 1.5. So we can assume that $\beta = 1$ too. We have then

$$G = \langle x, y, z \mid x^{e_1} = y^{e_2} = z^{e_3} = (xy)^3 = (x^\gamma z^\delta)^{f_2} = (y^\varepsilon z^\eta)^{f_3} = 1\rangle$$

with

$$\frac{1}{e_1} + \frac{1}{e_2} + \frac{1}{3} \ge 1$$

and $(f_2, f_3) = (2, 2)$, $(2, 3)$ or $(3, 2)$.

If $e_1 = 3$ then $e_2 = 3$ and G_1 admits an essential infinite metabelian image in $\mathrm{PSL}(2, \mathbb{C})$, and therefore G is infinite if $(e_3, f_2, f_3) \ne (2, 2, 2)$ by Theorem 1.5. For $e_1 = 3 = e_2$ and $(e_3, f_2, f_3) = (2, 2, 2)$, G is equivalent to an ordinary tetrahedron group. So we have now that $e_1 = 2$ and, further, that $\gamma = 1$. This argument is also valid for the case $e_1 = 2$ and $e_2 = 6$, i.e. if $e_2 = 6$ then G is infinite or equivalent to an ordinary tetrahedron group. So we have that $e_1 = 2$ and $2 \le e_2 \le 5$.

I) Let $e_2 = 4$. If

$$\frac{1}{e_2} + \frac{1}{e_3} + \frac{1}{f_3} < 1,$$

then G is infinite. Hence let $1/e_2 + 1/e_3 + 1/f_3 \ge 1$.

A) Assume first that $\varepsilon = 2$. If $f_3 = 3$, then $e_3 = 2$ and G_3 admits an essential non-elementary image in $\mathrm{PSL}(2, \mathbb{C})$, i.e. G is infinite.
So let $f_3 = 2$. If $e_3 = 3$, then G_3 admits an essential non-elementary image in $\mathrm{PSL}(2, \mathbb{C})$ and therefore, G is infinite. If $e_3 = 2$, we introduce the relation $y^2 = 1$ to get an infinite factor group of G. So let $e_3 = 4$. Without loss of generality $1 \le \eta \le 2$. If $\eta = 2$, we introduce the relation $y^2 = 1$ to get an infinite factor group of G. So let $\eta = 1$. If $\delta = 2$ and

$f_2 = 3$, then G_2 admits a non-elementary image in $\mathrm{PSL}(2,\mathbb{C})$; hence G is infinite.
If $\delta = 2$ and $f_2 = 2$, we introduce the relations $y^2 = z^2 = 1$ to get an infinite factor group of G. So assume (without loss of generality) that $\delta = 1$. Then we introduce again the relations $y^2 = z^2 = 1$ to get an infinite factor group of G. So we have that $\varepsilon \neq 2$.

B) Assume now that $\varepsilon = 1$ or 3, and so (without loss of generality) $\varepsilon = 1$. If $e_3 = 4$ and $\delta = 2$ or $\eta = 2$, we can use the same argument as above to see that G is infinite. So assume (without loss of generality) that $\delta = 1 = \eta$ if $e_3 = 4$.
If $e_2 = e_3 = 4$ then $f_3 = 2$, and G_3 admits an essential infinite metabelian image in $\mathrm{PSL}(2,\mathbb{C})$; in particular G is infinite by Theorem 1.5 since $f_3 = 3$. So let $e_3 \neq 4$. Since

$$\frac{1}{e_2} + \frac{1}{e_3} + \frac{1}{f_3} \geq 1$$

we have then that $e_3 \leq 3$, and G is equivalent to an ordinary tetrahedron group.

II) Let $e_2 \neq 4$, $e_3 = 4$. For

$$\frac{1}{e_2} + \frac{1}{e_3} + \frac{1}{f_3} < 1,$$

G is infinite. So let

$$\frac{1}{e_2} + \frac{1}{e_3} + \frac{1}{f_3} \geq 1.$$

Then, necessarily, $e_2 \leq 3$. Analogously to I), G is infinite, if $\delta = 2$ or $\eta = 2$. So let $\delta \neq 2 \neq \eta$. Then G is equivalent to an ordinary tetrahedron group.

III) $e_2 \neq 4 \neq e_3$. If

$$\frac{1}{e_2} + \frac{1}{e_3} + \frac{1}{f_3} < 1 \text{ or } \frac{1}{e_1} + \frac{1}{e_3} + \frac{1}{f_3} < 1,$$

G is infinite. So assume now that

$$\frac{1}{e_2} + \frac{1}{e_3} + \frac{1}{f_3} \geq 1 \text{ and } \frac{1}{e_1} + \frac{1}{e_3} + \frac{1}{f_3} \geq 1.$$

A) Let $e_3 \geq 7$. Then, necessarily, $e_2 = f_2 = f_3 = 2$.
If we set $d := \gcd(\delta, e_3) \geq 2$ and introduce the relation $z^d = 1$, we get an infinite factor group of G. Analogously, G is infinite for $\gcd(\eta, e_3) \geq 2$. If we have $\gcd(\eta, e_3) = \gcd(\delta, e_3) = 1$, then G is equivalent to an ordinary tetrahedron group.

B) Let $e_3 = 6$. Here $(e_2, f_2, f_3) \neq (2,2,2)$ is possible.
If $(e_2, f_2, f_3) = (2,2,2)$, we get, analogously to the case $e_3 \geq 7$, either an infinite factor group, or G is equivalent to an ordinary tetrahedron group.

If $e_2 = 3$, then $f_3 = 2$, and G_3 admits an essential non-elementary image in $\mathrm{PSL}(2,\mathbb{C})$ if $\eta = 2$, 3 or 4, and G_3 admits an essential infinite metabelian image if $\eta = 1$ or 5. In both cases, G is infinite, since $f_1 = 3$. If $f_3 = 3$ then $e_2 = 2$ and we have, analogously, that G is infinite. So let $e_2 = f_3 = 2$ and $f_2 = 3$. If $\delta = 2$, 3 or 4 then G_2 admits an essential non-elementary image in $\mathrm{PSL}(2,\mathbb{C})$, and, if $\delta = 1$ or 5, then G_2 admits an essential infinite metabelian image; G is infinite in both cases, since $f_1 = 3$.
So, in the following, let $e_3 \leq 5$.

C) $e_2 = 2$. Then

$$G = \langle x, y, z \mid x^2 = y^2 = z^{e_3} = (xy)^3 = (xz^\delta)^{f_2} = (yz^\eta)^{f_3} = 1\rangle$$

with $2 \leq e_3 \leq 5$, $e_3 \neq 4$ and $(f_2, f_3) = (2,2)$, $(2,3)$ or $(3,2)$.
If $e_3 = 2$ or 3, then G is equivalent to an ordinary tetrahedron group. Assume now that $e_3 = 5$. If $f_2 = f_3 = 2$, then G is equivalent to an ordinary tetrahedron group. So let $f_2 = 3$; then $f_3 = 2$. We can assume that $\delta = 1$ (replace z with a power) and, afterwards, we can assume that $\eta = 1$ (use the dihedral relation). So, again, G is equivalent to an ordinary tetrahedron group.
The same holds for $(f_2, f_3) = (2,3)$.

D) Here we have

$$G = \langle x, y, z \mid x^2 = y^3 = z^{e_3} = (xy)^3 = (xz^\delta)^{f_2} = (yz^\eta)^{f_3} = 1\rangle$$

with

$$2 \leq e_3 \leq 5,\ e_3 \neq 4,\ 1 \leq \delta, \eta < e_3 \text{ and } (f_2, f_3) = (2,2),\ (2,3) \text{ or } (3,2).$$

If $2 \leq e_3 \leq 3$ then G is equivalent to an ordinary tetrahedron group. So let $e_3 = 5$. Then $f_3 = 2$.
If $f_2 = 2$, then G is equivalent to an ordinary tetrahedron group (use the dihedral relation). So assume now that $f_2 = 3$. We can assume that $\eta = 1$ and $e_1 = 2$, since $1 \leq \delta \leq 2$. For $\eta = \delta = 1$, G is infinite by the Coxeter criterion. For $\eta = 1$, $\delta = 2$, G is infinite by GAP.

E) $e_2 = 5$. Since

$$\frac{1}{5} + \frac{1}{e_3} + \frac{1}{f_3} \geq 1,$$

we have, necessarily, that $2 \leq e_3 \leq 3$ and $f_3 = 2$ if $e_3 = 3$. Then, without loss of generality,

$$G = \langle x, y, z \mid x^2 = y^5 = z^{e_3} = (xy)^3 = (xz)^{f_2} = (y^\beta z)^{f_3} = 1\rangle$$

with $1 \leq \beta \leq 2$.
If $\beta = 1$, then G is equivalent to an ordinary tetrahedron group; so let $\beta = 2$. Assume first that $e_3 = 2$. If $f_3 = 2$, then G is equivalent to

an ordinary tetrahedron group (use the dihedral relation). Assume now that $e_3 = 2$, $f_3 = 3$. Then $f_2 = 2$ and

$$G = \langle x, y, z \mid x^2 = y^5 = z^2 = (xy)^3 = (xz)^2 = (y^2z)^3 = 1\rangle.$$

Here G is infinite by GAP.

So finally assume that $e_3 = 3$. Then $f_3 = 2$, and we have, without loss of generality, up to equivalence,

$$G = \langle x, y, z \mid x^2 = y^5 = z^3 = (xy)^3 = (xz)^{f_2} = (y^2z)^2 = 1\rangle.$$

For $f_2 = 3$, G is infinite by GAP. For $f_2 = 2$ we have the desired presentation with order 7200.

This completes the proof. □

Acknowledgements. The third author would like to thank Hilary Craig for all her help and encouragement.

References

[1] M. Conder, Three-relator quotients of the modular group, *Quart. J. Math. Oxford* **38** (1987), 427–447.
[2] H. S. M. Coxeter, The polytopes with regular-prismatic vertex figures (part 2), *Proc. London Math. Soc.* **34** (1932), 126–189.
[3] H. S. M. Coxeter, Discrete groups generated by reflections, *Ann. Math.* **35** (1934), 588–621.
[4] H. S. M. Coxeter, The complete enumaration of finite groups of the form $R_i^2 = (R_i, R_j)^{k_{ij}} = 1$, *J. London Math. Soc.* **10** (1935), 21–25.
[5] M. Edjvet, J. Howie, G. Rosenberger and R. M. Thomas, Finite generalized tetrahedron groups with a high-power relator, *Geometriae Dedicata* (to appear).
[6] M. Edjvet, G. Rosenberger, M. Stille and R. M. Thomas, On certain finite generalized tetrahedron groups, *in* M. D. Atkinson, N. D. Gilbert, J. Howie, S. A. Linton and E. F. Robertson (eds.), *Computational and Geometric Aspects of Modern Algebra*, London Math. Soc. Lecture Note Series **275**, Cambridge University Press (2000), 54–65.
[7] B. Fine, J. Howie and G. Rosenberger, One-relator quotients and free products of cyclics, *Proc. Amer. Math. Soc.* **102** (1988), 249-254.
[8] B. Fine, F. Levin, F. Roehl and G. Rosenberger, The generalized tetrahedron groups, *in* R. Charney, M. Davis and M. Shapiro (eds.), *Geometric Group Theory*, Ohio State University, Mathematical Research Institute Publications **3**, de Gruyter (1995), 99–119.
[9] B. Fine and G. Rosenberger, *Algebraic Generalizations of Discrete Groups*, Marcel Dekker (1999).
[10] J. Howie, V. Metaftsis and R. M. Thomas, Finite generalized triangle groups, *Trans. Amer. Math. Soc.* **347** (1995), 3613–3623.
[11] F. Levin and G. Rosenberger, On free subgroups of generalized triangle groups, part II, *in* S. Seghal et al. (eds.), *Proceedings of the Ohio State-Denison Conference on Group Theory*, World Scientific (1993), 206–222.
[12] G. Rosenberger, On free subgroups of generalized triangle groups, *Algebra i Logika* **28** (1989), 227–240.
[13] G. Rosenberger and M. Scheer, Classification of the finite generalized tetrahedron groups, *N.Y./Hoboken Proceedings 2001*, to appear.

[14] M. Scheer, *Darstellungen verallgemeinerter Dreiecks- und Tetraedergruppen in Gruppen von* 2×2 *- Matrizen uber Ringen*, Diplomarbeit, 2000.
[15] M. Schonert et al., *GAP - Groups, Algorithms and Programming*, Lehrstuhl D Mathematik RTWH Aachen, Germany (1995).
[16] J. Stallings, Non-positively curved triangles of groups, *in* E. Ghys, A. Haefliger and A. Verjovsky (eds.), *Group Theory from a Geometrical Viewpoint*, World Scientific (1991), 491-503.
[17] S. V. Tsaranov, On a generalization of Coxeter groups, *Algebras, Groups and Geometries* **6** (1989), 281-318.
[18] S. V. Tsaranov, Finite generalized Coxeter groups, *Algebras, Groups and Geometries* **6** (1989), 421-457.

CHARACTER DEGREES OF THE SYLOW p-SUBGROUPS OF CLASSICAL GROUPS

JOSU SANGRONIZ[1]

Universidad del País Vasco, Facultad de Ciencias, Departamento de Matemáticas,
Apartado 644, 48080 Bilbao, Spain
E-mail: mtpsagoj@lg.ehu.es

Abstract

If G is a finite group and q is an integer such that the degree of every irreducible character of G is a power of q, we say that G is a q-power-degree group. The aim of this note is to solve the following problem: Which are the classical groups defined over the finite field $\mathbb{F}_q$ ($\mathbb{F}_{q^2}$ in the unitary case) of characteristic p whose Sylow p-subgroups are q-power-degree groups?

1 Introduction

Throughout this paper p will denote the characteristic of the finite field $\mathbb{F}_q$ with q elements. It is an elementary exercise to check that the set $T_n(q)$ of unitriangular matrices (i. e., upper triangular matrices with ones in the diagonal) over $\mathbb{F}_q$ is a Sylow p-subgroup of the full linear group $GL_n(q)$. Isaacs proved in [4] that the degree of any irreducible character of this group is not only a power of p (as one should expect because we are dealing with a p-group) but actually, a power of q. For the sake of completeness we shall give later a simplified proof of this result.

Previtali [6], using techniques of Isaacs, showed that the same property holds for the Sylow p-subgroups of other families of classical groups (symplectic, unitary and orthogonal of maximal Witt index, always in odd characteristic and even dimension). Our goal is to settle the question of which are exactly the classical groups for which this result of Isaacs can be extended, and the answer is given in the following theorem.

Theorem 1.1 (Main Theorem) *Let $\mathcal{G}$ be any classical group, except symplectic or orthogonal in even characteristic. Then the degrees of the irreducible characters of a Sylow p-subgroup of $\mathcal{G}$ are powers of q. Moreover, this is not true with complete generality in the symplectic and orthogonal cases in even characteristic.*

It deserves to be noticed that the same result holds word for word if we replace 'degrees of the irreducible characters' by 'conjugacy class sizes' (see [7]).

A couple of works have appeared recently related to our theorem. For odd characteristic the same result has been established by Szegedy in [8] and, for even

[1]Research supported by the University of the Basque Country, grant 1/UPV 00127.310-E-13817/2001, and by the Plan Nacional de Investigación Científica, Desarrollo e Innovación Tecnológica (I+D+I) of the Ministerio de Ciencia y Tecnología, grant BFM2001-0180.

characteristic, the behaviour of the Sylow 2-subgroups of the symplectic and orthogonal groups of maximal Witt index has been clarified in [1].

A direct approach to obtain the desired result via Clifford theory fails (even for the unitriangular groups) because it is not clear how to relate the structure of the inertia groups that will arise in the analysis with the structure of the whole group. However, this procedure is useful to see for instance, that a given number (a power of q, of course!) is the degree of some irreducible character, which eventually may lead to the determination of the complete set of degrees of the irreducible characters, taking for granted that all these numbers are powers of q (see [2, 5]).

To circumvent this problem Isaacs enlarged the family of groups under consideration including all the groups of the form $G = 1+J$, where J is a finite dimensional nilpotent $\mathbb{F}_q$-algebra. He called these groups *algebra groups* and he also introduced the term *q-power-degree groups* for the groups whose irreducible characters have degrees that are powers of q. What Isaacs proved is that algebra groups, as well as certain subgroups of them, are q-power-degree groups. The subgroups we have in mind are the so-called *strong* subgroups: a subgroup $S \leq G = 1 + J$ is *strong* if for any subalgebra $\mathcal{H}$ of J, the order of the intersection $S \cap (1 + \mathcal{H})$ is a power of q. Our main theorem will follow from the fact that the groups that appear in it can be realized as strong subgroups of unitriangular groups.

2 Isaacs's result on algebra groups

We present in this section a shortened proof of the key lemma in Isaacs' result.

Lemma 2.1 *Suppose that a group G has normal subgroups M, H_i, $1 \leq i \leq r$, such that the following conditions are satisfied:*

(i) $|G : H_i| = q$ *for all* $1 \leq i \leq r$ *and* $|G : M| = q^2$.

(ii) *The intersection of any two subgroups H_i and H_j is M and the union of all of them is the whole G.*

(iii) *All subgroups H_i and M are q-power-degree groups.*

Then G is also a q-power-degree group.

Proof Some immediate consequences of the hypotheses are that the number of the subgroups H_i is $r = q + 1$ and

$$G = H_i H_j \quad \text{for any} \quad i \neq j. \tag{2.1}$$

Also, condition (iii) together with the fact that $|H_i : M| = q$ imply that for any irreducible character φ of M, either φ^{H_i} is irreducible or it is the sum of the q different extensions of φ to H_i. If R is the inertia group of φ in G, the former case holds when $R \cap H_i = M$ and the latter, when $H_i \subseteq R$. Combining this with (ii) and (2.1), we conclude that the inertia group R must be M, one of the H_i or the full group G.

Suppose now that χ is an irreducible character of G and φ is an irreducible constituent of χ_M. By Clifford's correspondence, $\chi = \eta^G$ for some irreducible character η of R, the inertia group of φ, and $\chi(1) = |G : R|\eta(1)$. At this point, it is clear that $\chi(1)$ is a power of q if R is M or one of the H_i, so we only need to consider the case $R = G$.

Then we have $\chi_M = e\varphi$ for some $1 \leq e \leq q$ and we'll be done once we prove that $e = 1$ or q. By restricting χ to H_i, we obtain that $\chi(1) = f_i|G : R_i|\varphi_i(1)$, where φ_i is an irreducible component of χ_{H_i} and R_i is its inertia group. Moreover, φ_i is an extension of φ, thus their degrees coincide and $e = f_i|G : R_i|$, whence

$$(\chi_{H_i}, \chi_{H_i}) = f_i^2|G : R_i| = ef_i.$$

Now, expanding the right hand side of the equality $1 = (\chi, \chi)$ as a sum running over G and grouping the terms according to the decomposition of G as the disjoint union of M and the sets $H_i \backslash M$, we obtain, after some simplifications, the relation

$$q = -e^2 + e \sum_{i=1}^{q+1} f_i,$$

so $q \geq -e^2 + e(q+1)$ and $(e-q)(e-1) \geq 0$. Since e is between 1 and q, only the cases $e = 1$ or $e = q$ can occur, which finishes the proof. □

Once the spadework has been done in the previous lemma, the results follow quite straightforwardly (see [4, Sections 3 and 4] for the details).

Corollary 2.1 *Algebra groups over $\mathbb{F}_q$ are q-power-degree groups.*

Proof Induct on the order of the group, the abelian case being trivial. If $G = 1 + J$ is not abelian, the dimension of J/J^2 is at least 2. Take a subspace U with $J^2 \subseteq U \subset J$ of codimension 2 and $\mathcal{H}_i$ the hyperplanes of J containing U. Then apply the lemma with $M = 1 + U$ and $H_i = 1 + \mathcal{H}_i$. □

Corollary 2.2 *Let G be an algebra group and $S \leq G$ a strong subgroup. Then S is a q-power-degree group.*

Proof Argue again by induction on the order of G, which can be supposed to be non-abelian. Let M and H_i be as in the proof of the previous corollary. If S is contained in one of the H_i, the result follows from the inductive hypothesis; otherwise apply the lemma in S with the subgroups $S \cap M$ and $S \cap H_i$. □

3 Sylow p-subgroups of the classical groups as strong subgroups

Suppose that $f \colon J \longrightarrow 1 + J$ is a bijection between a nilpotent algebra J and its associated algebra group $G = 1 + J$ and assume that f maps each subalgebra $\mathcal{H}$ of J into its corresponding algebra subgroup $1 + \mathcal{H}$. Note that this condition is not as restrictive as one might think, any polynomial bijection of the type $f(u) = 1 + a_1u + a_2u^2 + \cdots + a_mu^m$ will do. Now, if U is a subspace of J, its image $f(U)$ satisfies that

$$|f(U) \cap (1 + \mathcal{H})| = |f(U) \cap f(\mathcal{H})| = |f(U \cap \mathcal{H})| = |U \cap \mathcal{H}|$$

is a power of q ($U \cap \mathcal{H}$ being also a subspace of J), so whenever $f(U)$ is a subgroup of G, it is in fact a strong subgroup.

On the other hand, almost all polynomial functions from J to $1+J$ are bijective: if f is as before, the only requirement is that $a_1 = f'(0) \neq 0$ (this is so because any formal power series $a_1x + a_2x^2 + \cdots$ with $a_1 \neq 0$ is invertible in the ring of formal power series). In particular, any rational function $f(x) = r(x)/s(x)$ with $s(0) \neq 0$ (which ensures it can be expanded as a power series of x) satisfying $f(0) = 1$ and $f'(0) \neq 0$ can be used to prove that a given subgroup of G is strong.

Now we turn to classical groups. Symplectic groups and orthogonal groups in odd characteristic can be realized as groups of the form

$$\mathcal{G}_B = \{X \in GL_n(q) \mid XBX^t = B\}$$

(here X^t is the transposed matrix of X) for suitable fixed $n \times n$ matrices B. Conjugation by an invertible matrix P has the effect of replacing B by $P^{-1}B(P^{-1})^t$, so since $T_n(q)$ is a Sylow p-subgroup of $GL_n(q)$, there is no loss of generality if we assume that

$$T_B = \{X \in T_n(q) \mid XBX^t = B\}$$

is a Sylow p-subgroup of $\mathcal{G}_B$. To see that T_B is a strong subgroup of the algebra group $T_n(q)$ all we have to do is to find a rational function f and a subspace $\mathcal{T}_B$ of the algebra $\mathcal{T}_n(q)$ associated to $T_n(q)$ (that is, $\mathcal{T}_n(q)$ consists of the strictly upper triangular matrices) such that $T_B = f(\mathcal{T}_B)$. Routine calculations show that, if the characteristic is odd, we can take f to be the rational function $f(x) = (1-x)/(1+x)$ and

$$\mathcal{T}_B = \{X_0 \in \mathcal{T}_n(q) \mid X_0B + BX_0^t = 0\},$$

which is clearly a subspace of $\mathcal{T}_n(q)$. The map f is called the Cayley parametrization of the classical group $\mathcal{G}_B$ (see [9, Lemma 2.10A]).

Similar remarks can be made in the unitary case (in any characteristic): we fix an $n \times n$ matrix B over $\mathbb{F}_{q^2}$ and set ${}^2\mathcal{G}_B = \{X \in GL_n(q^2) \mid XB\overline{X}^t = B\}$ (the bar indicates the automorphism of the group $GL_n(q^2)$ induced by the automorphism of the base field $x \mapsto x^q$). Again, we can suppose that

$${}^2T_B = \{X \in T_n(q^2) \mid XB\overline{X}^t = B\}$$

is a Sylow p-subgroup of ${}^2\mathcal{G}_B$. To define f we take an element $a \in \mathbb{F}_{q^2}$ such that $a\bar{a} = a^{1+q} = 1$ and $a \neq -1$ (such an element always exists, in odd characteristic the obvious choice is $a = 1$). Then we set $f(x) = (1-x)/(1+ax)$. One can check that f satisfies the conditions required and that ${}^2T_B = f({}^2\mathcal{T}_B)$, where ${}^2\mathcal{T}_B$ is the $\mathbb{F}_q$-subspace

$${}^2\mathcal{T}_B = \{X_0 \in \mathcal{T}_n(q^2) \mid aX_0B + B\overline{X}_0^t = 0\}.$$

The following result is now clear from the preceding discussion.

Theorem 3.1 *The groups T_B (for q odd) and ${}^2T_B$ are q-power-degree groups. In particular this happens for the Sylow p-subgroups of the classical groups (except in the symplectic and orthogonal cases in even characteristic).*

4 Classical groups in even characteristic

In [4] (see also [1]) Isaacs proved that if q is a power of 2, the Sylow 2-subgroups of the symplectic group $Sp_6(q)$ have an irreducible character of degree $q/2$, so that Theorem 3.1 can not be extended to the symplectic case when the characteristic is 2. Thus, only the orthogonal cases in characteristic 2 remain to be considered. For a description of the structure of these groups we refer the reader to [7].

The Sylow 2-subgroups of $\Omega^+_{2n}(q)$, the orthogonal group of maximal Witt index and dimension $2n$, can be constructed as the semidirect product G of the additive group $\mathcal{A}$ of the $n \times n$ symmetric matrices with zero diagonal and the unitriangular group $T_n(q)$, where the action is given by $a^X = X^{-1}a(X^{-1})^t$ for $a \in \mathcal{A}$ and $X \in T_n(q)$.

Any $n \times n$ matrix m gives an irreducible character of $\mathcal{A}$ by means of $\zeta_m(a) = \psi(\mathrm{tr}(am))$, where $\psi\colon \mathbb{F}_q \longrightarrow \mathbb{C}^*$ is a fixed non-trivial linear character of the additive group of the field $\mathbb{F}_q$ and $\mathrm{tr}(x)$ gives the trace of the matrix x. In fact, all the irreducible characters of $\mathcal{A}$ arise in this way. Of course, repetitions may occur, to be more precise, $\zeta_m = \zeta_{m'}$ if and only if $m - m'$ is symmetric, that is $m + m^t = m' + (m')^t$. On the other hand, for $X \in T_n(q)$, the conjugate character $(\zeta_m)^X$ equals $\zeta_{X^t m X}$, thus X is in the inertia group of ζ_m if and only if $X^t(m+m^t)X = m+m^t$.

To simplify notation we set $n = 4$, although our construction works as well for any $n \geq 4$ (for $n \leq 3$, the groups are q-power-degree), and we take

$$m = \begin{pmatrix} 0 & 0 & 0 & 1 \\ 0 & 0 & 1 & 0 \\ 0 & 0 & 0 & 0 \\ 0 & 0 & 0 & 0 \end{pmatrix}$$

and $\zeta = \zeta_m$ (for a general n, just add the necessary zero rows and columns). Some computations show that the inertia group of ζ in $T_4(q)$ is

$$I = \left\{ \begin{pmatrix} 1 & x & xu+v & z \\ 0 & 1 & u & v \\ 0 & 0 & 1 & x \\ 0 & 0 & 0 & 1 \end{pmatrix} \mid x, u, v, z \in \mathbb{F}_q \right\} \tag{4.2}$$

and the commutator of the matrices in I corresponding to the parametres x, u, v, z and x', u', v', z' is

$$\begin{pmatrix} 1 & 0 & xu' + ux' & x^2u' + ux'^2 \\ 0 & 1 & 0 & xu' + ux' \\ 0 & 0 & 1 & 0 \\ 0 & 0 & 0 & 1 \end{pmatrix}.$$

From here it is easy to conclude that, for $q > 2$,

$$I' = Z(I) = \left\{ \begin{pmatrix} 1 & 0 & v & z \\ 0 & 1 & 0 & v \\ 0 & 0 & 1 & 0 \\ 0 & 0 & 0 & 1 \end{pmatrix} \mid v, z \in \mathbb{F}_q \right\}.$$

Since I has nilpotence class 2, for any irreducible character η of I, we have $\eta^2(1) = |I : Z(\eta)|$ (see [3, Theorem 2.31]). Now we make a suitable choice for η. The set $L = \{x + x^2 \mid x \in \mathbb{F}_q\}$ is an additive subgroup of $\mathbb{F}_q$ of index 2, so the matrices in $Z(I)$ for which $v + z \in L$ make up a subgroup K of $Z(I)$ of index 2 and there exists a linear character φ of $Z(I)$ with kernel K. We take η to be an irreducible constituent of φ^I. The condition for an element $X \in I$ to be in the center of η is that $[X, I] \subseteq \operatorname{Ker} \eta$, but $[X, I] \subseteq I' = Z(I)$, so this is equivalent to $[X, I]$ being inside $\operatorname{Ker} \varphi = K$. We conclude that a matrix like in (4.2) is in $Z(\eta)$ if and only if $(x + x^2)u' + u(x' + x'^2) \in L$ for all $x', u' \in \mathbb{F}_q$, which means that $x = 0$ or 1 and $uL \subseteq L$. Note that this last condition forces $u = 0$ or 1: if $u \neq 0$, we can regard the map $x \mapsto ux$ as an $\mathbb{F}_2$-linear semisimple operator of $\mathbb{F}_q$, so a one $\mathbb{F}_2$-dimensional invariant subspace must exist (to complement L) and this implies $u = 1$. To summarize, the center of η consists of the matrices in (4.2) with $u, x = 0$ or 1. In particular, $|Z(\eta)| = 4q^2$ and $\eta(1) = |I : Z(\eta)|^{1/2} = q/2$.

Now we use Clifford theory to obtain the desired character of our group G. The inertia group of ζ in G is the (semidirect) product $R = I\mathcal{A}$ and ζ has an obvious extension $\widehat{\zeta}$ to R. By Gallagher's theorem [3, Corollary 6.17], the product $\eta\widehat{\zeta}$ is an irreducible constituent of ζ^R so, by Clifford's correspondence, it induces irreducibly to G. The degree of the resulting character is $|G : R|\eta(1) = q^3/2$, which is not a power of q if $q > 2$. (For a general $n \geq 4$, one obtains a character of degree $q^{4n-13}/2$.)

We finally consider the orthogonal groups $\Omega^-_{2n+2}(q)$ of non-maximal Witt index. A Sylow 2-subgroup G of this group can be described as follows. We take the matrices

$$S = \begin{pmatrix} 0 & 1 \\ 1 & 0 \end{pmatrix} \quad \text{and} \quad C = \begin{pmatrix} \alpha & 0 \\ 1 & \alpha \end{pmatrix},$$

where $\alpha \in \mathbb{F}_q$ is such that the polynomial $\alpha x^2 + x + \alpha$ is irreducible and define M to be the set of matrices with a block decomposition of the type

$$\begin{pmatrix} I_n & u & a \\ 0 & I_2 & Su^t \\ 0 & 0 & I_n \end{pmatrix}, \tag{4.3}$$

where I_n denotes the $n \times n$ identity matrix and u and a are blocks of size $n \times 2$ and $n \times n$, respectively, which satisfy the condition $a + uCu^t \in \mathcal{A}$ ($\mathcal{A}$ is as before). As one can quickly check, M is a subgroup of $T_{2n+2}(q)$ and $\mathcal{A}$ is naturally embedded in it. The commutator of the matrices in M with blocks u, a and u', a' is

$$\begin{pmatrix} I_n & 0 & uSu'^t + u'Su^t \\ 0 & I_2 & 0 \\ 0 & 0 & I_n \end{pmatrix}$$

and we deduce form this that $M' = \mathcal{A} = Z(M)$. Now, conjugation by

$$\begin{pmatrix} X & 0 & 0 \\ 0 & I_2 & 0 \\ 0 & 0 & (X^{-1})^t \end{pmatrix}, \quad X \in T_n(q),$$

on (4.3) simply replaces u by $X^{-1}u$ and a by $X^{-1}a(X^{-1})^t$, thus $T_n(q)$ acts on M and we can consider the corresponding semidirect product, which turns out to be (isomorphic to) our group G.

Again, for the sake of simplicity, we take $n = 4$ although our argument can be extended to any $n \geq 4$. We fix m, ζ and I as in the previous case. Since $\mathcal{A}$ is central in M, it is clear that the inertia group of ζ in G is $R = IM$. We introduce the subgroup H of M which consists of the matrices (4.3) for which the last two rows of u are zero. Multiplication by elements in $T_4(q)$ preserves this property, so $T_4(q)$ normalizes H, that is, H is a normal subgroup of G. If $h \in H$ is like in (4.3) and $X \in I$, we define

$$\widehat{\zeta}(Xh) = \psi(\mathrm{tr}(am)).$$

Direct computations show that $\widehat{\zeta}$ is a character of $L = IH$, which obviously extends ζ. We put $\varphi = \widehat{\zeta}^R$. Note that $R = LM$ and $L \cap M = H$, so $\varphi_M = (\widehat{\zeta}^R)_M = (\widehat{\zeta}_H)^M$. An element $y \in M$ fixes $\widehat{\zeta}_H$ if and only if $[y, H] \subseteq \mathrm{Ker}\,\widehat{\zeta}$, a condition equivalent to $[y, H] \subseteq \mathrm{Ker}\,\zeta$ because $[y, H] \subseteq \mathcal{A}$. Moreover, $[y, H]$, viewed inside the vector space $\mathcal{A}$, is a subspace, so the last containment amounts to saying that $\mathrm{tr}([y, h]m) = 0$ for all $h \in H$ and it follows from here that y itself is in H. We have proved that the inertia group of $\widehat{\zeta}_H$ in M is H, whence $\varphi_M = (\widehat{\zeta}_H)^M$ is irreducible and so is φ. By Gallagher's theorem the characters $\eta\varphi$, for η any irreducible character of R/M, are the irreducible constituents of $(\varphi_M)^R$. But, as we know, $R/M \cong I$ has irreducible characters of degree $q/2$, so one of these constituents has degree $q^5/2$. By inducing it to G we obtain an irreducible character of G of degree $q^7/2$ (irreducibility is guaranteed by Clifford's correspondence). (For a general $n \geq 4$, one can obtain an irreducible character of degree $q^{4n-9}/2$.)

References

[1] R. Gow, M. Marjoram, A. Previtali, On the irreducible characters of a Sylow 2-subgroup of the finite symplectic group in characteristic 2, *J. Algebra* **241** (2001), 393-409.

[2] B. Huppert, A remark on the character-degrees of some p-groups, *Arch. Math.* **59** (1992), 313-318.

[3] I. M. Isaacs, *Character theory of finite groups*, Dover, New York, 1994.

[4] I. M. Isaacs, Characters of groups associated with finite algebras, *J. Algebra* **177** (1995), 708-730.

[5] A. Moretó, An elementary method for the calculation of the set of character degrees of some p-groups, *Rend. Circ. Mat. Palermo* Serie II, Tomo L (2001), 329-333.

[6] A. Previtali, On a conjecture concerning character degrees of some p-groups, *Arch. Math.* **65** (1995), 375-378.

[7] A. Previtali, Maps behaving like exponentials and maximal unipotent subgroups of groups of Lie type, *Comm. Algebra* **27** (1999), 2511-2519.

[8] B. Szegedy, Characters of the Borel and Sylow subgroups of the classical groups, Durham conference, July 2001.

[9] H. Weyl, *The classical groups*, Princeton University Press, Princeton, 1966.

CHARACTER CORRESPONDENCES AND PERFECT ISOMETRIES

LUCÍA SANUS[1]

Departament d'Àlgebra, Facultat de Matemàtiques, Universitat de València,
46100 Burjassot, València, Spain
E-mail: *lucia.sanus@uv.es*

Suppose that S acts coprimely on G. Write $C = \mathrm{C}_G(S)$ for the fixed points subgroup and $\mathrm{Irr}_S(G)$ for the set of S-invariant irreducible complex characters of G. It is well known that under these hypothesis there is a natural bijection

$$\hat{}: \mathrm{Irr}_S(G) \to \mathrm{Irr}(C)\,.$$

This map (the **Glauberman-Isaacs correspondence**) was constructed by G. Glauberman (when S is solvable [2]), and by I. M. Isaacs (when S is nonsolvable, and therefore $|G|$ is odd). Later on T. Wolf proved in [8] that both bijections were the same when both were defined.

Now, fix a prime p, and notice that S permutes the set of Brauer p-blocks of G. Although the Glauberman-Isaacs correspondence preserves many properties of characters, it does not preserve S-invariant characters in the same block: if $\chi, \psi \in \mathrm{Irr}_S(G)$ lie in the block B, it is not necessarily true that $\hat{\chi}$ and $\hat{\psi}$ lie in the same p-block of C (see [9] for more details). This anomalous behaviour does not occur when S centralizes some defect group of B by a remarkable result of A. Watanabe. In fact, she proves much more.

Theorem 0.1 (A. Watanabe) *Suppose that S solvable acts coprimely on G and let B be an S-invariant p-block of G. If S centralizes a defect group D of B, then all $\mathrm{Irr}(B)$ are S-invariant and there is a block b of C with defect group D such that $\{\hat{\chi} \mid \chi \in \mathrm{Irr}(B)\} = \mathrm{Irr}(b)$. Furthermore, $\chi \mapsto \pm\hat{\chi}$ defines a perfect isometry.*

(The important notion of a perfect isometry is due to M. Broué and we refer the reader to [1] for the definition and main properties.)

Of course, Watanabe's theorem left one case open: when $|G|$ is odd (and $\hat{}$ is the Isaacs correspondence). This case was proven by H. Horimoto in [3].

Another well-known character correspondence of characters was discovered by M. Isaacs in [4]. If G is a group and q is a prime, we write $\mathrm{Irr}_{q'}(G)$ for the irreducible characters of G of degree not divisible by q.

[1] Research partially supported by DGICYT.

Let $Q \in \mathrm{Syl}_q(G)$ and let G be a group of odd order. Isaacs proved in [4] that there is a natural bijection

$$^* : \mathrm{Irr}_{q'}(G) \to \mathrm{Irr}_{q'}(\mathrm{N}_G(Q))$$

(proving a stronger form of the McKay conjecture for groups of odd order).

Again, this correspondence does not preserve q'-degree irreducible characters in the same p-block. However, in our new situation, we can find the analogous condition to Watanabe's that guarantees that the Isaacs correspondence is a perfect isometry. This is the result that we present in this note.

Theorem 0.2 *Let B be a p-block of a group G of odd order and assume that* $\mathrm{Irr}(B) \subseteq \mathrm{Irr}_{q'}(G)$. *Then there is a unique p-block of* $\mathrm{N}_G(Q)$ *such that* $\{\chi^* \mid \chi \in \mathrm{Irr}\, B\} = \mathrm{Irr}(b)$. *The defect group of b is a defect group of B and the map $\chi \mapsto \chi^*$ is a perfect isometry.*

Proof This is the main result of [6]. □

Implicit in Theorem 2, we notice the fact that if q does not divide $\chi(1)$ for all $\chi \in \mathrm{Irr}(B)$, then a defect group of B normalizes a Sylow q-subgroup of G. This result, which is highly non-trivial, is the main result of [5]. Since it is not difficult to show that if q does not divide $\chi(1)$ for some $\chi \in \mathrm{Irr}(B)$ and a defect group of B normalizes a Sylow q-subgroup of G, then all $\mathrm{Irr}(B)$ have q'-degree we have that the above theorem is the analogous condition to Watanabe's in Theorem 1.

References

[1] M. BROUÉ, Isométries parfaites, types de blocks, catégories dérivées, *Astérisque.* **181-182** (1990), 61-92.

[2] G. GLAUBERMAN, Correspondence of characters for relatively prime operator groups, *Canad. J. Math.* **20** (1968), 1465-1488.

[3] H. HORIMOTO, On correspondence between blocks of finite groups induced from the Isaacs correpondence, *Hokkaido Math. J.* **30** (2001),65-74.

[4] I. M. ISAACS, Characters of solvable and symplectic groups, *Amer. J. Math.* **95** (1973), 594-635.

[5] G. NAVARRO, T. WOLF, Characters degrees and blocks of finite groups, J. reine angew. Math **531** (2001), 141-146.

[6] L. SANUS, Perfect isometries and the Isaacs correspondence, (submitted).

[7] A. WATANABE, The Glauberman character correspondence and perfect isometries for blocks of finite groups, *J. Algebra* **216** (1999), 548-565.

[8] T. WOLF, Character correspondences in solvable groups, *Illinois J. Math.* **22** (1978), 327-340.

[9] T. WOLF, Character correspondences and π-special characters in π-separable groups, *Can. J. Math.* **4** (1987), 920-937.

THE CHARACTERS OF FINITE PROJECTIVE SYMPLECTIC GROUP PSp$(4, q)$

M. A. SHAHABI* and H. MOHTADIFAR†

*Department of Mathematics, Tabriz University Tabriz, Iran
E-mail: Shahabi@tbrizu.ac.ir
† Department of Mathematics Tabriz University Tabriz, Iran.
E-mail: h_mohtadifar@hotmail.com

Abstract

In this article we obtain all the complex irreducible characters and the conjugacy classes of the group PSp$(4, q)$, where q is a power of an odd prime. [1]

1 Introduction

The complex irreducible characters of the symplectic group Sp$(4, q)$ with q odd prime power , has been calculated by B. Srinivasan [3]. But character tables of [3] have some misprints. Fortunately Professor B. Srinivasan and Dr Frank Luebeck provided us the corrections of these misprints. Now you can see in appendix I all of the necessary corrections. By using [3], we investigate the collection of the conjugacy classes of the group Sp$(4, q)$, one by one, to find conjugacy classes of PSp$(4, q)$ and the size of these classes. Then, we will able to obtain the character table of PSp$(4, q)$.

2 Notation

We will use the notation of [3] throughout this paper. The symplectic group Sp$(4, q)$ of dimension 4 over finite fields of odd order, $F_q = GF(q)$, is the set of all 4×4 matrices X which satisfies $XJX' = J$, where

$$J = \begin{pmatrix} 0 & 1 & 0 & 0 \\ -1 & 0 & 0 & 0 \\ 0 & 0 & 0 & 1 \\ 0 & 0 & -1 & 0 \end{pmatrix}.$$

The order of Sp$(4, q)$ is $q^4(q^2 - 1)(q^4 - 1)$, where q is a power of an odd prime p, and its center is $Z = \{I, -I\}$, where I is the identity matrix. The quotient group Sp$(4, q)/Z$ is called projective symplectic group. It is denoted by PSp$(4, q)$.

Let κ be a generator of the multiplicative group of $GF(q^4)$, and let $\zeta = \kappa^{q^2-1}$, $\theta = \kappa^{q^2+1}$, $\eta = \theta^{q-1}$ and $\gamma = \theta^{q+1}$. We choose a fixed monomorphism from the multiplicative group of the $GF(q^4)$ into the multiplicative group of the complex

[1] We would like to thank Professor B. Srinivasan and Dr Frank Luebeck for providing us, the corrections of the character table of Sp$(4, q)$ and remarks.

numbers, and let $\tilde{\zeta}$, $\tilde{\theta}$, $\tilde{\eta}$ and $\tilde{\gamma}$ be the images of ζ, θ, η and γ respectively under this monomorphism. Let R_1 be the set of $\frac{1}{4}(q^2-1)$ distinct positive integers i, such that all of the scalers ζ^i, ζ^{-i}, ζ^{qi} and ζ^{-qi} be distinct, and also let R_2 be the set of $\frac{1}{4}(q-1)^2$ distinct positive integers i, such that all of the scalers θ^i, θ^{-i}, θ^{qi} and θ^{-qi} be distinct. Assume $T_1 = \{1, 2, ..., \frac{1}{2}(q-3)\}$, $T_2 = \{1, 2, ..., \frac{1}{2}(q-1)\}$, $\alpha_j = \tilde{\gamma}^j + \tilde{\gamma}^{-j}$, $\beta_j = \tilde{\eta}^j + \tilde{\eta}^{-j}$, $\tilde{\epsilon} = -s(s + \sqrt{sq})$ and $\tilde{\epsilon}' = -s(s - \sqrt{sq})$, where $s = (-1)^{(q-1)/2}$.

First we calculate conjugacy classes of PSp$(4, q)$.

3 Conjugacy classes of PSp$(4, q)$

In this section we will determine the conjugacy classes of PSp$(4, q)$. To do this we will use conjugacy classes and character table of the group Sp$(4, q)$. One can see both the character table and the conjugacy classes of the group Sp$(4, q)$ in [3]. (see appendix I the corrected table.)

We choose a representative X of a conjugacy class A and then multiply it by $-I$. Then we search to see that, $-X$ lies in which conjugacy classes of Sp$(4, q)$. If both X and $-X$ lie in A, then the canonical image of A in PSp$(4, q)$ will form an independent conjugacy class, where its size is half of the size of the conjugacy class A. If X and $-X$ lie in the distinct classes A and A' respectively, then any element of A' will be the additive inverse of some element in A, and so the canonical image of A and A' coincide in PSp$(4, q)$. Therefore, they make only one class, $\bar{A}$, in PSp$(4, q)$, where its size is the same as A. But this method does not work for any class of Sp$(4, q)$. For, in some classes it needs lots of work to see that $-X$ lies in which classes of Sp$(4, q)$. Therefore, we have to use the character table of the group Sp$(4, q)$. We consider those irreducible characters of Sp$(4, q)$, which their values in I and $-I$ are equal. Then these irreducible characters also will be the irreducible characters of PSp$(4, q)$ [See I, page 24]. Therefore, if the values of these irreducible characters are the same in some conjugacy classes of Sp$(4, q)$, then the canonical images of these classes form an independent class in PSp$(4, q)$ and the size of this new conjugacy class is half of the sum of the sizes of these merged classes. But, also this method can not work for any classes in Sp$(4, q)$. For, the existence of a large numbers of the irreducible characters of Sp$(4, q)$ and the n-th roots of unity in their values is an obstacle, so we need to do more work in some classes.

Now by using above methods and notations of [3], we will obtain the conjugacy classes of PSp$(4, q)$. Note that we will use only one notation for a class and its representative.

3.1 The classes A and A'

By [3], it is easy to see that, any conjugacy class A' with any index is the additive inverse of A with the same index. Therefore, from two classes A and A' with the same index , we obtain a class $\bar{A}$ with the same index in PSp$(4, q)$, where its size is equal to the size of class A.

3.2 Collections of the classes B

From the table of the conjugacy classes in [3], one can see that, the conjugacy classes of type B have been classified with respect to their indices. So we will consider each of them separately.

3-2.1 Collection of the classes B_1

By [3], we have

$$B_1(i) = \text{Diag}(\zeta^i, \zeta^{-i}, \zeta^{qi}, \zeta^{-qi});\ i \in R_1\ \&\ \zeta^{q^2+1} = 1.$$

Note : In [3], the set R_1 is misprinted. For, with this definition of R_1, the conjugacy classes $B_1(i)$ can not be distinct. But, by new definition of R_1(see section 2), the classes $B_1(i)$ with $i \in R_1$ are distinguished by the corresponding irreducible characters $\chi_1(i)$ with $i \in R_1$.
It is easy to see that each conjugacy class $B_1(i)$ with $i \in R_1$ is determined with the set

$$\{\pm i,\ \pm qi\}\ mod\ (q^2+1).\ eqno(1)$$

Since $\zeta^{\pm(q^2+1)/2} = -1$, if we multiply $B_1(i)$ by $-I$, then modulo (q^2+1), the set (1) is transformed to the set

$$\{\pm i \mp (q^2+1)/2,\ \pm qi \mp (q^2+1)/2\}. \tag{2}$$

But, $\frac{1}{2}(q^2+1)$ is odd, therefore, if i is odd(even), then $-i+\frac{1}{2}(q^2+1)$ is even(odd). It follows that, the elements of the classes $B_1(i)$ with $i \in R_1$ and i odd, and the elements of the classes $B_1(i)$ with $i \in R_1$ and i even, are additive inverses of each other. Hence their canonical images are merged in $\text{PSp}(4, q)$. So we obtain $\frac{1}{8}(q^2+1)$ classes $\bar{B}_1(i)$ with $i \in R_1$ and i even, where their sizes are is the same as $B_1(i)$.

3-2.2 Collection of the classes B_2

By [3], we have

$$B_2(i) = \text{Diag}(\theta^i,\ \theta^{-i},\ \theta^{qi},\ \theta^{-qi});\ i \in R_2\ \&\ \theta^{q^2-1} = 1.$$

Like $B_1(i)$ each class $B_2(i)$ with $i \in R_2$ is determined with the set

$$\{\pm i,\ \pm qi\}\ mod\ (q^2-1). \tag{3}$$

Since $\theta^{\pm(q^2-1)/2} = -1$, if we multiply $B_2(i)$ by $-I$, then modulo (q^2-1), the set (3) is transformed to the set

$$\{\pm i \mp (q^2-1)/2,\ \pm qi \mp (q^2-1)/2\}. \tag{4}$$

Now we consider two cases :
a) If $i \in R_2$ satisfies the following equation

$$i - (q^2-1)/2 \equiv \pm qi\ mod\ (q^2-1), \tag{5}$$

then the conjugacy class $B_2(i)$ contains the additive inverse of each of its elements. Hence the size of its canonical image $\bar{B}_2(i)$ is half of the size of the class $B_2(i)$.

The congruent equation (5) have $\frac{1}{2}(q-1)$ distinct solution in R_2. If we denote this set of solutions by R'_2, then in PSp$(4,q)$ we have $\frac{1}{2}(q-1)$ conjugacy classes $\bar{B}_2(i)$ with $i \in R'_2$ such that their sizes are $\frac{1}{2}q^4(q^4-1)$.
b) If $i \in R_2$ dose not satisfy the equation (5), then $-B_2(i)$ is another class from the remaining class of this collection of classes. In other word, there is a $j \in R_2 - R'_2$ with $j \neq i$, such that $-B_2(i)$=$B_2(j)$. Thus, these two classes are merged in PSp$(4,q)$ under the canonical homomorphism. Therefore, we have $\frac{1}{8}(q-1)(q-3)$ classes $\bar{B}_2(i)$, $i \in R''$ with the size $q^4(q^4-1)$, where R'' is the set of $\frac{1}{8}(q-1)(q-3)$ positive integers $i \in R - R'$, such that all of the scalers $\theta^i, \theta^{-i}, \theta^{qi}$ *and* θ^{-qi} are distinct, without considering their signs.

3-2.3 Collection of the classes B_3

By [3], we have

$$B_3(i,j) = \mathrm{Diag}(\gamma^i, \gamma^{-i}, \gamma^j, \gamma^{-j});\ \gamma^{q-1} = 1,\ i,j \in T_1\ \&\ \ i < j.$$

Each class $B_3(i,j)$ is determined by the set

$$\{\pm i,\ \pm j\}\ mod\ (q-1). \tag{6}$$

If we multiply $B_3(i,j)$ by $-I$, then the set (6) is transformed to the set

$$\{\pm i + \frac{q-1}{2},\ \pm j + \frac{q-1}{2}\}, \tag{7}$$

since $\gamma^{(q-1)/2} = -1$. Now modulo (q-1), the set (7) will be

$$\{\pm\frac{q-1}{2} \mp i,\ \pm\frac{q-1}{2} \mp j\}. \tag{8}$$

Therefore the classes $B_3(i,j)$ and $-B_3(i,j)$ are equal, if and only if $\frac{1}{2}(q-1) - i = j$, or if and only if $\frac{1}{2}(q-1) = i + j$. Note that the classes $B_3(i,j)$ and $B_3(j,i)$ are equal. For, $EB_3(i,j)E = B_3(i,j)$, where

$$E = \begin{pmatrix} 0 & 0 & 1 & 0 \\ 0 & 0 & 0 & 1 \\ 1 & 0 & 0 & 0 \\ 0 & 1 & 0 & 0 \end{pmatrix}.$$

Therefore, we have two cases :
a) For q=4k+1, the classes $B_3(i,j)$ and $-B_3(i,j)$ are equal if and only if $i = 1, 2, ..., \frac{1}{4}(q-5)$ and $j = \frac{1}{2}(q-1) - i$. Thus each of the $\frac{1}{4}(q-5)$ classes

$$B_3(i,j);\ i = 1, 2, ..., \frac{1}{4}(q-5)\ ,\ j = \frac{1}{2}(q-1) - i,$$

contains the additive inverse of each element of itself. Hence the canonical images of these classes in PSp$(4,q)$ form an independent class $\bar{B}_3(i,j)$, where its size equals to half of the size of $B_3(i,j)$. On the other hand, in the remaining classes of the collection B_3, we have the following properties. For any $B_3(i,j)$ with $i+j<\frac{1}{2}(q+1)$, if we multiply it by $-I$, then we obtain the class $B_3(i',j')$ with $i'+j'>\frac{1}{2}(q-1)$ where $i'=\frac{1}{2}(q-1)-j$ *and* $j'=\frac{1}{2}(q-1)-i$. Hence, the canonical images of these two classes coincide in $\mathrm{PSp}(4,q)$ and form a class with the same size as $B_3(i,j)$. Consequently, from the remaining classes, we obtain $\frac{1}{16}(q-1)^2$ distinct classes

$$\bar{B}_3(i,j);\ 1\leq i\leq \frac{q-5}{4}\ \&\ i<j\leq \frac{q-3}{4}-i,$$

in $\mathrm{PSp}(4,q)$ with the same size as $B_3(i,j)$.
b)For q=4k+3, two classes $B_3(i,j)$ and $-B_3(i,j)$ are equal, if and only if $i=1,2,...,k$ and $j=\frac{1}{2}(q-1)-i$. So with the similar argument as in case (a), in $\mathrm{PSp}(4,q)$ we obtain $k=\frac{1}{4}(q-3)$ classes $\bar{B}_3(i,j)$ with the size equal to the half of the size of $B_3(i,j)$, where $i=1,2,...,\frac{1}{4}(q-3)$ and $j=\frac{1}{2}(q-1)-i$ and also $\frac{1}{16}(q-3)(q-7)$ classes $\bar{B}_3(i,j)$ with the same size as $B_3(i,j)$, where $1\leq i\leq\frac{1}{4}(q-7)$ and $i<j\leq\frac{1}{4}(q-3)-i$.

3-2.4 Collection of the classes B_4

By [3], we have

$$B_4(i,j)=\mathrm{Diag}(\eta^i,\eta^{-i},\eta^j,\eta^{-j});\ \eta^{q+1}=1,\ i,j\in T_2\ \&\ i<j.$$

With the similar argument as in section 3-2.2, we see that, if $q=4k+1$ ($q=4k+3$), we have $k=\frac{1}{4}(q-1)$ ($k=\frac{1}{4}(q-3)$) classes

$$B_4(i,j);\ i=1,2,...,k\ \&\ j=\frac{1}{4}(q+1)-i,$$

where each class $B_4(i,j)$ contains the additive inverses of its elements. So the canonical images of these classes are distinct and their sizes are divided by 2. But, from the remaining classes of this collection we obtain the following classes in $\mathrm{PSp}(4,q)$. If $q=4k+1$, then, we have $\frac{1}{16}(q-1)(q-5)$ classes $\bar{B}_4(i,j)$; $1\leq i\leq\frac{1}{4}(q-5)$ and $i<j\leq\frac{1}{2}(q-1)-i$. If $q=4k+3$, then we have $\frac{1}{16}(q-3)^2$ classes $\bar{B}_4(i,j)$; $1\leq i\leq\frac{1}{4}(q-3)$ and $i<j<\frac{1}{2}(q-1)-i$, where in both cases the size of $\bar{B}_4(i,j)$ is the same as $B_4(i,j)$.

3-2.5 Collection of classes B_5

We have,

$$B_5(i,j)=\mathrm{Diag}(\eta^i,\eta^{-i},\gamma^j,\gamma^{-j});\ \eta^{q+1}=1=\gamma^{q-1},\ i\in T_2\ \&\ j\in T_1$$

By multiplying $B_5(i,j)$ by $-I$, we obtain the class $B_5(\frac{q+1}{2}-i,\ \frac{q-1}{2}-j)$. It is easy to see that these two classes are distinct. (Note that, if equality occurs, then $i=\frac{1}{4}(q-1)$ and $j=\frac{1}{4}(q-3)$. But, for such an i and j there is no integral solution for q). Consequently, this collection of classes contains $B_5(i,j)$ and $-B_5(i,j)$, for each $i\in T_2$ and $j\in T_1$. Hence, the canonical image of them coincide in $\mathrm{PSp}(4,q)$. Thus, in $\mathrm{PSp}(4,q)$ there are a number of $\frac{1}{8}(q-1)(q-3)$ distinct classes $\bar{B}_5(i,j)$ with size $q^4(q^4-1)$ where; if $q=4k+1$, then $1\leq i\leq\frac{1}{4}(q-1)$ and $1\leq j\leq\frac{1}{2}(q-3)$, and if $q=4k+3$, then $1\leq i\leq\frac{1}{2}(q-1)$ and $1\leq j\leq\frac{1}{4}(q-3)$.

3-2.6 Collection of classes B_6

By [3], we have

$$B_6(i) = \text{Diag}(\eta^i, \eta^{-i}, \eta^i, \eta^{-i});\ \eta^{q+1} = 1\ \&\ \in T_2.$$

We determine each class $B_6(i)$ in this collection by the set

$$\{\pm i\}\ mod\ (q+1). \tag{9}$$

While the class $-B_6(i)$ is determined by the set

$$\{\pm\frac{q+1}{2} \mp i\}\ mod\ (q+1). \tag{10}$$

We see that modulo $(q+1)$, the sets (9) and (10) are distinct, unless, $i = \frac{1}{4}(q+1)$ and $q = 4k+1$. Therefore, we have two cases :
a) If $q = 4k+1$, then under the canonical homomorphism, the classes $B_6(i)$, $i \in T_2$, will give $\frac{1}{4}(q-1)$ classes $\bar{B}_6(i)$, $i = 1, 2, ..., \frac{1}{4}(q-1)$ with the sizes $q^3(q-1)(q^2+1)$ in PSp(4, q).
b) If $q = 4k+3$, then the canonical image $\bar{B}_6(\frac{q+1}{4})$ of the class $B_6(\frac{q+1}{4})$ has the size $\frac{1}{2}q^3(q-1)(q^2+1)$. Under the canonical homomorphism, from $\frac{1}{2}(q-3)$ remaining classes, half of them merge with another half, and we obtain $\frac{1}{4}(q-3)$ classes $\bar{B}_6(i)$, $i = 1, 2, ..., \frac{1}{4}(q-3)$ with the sizes $q^3(q-1)(q^2+1)$ in PSp(4, q).

We postpone $B_7(i)$ to be after $B_8(i)$, since in the investigation of this class and $B_9(i)$ we use method 2.

3-2.7 Collection of the classes B_8

By [3], we have

$$B_8(i) = \text{Diag}(\gamma^i, \gamma^{-i}, \gamma^i, \gamma^{-i});\ \gamma^{q-1} = 1\ \&\ i \in T_1$$

By the Similar argument of $B_6(i)$, we have :
a) If $q = 4k+1$, then the classes $B_8(\frac{q-1}{4})$ causes a class $\bar{B}_8(\frac{q-1}{4})$ with the size $\frac{1}{2}q^3(q+1)(q^2+1)$ in PSp(4, q). From the remaining classes, we obtain classes $\bar{B}_8(i), i = 1, 2, ..., \frac{1}{4}(q-5)$ with the size $q^3(q+1)(q^2+1)$ in PSp(4, q).
b) If $q = 4k+3$, then from the collection of the classes $B_8(i)$, we obtain $\frac{1}{4}(q-3)$ classes $\bar{B}_8(i)$; $i = 1, 2, ..., \frac{1}{4}(q-3)$ with the size $q^3(q+1)(q^2+1)$ in PSp(4, q).

3-2.8 Collection of classes B_7

To investigate the situation of the canonical images of $B_7(i)$ with $i \in T_2$, in PSp(4, q), we consider the values of those irreducible characters of $Sp(4, q)$ in $B_7(i)$ which have equal value in $\pm I$. For, these are also irreducible character of PSp(4, q). Again, we have two cases :
a) If $q = 4k+1$, then the irreducible characters which have equal values in $\pm I$, also have equal values in $B_7(i)$ and $B_7(\frac{q+1}{2} - i)$. For, the values of these irreducible characters in this collection are determined by $\beta_{ik} = \tilde{\eta}^{ik} + \tilde{\eta}^{-ik}$. Since $\eta^{(q+1)/2} = -1$,

we have $\beta_{(\frac{q+1}{2}-i)k} = (-1)^k\beta_{ik}$ and in our above mentioned characters $(-1)^k$ is cancelled. For example, the value of the irreducible character $\xi'_{21}(k)(k$ is odd) in $B_7(i)$ is equal to $(-1)^i\beta_{ik}$ and in $B_7(\frac{q+1}{2} - i)$ is equal to $(-1)^{\frac{q+1}{2}-i}\beta_{(\frac{q+1}{2}-i)k} = (-1)^{\frac{q+1}{2}}(-1)^k(-1)^i\beta_{ik} = (-1)^i\beta_{ik}$, since, k and $\frac{1}{2}(q+1)$ are odd. Therefore, the canonical images of these classes coincide in $\mathrm{PSp}(4,q)$. Thus, in $\mathrm{PSp}(4,q)$, we have $\frac{1}{4}(q-1)$ classes $\bar{B}_7(i)$ with the size $q^3(q-1)(q^4-1)$, where $i = 1, 2, ..., \frac{1}{4}(q-1)$.
b) If $q = 4k+3$, then all of the classes $B_7(i)$ are distinguished by some of the above mentioned irreducible characters. For example, according to (a), the values of $\xi'_{21}(k)$ in $B_7(i)$ and $B_7(\frac{q+1}{7} - i)$ are $(-1)^i\beta_{ik}$ and $-(-1)^i\beta_{ik}$ respectively. Therefore, there are $\frac{1}{2}(q-1)$ classes $\bar{B}_7(i)$ with the size $\frac{1}{2}q^3(q-1)(q^4-1)$ in $\mathrm{PSp}(4,q)$.

3-2.9 Collection of classes B_9

By the similar argument of $B_7(i)$, if we consider the character table of $Sp(4,q)$, we can see that for any $i \in T_1$ the classes $B_9(i)$ and $B_9(\frac{q-1}{2} - i)$ have the same values in all of the corresponding irreducible characters of $Sp(4,q)$, so their canonical image coincide in $\mathrm{PSp}(4,q)$. Therefore,
a) If $q = 4k+1$, then we obtain $\frac{1}{4}(q-5$ classes $\bar{B}_9(i)$; $i = 1, 2, ..., \frac{1}{4}(q-5)$ with the size $q^3(q+1)(q^4-1)$, and one class $\bar{B}_9(\frac{q-1}{4})$ with the size $\frac{1}{2}q^3(q+1)(q^4-1)$ in $\mathrm{PSp}(4,q)$.
b) If $q = 4k+3$, then in $\mathrm{PSp}(4,q)$ we have $\frac{q-3}{2}$ classes $\bar{B}_9(i)$; $i = 1, 2, ..., \frac{1}{4}(q-3)$ with the size $q^3(q+1)(q^4-1)$.

3-3 Collection of the classes C and C'

By [3], it is easy to see that for any classes $C_1(i)$, $i \in T_2$ of the collection C_1, the classes $-C_1(i)$ are classes $C'_1(j)$, $j \in T_2$ of the collection C'_1, where i and j are not necessarily equal. Therefore, the canonical images of the elements of these collection of classes are merged pair wise in $\mathrm{PSp}(4,q)$. Therefore, from these collection of classes, we obtain $\frac{q-1}{2}$ classes $\bar{C}_1(i)$, $i \in T_2$ with the size $q^3(q-1)(q^2+1)$ in $\mathrm{PSp}(4,q)$.

By the similar argument we also see that, there are the following collection of classes classes in $\mathrm{PSp}(4,q)$:

$$\{\bar{C}_{21}(i) : i \in T_2\}\ ;\quad |\bar{C}_{21}(i)| = q^3(q-1)(q^4-1)/2$$

$$\{\bar{C}_{22}(i) : i \in T_2\}\ ;\quad |\bar{C}_{22}(i)| = q^3(q-1)(q^4-1)/2$$

$$\{\bar{C}_{3}(i) : i \in T_2\}\ ;\quad |\bar{C}_{3}(i)| = q^3(q+1)(q^2+1)/2$$

$$\{\bar{C}_{41}(i) : i \in T_2\}\ ;\quad |\bar{C}_{41}(i)| = q^3(q+1)(q^4-1)/2$$

$$\{\bar{C}_{41}(i) : i \in T_2\}\ ;\quad |\bar{C}_{41}(i)| = q^3(q+1)(q^4-1)/2$$

3-4 The classes D

By [3], we can see that additive inverse of each element of D_1 is in D_1. For, the representative of D_1 in $Sp(4,q)$ is Diag$(1,1,-1,-1)$, and $ED_1E = -D_1$, where

$$E = \begin{pmatrix} 0 & 0 & 1 & 0 \\ 0 & 0 & 0 & 1 \\ 1 & 0 & 0 & 0 \\ 0 & 1 & 0 & 0 \end{pmatrix}.$$

Again, by considering the character table and the conjugacy classes of $Sp(4,q)$, in [3], one can see that, the elements of the pair of the classes (D_{21}, D_{23}) and (D_{22}, D_{24}) are additive inverse of each other. The irreducible characters of Sp$(4,q)$ corresponding to the irreducible characters of PSp$(4,q)$, have equal values in D_{32} and D_{33}, but on the classes D_{31} and D_{34} these characters are distinct. (The differences are in the ϵ and ϵ' which occurs in the values of some characters.) Therefore, the following classes also must be added to the conjugacy classes of PSp$(4,q)$:
$\bar{D}_1$ with the size $\frac{1}{2}q^2(q^2+1)$, $\bar{D}_{21}$ with the size $\frac{1}{2}q^2(q^4-1)$, $\bar{D}_{22}$ with the size $\frac{1}{2}q^2(q^4-1)$, $\bar{D}_{31}$ with the size $\frac{1}{8}q^2(q^2-1)(q^4-1)$, $\bar{D}_{32}$ with the size $\frac{1}{4}q^2(q^2-1)(q^4-1)$ and $\bar{D}_{34}$ with the size $\frac{1}{8}q^2(q^2-1)(q^4-1)$.

4 Table of conjugacy classes and characters

Now by summarizing the above discussion, we can state the following theorems.

Theorem 4.1 *Tables with numbers 1 and 2 state the conjugacy classes of PSp$(4,q)$ with their sizes for, $q = 4k+1$ and $q = 4k+3$ respectively.*

Like [3] in the first column of these tables, we have given as class representatives not necessarily elements of PSp$(4,q)$, but their canonical forms in an extension filed of $GF(q)$, and also we use A rather than $A\{I,-I\}$.

Theorem 4.2 *Tables with numbers 3 and 4 state the character table of PSp$(4,q)$ for, $q = 4k+1$ and $q = 4k+3$ respectively.*

In the tables 3 and 4 the values of the characters at classes $\bar{A}_{22}$,$\bar{A}_{42}$, $\bar{C}_{22}(i)$, $\bar{C}_{42}(i)$, $\bar{D}_{22}$, $\bar{D}_{34}$,are omitted, since they can be obtained from the values at $\bar{A}_{21}$,$\bar{A}_{41}$, $\bar{C}_{21}(i)$, $\bar{C}_{41}(i)$, $\bar{D}_{21}$, $\bar{D}_{31}$ by replacing $\tilde{\epsilon}$ by $\tilde{\epsilon'}$ and $\tilde{\epsilon'}$ by $\tilde{\epsilon}$ respectively.
It is sufficient to give the values of one character from the pair $\{\xi_{21}(k), \xi_{22}(k)\}$. For, the values of the others are then obtained by replacing $\tilde{\epsilon}$ by $\tilde{\epsilon'}$ and $\tilde{\epsilon'}$ by $\tilde{\epsilon}$. A similar statement holds for the pairs $\{\xi'_{21}(k), \xi'_{22}(k)\}$, $\{\xi_{41}(k), \xi_{42}(k)\}$, $\{\xi'_{41}(k), \xi'_{42}(k)\}$, $\{\Phi_1\Phi_2\}$, $\{\Phi_3\Phi_4\}$, $\{\Phi_5\Phi_6\}$, $\{\Phi_7\Phi_8\}$, $\{\theta_1\theta_2\}$, and $\{\theta_3\theta_4\}$. The absence of entry in the table indicates that the corresponding value is zero. In some cases it is the negative of a character that is irreducible, and this will not be mentioned explicitly in the tables.

Table 1-1 The Conjugacy Classes of $\mathrm{PSp}(4,q)$, $q=4k+1$

Class Representative	No. classes	Order of classes	notation
$\begin{pmatrix}1&&&\\&1&&\\&&1&\\&&&1\end{pmatrix}$	1	1	$\overline{A}_1$
$\begin{pmatrix}1&1&&\\&1&&\\&&1&\\&&&1\end{pmatrix}$	1	$\frac{1}{2}(q^4-1)$	$\overline{A}_{21}$
$\begin{pmatrix}1&\gamma&&\\&1&&\\&&1&\\&&&1\end{pmatrix}$	1	$\frac{1}{2}(q^4-1)$	$\overline{A}_{22}$
$\begin{pmatrix}1&1&&\\&1&&\\&&1&-1\\&&&1\end{pmatrix}$	1	$\frac{1}{2}q(q+1)(q^4-1)$	$\overline{A}_{31}$
$\begin{pmatrix}1&1&&\\&1&&\\&&1&-\gamma\\&&&1\end{pmatrix}$	1	$\frac{1}{2}q(q-1)(q^4-1)$	$\overline{A}_{32}$
$\begin{pmatrix}1&1&&\\&1&&1\\-1&&1&\\&&&1\end{pmatrix}$	1	$\frac{1}{2}q^2(q^2-1)(q^4-1)$	$\overline{A}_{41}$
$\begin{pmatrix}1&\gamma&&\\&1&&1\\-1&&1&\\&&&1\end{pmatrix}$	1	$\frac{1}{2}q^2(q^2-1)(q^4-1)$	$\overline{A}_{42}$
$\begin{pmatrix}\zeta^i&&&\\&\zeta^{-i}&&\\&&\zeta^{qi}&\\&&&\zeta^{-qi}\end{pmatrix}$	$\frac{1}{8}(q^2-1)$ $i\in R_1$ & *i even*	$q^4(q^2-1)^2$	$\overline{B}_1(i)$

Table 1-2 The Conjugacy Classes of PSp(4, q), q=4k+1

Class Representative	No. classes	Order of classes	notation
$\begin{pmatrix} \theta^i & & & \\ & \theta^{-i} & & \\ & & \theta^{qi} & \\ & & & \theta^{-qi} \end{pmatrix}$	$\frac{1}{2}(q-1)$ $i \in R_2'$	$\frac{1}{2}q^4(q^4-1)$	$\overline{B}_2(i)$
$''$	$\frac{1}{8}(q-1)(q-3)$ $i \in R_2''$	$q^4(q^4-1)$	$''$
$\begin{pmatrix} \gamma^i & & & \\ & \gamma^{-i} & & \\ & & \gamma^j & \\ & & & \gamma^{-j} \end{pmatrix}$	$\frac{1}{4}(q-5)$ $1 \leq i \leq \frac{q-5}{4}$ $j = \frac{q-1}{2} - i$	$\frac{1}{2}q^4(q^2+1)(q+1)^2$	$\overline{B}_3(i,j)$
$\begin{pmatrix} \gamma^i & & & \\ & \gamma^{-i} & & \\ & & \gamma^j & \\ & & & \gamma^{-j} \end{pmatrix}$	$\frac{1}{16}(q-5)^2$ $1 \leq i \leq \frac{q-5}{4}$ $i < j \leq \frac{q-3}{2} - i$	$q^4(q^2+1)(q+1)^2$	$\overline{B}_3(i,j)$
$\begin{pmatrix} \eta^i & & & \\ & \eta^{-i} & & \\ & & \eta^j & \\ & & & \eta^{-j} \end{pmatrix}$	$\frac{1}{4}(q-1)$ $1 \leq i \leq \frac{q-1}{4}$ $j = \frac{q+1}{2} - i$	$\frac{1}{2}q^4(q-1)^2(q^2+1)$	$\overline{B}_4(i,j)$
$''$	$\frac{1}{16}(q-1)(q-5)$ $1 \leq i \leq \frac{q-5}{4}$ $i < j \leq \frac{q-1}{2} - i$	$q^4(q-1)^2(q^2+1)$	$''$
$\begin{pmatrix} \eta^i & & & \\ & \eta^{-i} & & \\ & & \gamma^j & \\ & & & \gamma^{-j} \end{pmatrix}$	$\frac{1}{8}(q-1)(q-3)$ $1 \leq i \leq \frac{q-1}{4}$ $1 \leq j \leq \frac{q-3}{2}$	$q^4(q^4-1)$	$\overline{B}_5(i,j)$
$\begin{pmatrix} \eta^i & & & \\ & \eta^{-i} & & \\ & & \eta^i & \\ & & & \eta^{-i} \end{pmatrix}$	$\frac{1}{4}(q-1)$ $1 \leq i \leq \frac{q-1}{4}$	$q^3(q^2+1)(q-1)$	$\overline{B}_6(i)$
$\begin{pmatrix} \eta^i & & 1 & \\ & \eta^{-i} & & 1 \\ & & \eta^i & \\ & & & \eta^{-i} \end{pmatrix}$	$\frac{1}{4}(q-1)$ $1 \leq i \leq \frac{q-1}{4}$	$q^3(q^4-1)(q-1)$	$\overline{B}_7(i)$

Table 1-3 The Conjugacy Classes of PSp$(4,q)$, q=4k+1

Class Representative	No. classes	Order of classes	notation
$\begin{pmatrix}\gamma^i&&&\\&\gamma^{-i}&&\\&&\gamma^i&\\&&&\gamma^{-i}\end{pmatrix}$	$\frac{1}{4}(q-5)$ $1\le i\le \frac{q-5}{4}$	$q^3(q+1)(q^2+1)$	$\overline{B}_8(i)$
"	1 $i=\frac{q-1}{2}$	$\frac{1}{2}q^3(q+1)(q^2+1)$	"
$\begin{pmatrix}\gamma^i&&1&\\&\gamma^{-i}&&1\\&&\gamma^i&\\&&&\gamma^{-i}\end{pmatrix}$	$\frac{1}{4}(q-5)$ $1\le i\le \frac{q-5}{4}$	$q^3(q+1)(q^4-1)$	$\overline{B}_9(i)$
"	1 $i=\frac{q-1}{4}$	$\frac{1}{2}q^3(q+1)(q^4-1)$	"
$\begin{pmatrix}\eta^i&&&\\&\eta^{-i}&&\\&&1&\\&&&1\end{pmatrix}$	$\frac{1}{2}(q-1)$ $i\in T_2$	$q^3(q-1)(q^2+1)$	$\overline{C}_1(i)$
$\begin{pmatrix}\eta^i&&&\\&\eta^{-i}&&\\&&1&1\\&&&1\end{pmatrix}$	$\frac{1}{2}(q-1)$ $i\in T_2$	$\frac{1}{2}q^3(q-1)(q^4-1)$	$\overline{C}_{21}(i)$
$\begin{pmatrix}\eta^i&&&\\&\eta^{-i}&&\\&&1&\gamma\\&&&1\end{pmatrix}$	$\frac{1}{2}(q-1)$ $i\in T_2$	$\frac{1}{2}q^3(q-1)(q^4-1)$	$\overline{C}_{22}(i)$
$\begin{pmatrix}\gamma^i&&&\\&\gamma^{-i}&&\\&&1&\\&&&1\end{pmatrix}$	$\frac{1}{2}(q-3)$ $i\in T_1$	$q^3(q+1)(q^2+1)$	$\overline{C}_3(i)$
$\begin{pmatrix}\gamma^i&&&\\&\gamma^{-i}&&\\&&1&1\\&&&1\end{pmatrix}$	$\frac{1}{2}(q-3)$ $i\in T_1$	$\frac{1}{2}q^3(q+1)(q^4-1)$	$\overline{C}_{41}(i)$
$\begin{pmatrix}\gamma^i&&&\\&\gamma^{-i}&&\\&&1&\gamma\\&&&1\end{pmatrix}$	$\frac{1}{2}(q-3)$ $i\in T_1$	$\frac{1}{2}q^3(q+1)(q^4-1)$	$\overline{C}_{42}(i)$

Table 1-4 The Conjugacy Classes of PSp$(4, q)$, q=4k+1

Class Representative	No. classes	Order of classes	notation
$\begin{pmatrix} 1 & & & \\ & 1 & & \\ & & -1 & \\ & & & -1 \end{pmatrix}$	1	$\frac{1}{2}q^2(q^2+1)$	$\overline{D}_1$
$\begin{pmatrix} 1 & & & \\ & 1 & & \\ & & -1 & -1 \\ & & & -1 \end{pmatrix}$	1	$\frac{1}{2}q^2(q^4-1)$	$\overline{D}_{21}$
$\begin{pmatrix} 1 & & & \\ & 1 & & \\ & & -1 & -\gamma \\ & & & -1 \end{pmatrix}$	1	"	$\overline{D}_{22}$
$\begin{pmatrix} 1 & 1 & & \\ & 1 & & \\ & & -1 & -1 \\ & & & -1 \end{pmatrix}$	1	$\frac{1}{8}q^2(q^2-1)(q^4-1)$	$\overline{D}_{31}$
$\begin{pmatrix} 1 & 1 & & \\ & 1 & & \\ & & -1 & -\gamma \\ & & & -1 \end{pmatrix}$	1	$\frac{1}{4}q^2(q^2-1)(q^4-1)$	$\overline{D}_{32}$
$\begin{pmatrix} 1 & \gamma & & \\ & 1 & & \\ & & -1 & -\gamma \\ & & & -1 \end{pmatrix}$	1	$\frac{1}{8}q^2(q^2-1)(q^4-1)$	$\overline{D}_{34}$

Table 2-1 The Conjugacy Classes of PSp$(4, q)$, q=4k+3

Class Representative	No. classes	Order of classes	notation
$\begin{pmatrix} 1 & & & \\ & 1 & & \\ & & 1 & \\ & & & 1 \end{pmatrix}$	1	1	$\overline{A}_1$
$\begin{pmatrix} 1 & 1 & & \\ & 1 & & \\ & & 1 & \\ & & & 1 \end{pmatrix}$	1	$\frac{1}{2}(q^4-1)$	$\overline{A}_{21}$
$\begin{pmatrix} 1 & \gamma & & \\ & 1 & & \\ & & 1 & \\ & & & 1 \end{pmatrix}$	1	$\frac{1}{2}(q^4-1)$	$\overline{A}_{22}$
$\begin{pmatrix} 1 & 1 & & \\ & 1 & & \\ & & 1 & -1 \\ & & & 1 \end{pmatrix}$	1	$\frac{1}{2}q(q+1)(q^4-1)$	$\overline{A}_{31}$
$\begin{pmatrix} 1 & 1 & & \\ & 1 & & \\ & & 1 & -\gamma \\ & & & 1 \end{pmatrix}$	1	$\frac{1}{2}q(q-1)(q^4-1)$	$\overline{A}_{32}$
$\begin{pmatrix} 1 & 1 & & \\ & 1 & & 1 \\ -1 & & 1 & \\ & & & 1 \end{pmatrix}$	1	$\frac{1}{2}q^2(q^2-1)(q^4-1)$	$\overline{A}_{41}$
$\begin{pmatrix} 1 & \gamma & & \\ & 1 & & 1 \\ -1 & & 1 & \\ & & & 1 \end{pmatrix}$	1	$\frac{1}{2}q^2(q^2-1)(q^4-1)$	$\overline{A}_{42}$
$\begin{pmatrix} \zeta^i & & & \\ & \zeta^{-i} & & \\ & & \zeta^{qi} & \\ & & & \zeta^{-qi} \end{pmatrix}$	$\frac{1}{8}(q^2-1)$ $i \in R_1$ & *i even*	$q^4(q^2-1)^2$	$\overline{B}_1(i)$

Table 2-2 The Conjugacy Classes of PSp$(4, q)$, q=4k+3

Class Representative	No. classes	Order of classes	notation
$\begin{pmatrix} \theta^i &&& \\ & \theta^{-i} && \\ && \theta^{qi} & \\ &&& \theta^{-qi} \end{pmatrix}$	$\frac{1}{2}(q-1)$ $i \in R_2'$	$\frac{1}{2}q^4(q^4-1)$	$\overline{B}_2(i)$
"	$\frac{1}{8}(q-1)(q-3)$ $i \in R_2''$	$q^4(q^4-1)$	"
$\begin{pmatrix} \gamma^i &&& \\ & \gamma^{-i} && \\ && \gamma^{j} & \\ &&& \gamma^{-j} \end{pmatrix}$	$\frac{1}{4}(q-3)$ $1 \leq i \leq \frac{q-3}{4}$ $j = \frac{q-1}{2} - i$	$\frac{1}{2}q^4(q^2+1)(q+1)^2$	$\overline{B}_3(i,j)$
$\begin{pmatrix} \gamma^i &&& \\ & \gamma^{-i} && \\ && \gamma^{j} & \\ &&& \gamma^{-j} \end{pmatrix}$	$\frac{1}{16}(q-3)(q-7)$ $1 \leq i \leq \frac{q-7}{4}$ $i < j \leq \frac{q-3}{2} - i$	$q^4(q^2+1)(q+1)^2$	$\overline{B}_3(i,j)$
$\begin{pmatrix} \eta^i &&& \\ & \eta^{-i} && \\ && \eta^{j} & \\ &&& \eta^{-j} \end{pmatrix}$	$\frac{1}{4}(q-3)$ $1 \leq i \leq \frac{q-3}{4}$ $j = \frac{q+1}{2} - i$	$\frac{1}{2}q^4(q-1)^2(q^2+1)$	$\overline{B}_4(i,j)$
"	$\frac{1}{16}(q-3)^2$ $1 \leq i \leq \frac{q-3}{4}$ $i < j \leq \frac{q-1}{2} - i$	$q^4(q-1)^2(q^2+1)$	"
$\begin{pmatrix} \eta^i &&& \\ & \eta^{-i} && \\ && \gamma^{j} & \\ &&& \gamma^{-j} \end{pmatrix}$	$\frac{1}{8}(q-1)(q-3)$ $1 \leq i \leq \frac{q-1}{2}$ $1 \leq j \leq \frac{q-3}{4}$	$q^4(q^4-1)$	$\overline{B}_5(i,j)$
$\begin{pmatrix} \eta^i &&& \\ & \eta^{-i} && \\ && \eta^{i} & \\ &&& \eta^{-i} \end{pmatrix}$	1 $i = \frac{q+1}{4}$	$\frac{1}{2}q^3(q^2+1)(q-1)$	$\overline{B}_6(i)$
"	$\frac{1}{4}(q-3)$ $1 \leq i \leq \frac{q-3}{4}$	$q^3(q^2+1)(q-1)$	"

Table 2-3 The Conjugacy Classes of $\mathrm{PSp}(4,q)$, q=4k+3

Class Representative	No. classes	Order of classes	notation
$\begin{pmatrix} \eta^i & & 1 & \\ & \eta^{-i} & & 1 \\ & & \eta^i & \\ & & & \eta^{-i} \end{pmatrix}$	$\frac{1}{2}(q-1)$ $1 \leq i \leq \frac{q-1}{2}$	$\frac{1}{2}q^3(q^4-1)(q-1)$	$\overline{B}_7(i)$
$\begin{pmatrix} \gamma^i & & & \\ & \gamma^{-i} & & \\ & & \gamma^i & \\ & & & \gamma^{-i} \end{pmatrix}$	$\frac{1}{4}(q-3)$ $1 \leq i \leq \frac{q-3}{4}$	$q^3(q+1)(q^2+1)$	$\overline{B}_8(i)$
$\begin{pmatrix} \gamma^i & & 1 & \\ & \gamma^{-i} & & 1 \\ & & \gamma^i & \\ & & & \gamma^{-i} \end{pmatrix}$	$\frac{1}{4}(q-3)$ $1 \leq i \leq \frac{q-3}{4}$	$q^3(q+1)(q^4-1)$	$\overline{B}_9(i)$
$\begin{pmatrix} \eta^i & & & \\ & \eta^{-i} & & \\ & & 1 & \\ & & & 1 \end{pmatrix}$	$\frac{1}{2}(q-1)$ $i \in T_2$	$q^3(q-1)(q^2+1)$	$\overline{C}_1(i)$
$\begin{pmatrix} \eta^i & & & \\ & \eta^{-i} & & \\ & & 1 & 1 \\ & & & 1 \end{pmatrix}$	$\frac{1}{2}(q-1)$ $i \in T_2$	$\frac{1}{2}q^3(q-1)(q^4-1)$	$\overline{C}_{21}(i)$
$\begin{pmatrix} \eta^i & & & \\ & \eta^{-i} & & \\ & & 1 & \gamma \\ & & & 1 \end{pmatrix}$	$\frac{1}{2}(q-1)$ $i \in T_2$	$\frac{1}{2}q^3(q-1)(q^4-1)$	$\overline{C}_{22}(i)$
$\begin{pmatrix} \gamma^i & & & \\ & \gamma^{-i} & & \\ & & 1 & \\ & & & 1 \end{pmatrix}$	$\frac{1}{2}(q-3)$ $i \in T_1$	$q^3(q+1)(q^2+1)$	$\overline{C}_3(i)$
$\begin{pmatrix} \gamma^i & & & \\ & \gamma^{-i} & & \\ & & 1 & 1 \\ & & & 1 \end{pmatrix}$	$\frac{1}{2}(q-3)$ $i \in T_1$	$\frac{1}{2}q^3(q+1)(q^4-1)$	$\overline{C}_{41}(i)$

Table 2-4 The Conjugacy Classes of PSp$(4,q)$, q=4k+3

Class Representative	No. classes	Order of classes	notation
$\begin{pmatrix} \gamma^i & & & \\ & \gamma^{-i} & & \\ & & 1 & \gamma \\ & & & 1 \end{pmatrix}$	$\frac{1}{2}(q-3)$ $i \in T_1$	$\frac{1}{2}q^3(q+1)(q^4-1)$	$\overline{C}_{42}(i)$
$\begin{pmatrix} 1 & & & \\ & 1 & & \\ & & -1 & \\ & & & -1 \end{pmatrix}$	1	$\frac{1}{2}q^2(q^2+1)$	$\overline{D}_1$
$\begin{pmatrix} 1 & & & \\ & 1 & & \\ & & -1 & -1 \\ & & & -1 \end{pmatrix}$	1	$\frac{1}{2}q^2(q^4-1)$	$\overline{D}_{21}$
$\begin{pmatrix} 1 & & & \\ & 1 & & \\ & & -1 & -\gamma \\ & & & -1 \end{pmatrix}$	1	"	$\overline{D}_{22}$
$\begin{pmatrix} 1 & 1 & & \\ & 1 & & \\ & & -1 & -1 \\ & & & -1 \end{pmatrix}$	1	$\frac{1}{8}q^2(q^2-1)(q^4-1)$	$\overline{D}_{31}$
$\begin{pmatrix} 1 & 1 & & \\ & 1 & & \\ & & -1 & -\gamma \\ & & & -1 \end{pmatrix}$	1	$\frac{1}{4}q^2(q^2-1)(q^4-1)$	$\overline{D}_{32}$
$\begin{pmatrix} 1 & \gamma & & \\ & 1 & & \\ & & -1 & -\gamma \\ & & & -1 \end{pmatrix}$	1	$\frac{1}{8}q^2(q^2-1)(q^4-1)$	$\overline{D}_{34}$

Table 3-1 The Character table of PSp(4,q), q=4k+1

Character $\longrightarrow$	$\chi_1(j)$ $j \in R_1$, j *even*	$\chi_2(j)$ $j \in R_2$, j *even*	$\chi_3(k,r)$ $k,r \in T_1,\ k \neq r$ $k+r$ *even*	$\chi_4(k,r)$ $k,r \in T_2,\ k \neq r$ $k+r$ *even*	$\chi_5(k,r)$ $k \in T_2,\ r \in T_1$ $k+r$ *even*	$\chi_6(k)$ $k \in T_2$	$\chi_7(k)$ $k \in T_2$
class $\downarrow$ A_1	$(1-q^2)^2$	$1-q^4$	$(1+q)^2(1+q^2)$	$(1-q)^2(1+q^2)$	$1-q^4$	$(1-q)(1+q^2)$	$q(1-q)(1+q^2)$
A_{21}	$1-q^2$	$1-q^2$	$(1+q)^2$	$(1-q)^2$	$1+q^2$	1-q	q(1-q)
A_{31}	1-q	1+q	1+3q	1-q	1-q	1	-q
A_{32}	1+q	1-q	1+q	1-3q	1+q	1-2q	-q
A_{41}	1	1	1	1	1	1	
$B_1(i)$	$\tilde{\zeta}^{ij}+\tilde{\zeta}^{-ij}+$ $\tilde{\zeta}^{qij}+\tilde{\zeta}^{-qij}$						
$B_2(i)$		$\tilde{\theta}^{ij}+\tilde{\theta}^{-ij}+$ $\tilde{\theta}^{qij}+\tilde{\theta}^{-qij}$				β_{ik}	$-\beta_{ik}$
$B_3(i,\ j)$			$\alpha_{ik}\alpha_{jr}+\alpha_{jk}\alpha_{ir}$				
$B_4(i\ j)$				$\beta_{ik}\beta_{jr}+\beta_{jk}\beta_{ir}$		$\beta_{ik}\beta_{jk}$	$\beta_{ik}\beta_{jk}$
$B_5(i\ j)$					$\beta_{ik}\alpha_{jr}$		
$B_6(i)$		$(1+q)\beta_{ij}$		$(1-q)\beta_{ik}\beta_{ir}$		$\beta_{2ik}+1-q$	$-q\beta_{2ik}+1-q$
$B_7(i)$		β_{ij}		$\beta_{ik}\beta_{ir}$		$\beta_{2ik}+1$	1
$B_8(i)$		$(1-q)\alpha_{ij}$	$(1+q)\alpha_{ik}\alpha_{ir}$			1-q	-(1-q)
$B_9(i)$		α_{ij}	$\alpha_{ik}\alpha_{ir}$			1	-1
$C_1(i)$				$(1-q)(\beta_{ik}+\beta_{ir})$	$(1+q)\beta_{ik}$	$(1-q)\beta_{ik}$	$\beta_{ik}(1-q)$
$C_{21}(i)$				$(\beta_{ik}+\beta_{ir})$	β_{ik}	β_{ik}	β_{ik}
$C_3(i)$			$(1+q)(\alpha_{ik}+\alpha_{ir})$		$(1-q)\alpha_{ir}$		
$C_{41}(i)$			$(\alpha_{ik}+\alpha_{ir})$		α_{ir}		
D_1			$2(-1)^k(1+q)^2$	$2(-1)^k(1-q)^2$	$2(-1)^k(1-q^2)$	$(-1)^k(1-q)^2$	$(-1)^k(1-q)^2$
D_{21}			$2(-1)^k(1+q)$	$2(-1)^k(1-q)$	$2(-1)^k$	$(-1)^k(1-q)$	$(-1)^k(1-q)$
D_{31}			$2(-1)^k$	$2(-1)^k$	$2(-1)^k$	$(-1)^k$	$(-1)^k$
D_{32}			$2(-1)^k$	$2(-1)^k$	$2(-1)^k$	$(-1)^k$	$(-1)^k$

Table 3-2 The Character table of PSp(4,q), q=4k+1

Character $\longrightarrow$	$\chi_8(k)$ $k \in T_1$	$\chi_9(k)$ $k \in T_1$	$\xi_1(k)$ $k \in T_2,\ k$ *even*	$\xi_1'(k)$ $k \in T_2,\ k$ *even*	$\xi_3(k)$ $k \in T_1,\ k$ *even*	$\xi_3'(k)$ $k \in T_1,\ k$ *even*	$\xi_{21}(k)$ $k \in T_2,\ k$ *even*
class $\downarrow$ A_1	$(1+q)(1+q^2)$	$q(q+1)(q^2+1)$	$(1-q)(q^2+1)$	$q(1-q)(q^2+1)$	$(q+1)(q^2+1)$	$q(q+1)(q^2+1)$	$\frac{1}{2}(1-q^4)$
A_{21}	1+q	q(q+1)	$1+q^2-q$	q	$1+q+q^2$	q	$\frac{1}{2}(1+q)+$ $q(1-q)\tilde{\varepsilon}$
A_{31}	1+2q	q	1-q		1+q	2q	$\frac{1}{2}(1-q)$
A_{32}	1	q	1-q	2q	1+q		$\frac{1}{2}(1+q)$
A_{41}	1		1		1		$-\tilde{\varepsilon}'$
$B_1(i)$							
$B_2(i)$	α_{ik}	$-\alpha_{ik}$					
$B_3(i,j)$	$\alpha_{ik}\alpha_{jk}$	$\alpha_{ik}\alpha_{jk}$			$\alpha_{ik}+\alpha_{jk}$	$\alpha_{ik}+\alpha_{jk}$	
$B_4(i,j)$			$\beta_{ik}+\beta_{jk}$	$-\beta_{ik}-\beta_{jk}$			
$B_5(i,j)$			β_{ik}	β_{ik}	α_{jk}	$-\alpha_{jk}$	$(-1)^j\beta_{ik}$
$B_6(i)$	1+q	-(1+q)	$(1-q)\beta_{ik}$	$-(1-q)\beta_{ik}$			
$B_7(i)$	1	-1	β_{ik}	$-\beta_{ik}$			
$B_8(i)$	$\alpha_{2ik}+1+q$	$q\alpha_{2ik}+1+q$			$(1+q)\alpha_{ik}$	$(1+q)\alpha_{ik}$	
$B_9(i)$	$\alpha_{2ik}+1$	1			α_{ik}	α_{ik}	
$C_1(i)$			$1-q+\beta_{ik}$	$q-1+q\beta_{ik}$	1+q	-(1+q)	$\frac{1}{2}(1+q)\beta_{ik}$
$C_{21}(i)$			$1+\beta_{ik}$	-1	1	-1	$-\beta_{ik}\tilde{\varepsilon}$
$C_3(i)$	$(1+q)\alpha_{ik}$	$(1+q)\alpha_{ik}$	1-q	1-q	$1+q+\alpha_{ik}$	$1+q+q\alpha_{ik}$	$(-1)^i(1-q)$
$C_{41}(i)$	α_{ik}	α_{ik}	1	1	$1+\alpha_{ik}$	1	$(-1)^i$
D_1	$(-1)^k(1+q)^2$	$(-1)^k(1+q)^2$	$2(1-q)$	2q(1-q)	2(1+q)	2q(1+q)	$(1-q^2)$
D_{21}	$(-1)^k(1+q)$	$(-1)^k(1+q)$	2-q	q	2+q	q	$\frac{1}{2}(1+q)-$ $(1-q)\tilde{\varepsilon}$
D_{31}	$(-1)^k$	$(-1)^k$	2		2		$-2\tilde{\varepsilon}$
D_{32}	$(-1)^k$	$(-1)^k$	2		2		$-\tilde{\varepsilon}-\tilde{\varepsilon}'$

Table 3-3 The Character table of PSp(4,q), q=4k+1

Character $\longrightarrow$	$\xi'_{21}(k)$ $k \in T_2,\ k\ odd$	$\xi_{41}(k)$ $k \in T_1,\ k\ even$	$\xi'_{41}(k)$ $k \in T_1,\ k\ odd$	Φ_5	Φ_7	Φ_9
classes $\downarrow$						
A_1	$\frac{1}{2}(1-q)^2(1+q^2)$	$\frac{1}{2}(1+q)^2(1+q^2)$	$\frac{1}{2}(1-q^4)$	$\frac{1}{2}(1+q)(1+q^2)$	$\frac{1}{2}q(1+q)(1+q^2)$	$q(1+q^2)$
A_{21}	$\frac{1}{2}(1-q)+q(1-q)\tilde{\varepsilon}$	$\frac{1}{2}(1+q)-q(1+q)\tilde{\varepsilon}$	$\frac{1}{2}(1-q)-q(1+q)\tilde{\varepsilon}$	$\frac{1}{2}(1+q)^2+q\tilde{\varepsilon}'$	$\frac{1}{2}q(1+q)+q^2\tilde{\varepsilon}'$	q
A_{31}	$\frac{1}{2}(1-q)$	$\frac{1}{2}(1+3q)$	$\frac{1}{2}(1-q)$	$\frac{1}{2}(1+q)$	q	q
A_{32}	$\frac{1}{2}(1-3q)$	$\frac{1}{2}(1+q)$	$\frac{1}{2}(1+q)$	$\frac{1}{2}(1+q)$		q
A_{41}	$-\tilde{\varepsilon}'$	$-\tilde{\varepsilon}$	$-\tilde{\varepsilon}$	$-\tilde{\varepsilon}$		
$B_1(i)$						
$B_2(i)$						
$B_3(i,j)$		$(-1)^j\alpha_{ik}+(-1)^i\alpha_{jk}$		$(-1)^i+(-1)^j$	$(-1)^i+(-1)^j$	$2(-1)^{i+j}$
$B_4(i,j)$	$(-1)^j\beta_{ik}+(-1)^i\beta_{jk}$					$2(-1)^{i+j+1}$
$B_5(i)$			$(-1)^i\alpha_{jk}$	$(-1)^j$	$(-1)^{j+1}$	
$B_6(i)$	$(-1)^i(1-q)\beta_{ik}$					q-1
$B_7(i)$	$(-1)^i\beta_{ik}$					-1
$B_8(i)$		$(-1)^i(1+q)\alpha_{ik}$		$(-1)^i(1+q)$	$(-1)^i(1+q)$	q+1
$B_9(i)$		$(-1)^i\alpha_{ik}$		$(-1)^i$	$(-1)^i$	1
$C_1(i)$	$(-1)^i(1-q)+$ $\frac{1}{2}(1-q)\beta_{ik}$		$(-1)^i(1+q)$	$\frac{1}{2}(1+q)$	$-\frac{1}{2}(1+q)$	$(-1)^i(q-1)$
$C_{21}(i)$	$(-1)^i-\beta_{ik}\tilde{\varepsilon}$		$(-1)^i$	$-\tilde{\varepsilon}'$	$\tilde{\varepsilon}'$	$(-1)^{i+1}$
$C_3(i)$		$(-1)^i(1+q)+$ $\frac{1}{2}(1+q)\alpha_{ik}$	$\frac{1}{2}(1-q)\alpha_{ik}$	$(-1)^i+\frac{1}{2}(1-q)$	$q(-1)^i+\frac{1}{2}(1-q)$	$(q+1)(-1)^i$
$C_{41}(i)$		$(-1)^i-\alpha_{ik}\tilde{\varepsilon}$	$-\alpha_{ik}\tilde{\varepsilon}$	$(-1)^i-\tilde{\varepsilon}'$	$-\tilde{\varepsilon}$	$(-1)^i$
D_1	$-(1-q)^2$	$(1+q)^2$	$-(1-q)^2$	1+q	q(1+q)	q^2+1
D_{21}	$-\frac{1}{2}(1-q)+(1-q)\tilde{\varepsilon}$	$\frac{1}{2}(1+q)-(1+q)\tilde{\varepsilon}$	$-\frac{1}{2}(1-q)+(1+q)\tilde{\varepsilon}$	$(1+q)+q\tilde{\varepsilon}'$	$\frac{1}{2}(1+q)+\tilde{\varepsilon}'$	1
D_{31}	$2\tilde{\varepsilon}$	$-2\tilde{\varepsilon}$	$2\tilde{\varepsilon}$	1	$\tilde{\varepsilon}'-\tilde{\varepsilon}$	1
D_{32}	$\tilde{\varepsilon}+\tilde{\varepsilon}'$	$-\tilde{\varepsilon}-\tilde{\varepsilon}'$	$\tilde{\varepsilon}+\tilde{\varepsilon}'$	1		1

Table 3-4 The Character table of PSp(4,q), q=4k+1

Character $\longrightarrow$ classes$\downarrow$	θ_1	θ_3	θ_9	θ_{10}	θ_{11}	θ_{12}	θ_{13}
A_1	$\frac{1}{2}q^2(1+q^2)$	$\frac{1}{2}(1+q^2)$	$\frac{1}{2}q(1+q)^2$	$\frac{1}{2}q(1-q)^2$	$\frac{1}{2}q(1+q^2)$	$\frac{1}{2}q(1+q^2)$	q^4
A_{21}	$-q^2\tilde{\varepsilon}$	$\frac{1}{2}(1+q)+q\tilde{\varepsilon}$	$\frac{1}{2}q(1+q)$	$\frac{1}{2}q(1-q)$	$\frac{1}{2}q(1-q)$	$\frac{1}{2}q(1+q)$	
A_{31}		$\frac{1}{2}(1+q)$	q		q		
A_{32}		$\frac{1}{2}(1-q)$		q		q	
A_{41}		$-\tilde{\varepsilon}'$					
$B_1(i)$			-1	1			1
$B_2(i)$	$(-1)^{i+1}$	$(-1)^i$			1	-1	-1
$B_3(i,j)$	$(-1)^{i+j}$	$(-1)^{i+j}$	2		1	1	1
$B_4(i,j)$	$(-1)^{i+j}$	$(-1)^{i+j}$		-2	-1	-1	1
$B_5(i,j)$					-1	1	-1
$B_6(i)$	-q	1		q-1	q	-1	-q
$B_7(i)$		1		-1		-1	
$B_8(i)$	q	1	1+q		1	q	q
$B_9(i)$		1	1		1		
$C_1(i)$	$\frac{1}{2}(-1)^i(1-q)$	$\frac{1}{2}(-1)^i(1-q)$		q-1	-1	q	-q
$C_{21}(i)$	$(-1)^{i+1}\tilde{\varepsilon}$	$(-1)^{i+1}\tilde{\varepsilon}$		-1	-1		
$C_3(i)$	$\frac{1}{2}(-1)^i(1+q)$	$\frac{1}{2}(-1)^i(1+q)$	1+q		q	1	q
$C_{41}(i)$	$(-1)^{i+1}\tilde{\varepsilon}$	$(-1)^{i+1}\tilde{\varepsilon}'$	1			1	
D_1	q	q	$\frac{1}{2}(1+q)^2$	$-\frac{1}{2}(1-q)^2$	$\frac{1}{2}(q^2-1)+q$	$\frac{1}{2}(1+2q-q^2)$	q^2
D_{21}	$-q\tilde{\varepsilon}$	$\tilde{\varepsilon}+\frac{1}{2}(1+q)$	$\frac{1}{2}(1+q)$	$\frac{1}{2}(q-1)$	$\frac{1}{2}(q-1)$	$\frac{1}{2}(1+q)$	
D_{31}		$-\tilde{\varepsilon}'+\tilde{\varepsilon}$	$\frac{1}{2}(1+q)$	$\frac{1}{2}(1-q)$	$-\frac{1}{2}(1+q)$	$\frac{1}{2}(1-q)$	
D_{32}			$\frac{1}{2}(1-q)$	$-\frac{1}{2}(1+q)$	$\frac{1}{2}(q-1)$	$\frac{1}{2}(1+q)$	

Table 4-1 The Character table of PSp(4,q), q=4k+3

Character $\longrightarrow$	$\chi_1(j)$ $j \in R_1$, j *even*	$\chi_2(j)$ $j \in R_2$, j *even*	$\chi_3(k,r)$ $k,r \in T_1,\ k \neq r$ $k+r$ *even*	$\chi_4(k,r)$ $k,r \in T_2,\ k \neq r$ $k+r$ *even*	$\chi_5(k,r)$ $k \in T_2,\ r \in T_1$ $k+r$ *even*	$\chi_6(k)$ $k \in T_2$	$\chi_7(k)$ $k \in T_2$
class $\downarrow$ A_1	$(1-q^2)^2$	$1-q^4$	$(1+q)^2(1+q^2)$	$(1-q)^2(1+q^2)$	$1-q^4$	$(1-q)(1+q^2)$	$q(1-q)(1+q^2)$
A_{21}	$1-q^2$	$1-q^2$	$(1+q)^2$	$(1-q)^2$	$1+q^2$	1-q	q(1-q)
A_{31}	1-q	1+q	1+3q	1-q	1-q	1	-q
A_{32}	1+q	1-q	1+q	1-3q	1+q	1-2q	-q
A_{41}	1	1	1	1	1	1	
$B_1(i)$	$\tilde{\zeta}^{ij}+\tilde{\zeta}^{-ij}+\tilde{\zeta}^{qij}+\tilde{\zeta}^{-qij}$						
$B_2(i)$		$\tilde{\theta}^{ij}+\tilde{\theta}^{-ij}+\tilde{\theta}^{qij}+\tilde{\theta}^{-qij}$				β_{ik}	$-\beta_{ik}$
$B_3(i,\ j)$			$\alpha_{ik}\alpha_{jr}+\alpha_{jk}\alpha_{ir}$				
$B_4(i\ j)$				$\beta_{ik}\beta_{jr}+\beta_{jk}\beta_{ir}$		$\beta_{ik}\beta_{jk}$	$\beta_{ik}\beta_{jk}$
$B_5(i\ j)$					$\beta_{ik}\alpha_{jr}$		
$B_6(i)$		$(1+q)\beta_{ij}$		$(1-q)\beta_{ik}\beta_{ir}$		$\beta_{2ik}+1-q$	$-q\beta_{2ik}+1-q$
$B_7(i)$		β_{ij}		$\beta_{ik}\beta_{ir}$		$\beta_{2ik}+1$	1
$B_8(i)$		$(1-q)\alpha_{ij}$	$(1+q)\alpha_{ik}\alpha_{ir}$			1-q	-(1-q)
$B_9(i)$		α_{ij}	$\alpha_{ik}\alpha_{ir}$			1	-1
$C_1(i)$				$(1-q)(\beta_{ik}+\beta_{ir})$	$(1+q)\beta_{ik}$	$(1-q)\beta_{ik}$	$\beta_{ik}(1-q)$
$C_{21}(i)$				$(\beta_{ik}+\beta_{ir})$	β_{ik}	β_{ik}	β_{ik}
$C_3(i)$			$(1+q)(\alpha_{ik}+\alpha_{ir})$		$(1-q)\alpha_{ir}$		
$C_{41}(i)$			$(\alpha_{ik}+\alpha_{ir})$		α_{ir}		
D_1			$2(-1)^k(1+q)^2$	$2(-1)^k(1-q)^2$	$2(-1)^k(1-q^2)$	$(-1)^k(1-q)^2$	$(-1)^k(1-q)^2$
D_{21}			$2(-1)^k(1+q)$	$2(-1)^k(1-q)$	$2(-1)^k$	$(-1)^k(1-q)$	$(-1)^k(1-q)$
D_{31}			$2(-1)^k$	$2(-1)^k$	$2(-1)^k$	$(-1)^k$	$(-1)^k$
D_{32}			$2(-1)^k$	$2(-1)^k$	$2(-1)^k$	$(-1)^k$	$(-1)^k$

Table 4-2 The Character table of PSp(4,q), q=4k+3

Character $\longrightarrow$ class $\downarrow$	$\chi_8(k)$ $k \in T_1$	$\chi_9(k)$ $k \in T_1$	$\xi_1(k)$ $k \in T_2,\ k\ even$	$\xi_1'(k)$ $k \in T_2,\ k\ even$	$\xi_3(k)$ $k \in T_1,\ k\ even$	$\xi_3'(k)$ $k \in T_1,\ k\ even$	$\xi_{21}(k)$ $k \in T_2,\ k\ odd$
A_1	$(1+q)(1+q^2)$	$q(q+1)(q^2+1)$	$(1-q)(q^2+1)$	$q(1-q)(q^2+1)$	$(q+1)(q^2+1)$	$q(q+1)(q^2+1)$	$\frac{1}{2}(1-q^4)$
A_{21}	1+q	q(q+1)	$1+q^2-q$	q	$1+q+q^2$	q	$\frac{1}{2}(1+q)+$ $q(1-q)\tilde{\varepsilon}$
A_{31}	1+2q	q	1-q		1+q	2q	$\frac{1}{2}(1-q)$
A_{32}	1	q	1-q	2q	1+q		$\frac{1}{2}(1+q)$
A_{41}	1		1		1		$-\tilde{\varepsilon}'$
$B_1(i)$							
$B_2(i)$	α_{ik}	$-\alpha_{ik}$					
$B_3(i,j)$	$\alpha_{ik}\alpha_{jk}$	$\alpha_{ik}\alpha_{jk}$			$\alpha_{ik}+\alpha_{jk}$	$\alpha_{ik}+\alpha_{jk}$	
$B_4(i,j)$			$\beta_{ik}+\beta_{jk}$	$-\beta_{ik}-\beta_{jk}$			
$B_5(i,j)$			β_{ik}	β_{ik}	α_{jk}	$-\alpha_{jk}$	$(-1)^j\beta_{ik}$
$B_6(i)$	1+q	-(1+q)	$(1-q)\beta_{ik}$	$-(1-q)\beta_{ik}$			
$B_7(i)$	1	-1	β_{ik}	$-\beta_{ik}$			
$B_8(i)$	$\alpha_{2ik}+1+q$	$q\alpha_{2ik}+1+q$			$(1+q)\alpha_{ik}$	$(1+q)\alpha_{ik}$	
$B_9(i)$	$\alpha_{2ik}+1$	1			α_{ik}	α_{ik}	
$C_1(i)$			$1-q+\beta_{ik}$	$q-1+q\beta_{ik}$	1+q	-(1+q)	$\frac{1}{2}(1+q)\beta_{ik}$
$C_{21}(i)$			$1+\beta_{ik}$	-1	1	-1	$-\beta_{ik}\tilde{\varepsilon}$
$C_3(i)$	$(1+q)\alpha_{ik}$	$(1+q)\alpha_{ik}$	1-q	1-q	$1+q+\alpha_{ik}$	$1+q+q\alpha_{ik}$	$(-1)^i(1-q)$
$C_{41}(i)$	α_{ik}	α_{ik}	1	1	$1+\alpha_{ik}$	1	$(-1)^i$
D_1	$(-1)^k(1+q)^2$	$(-1)^k(1+q)^2$	$2(1-q)$	2q(1-q)	2(1+q)	2q(1+q)	q^2-1
D_{21}	$(-1)^k(1+q)$	$(-1)^k(1+q)$	2-q	q	2+q	q	$-\frac{1}{2}(1+q)+$ $(1-q)\tilde{\varepsilon}$
D_{31}	$(-1)^k$	$(-1)^k$	2		2		$2\tilde{\varepsilon}$
D_{32}	$(-1)^k$	$(-1)^k$	2		2		$\tilde{\varepsilon}+\tilde{\varepsilon}'$

Table 4-3 The Character table of PSp(4,q), q=4k+3

Character $\longrightarrow$	$\xi'_{21}(k)$ $k \in T_2,\ k\ odd$	$\xi_{41}(k)$ $k \in T_1,\ k\ even$	$\xi'_{41}(k)$ $k \in T_1,\ k\ odd$	Φ_1	Φ_3	Φ_9
classes $\downarrow$						
A_1	$\frac{1}{2}(1-q)^2(1+q^2)$	$\frac{1}{2}(1+q)^2(1+q^2)$	$\frac{1}{2}(1-q^4)$	$\frac{1}{2}(1+q)(1+q^2)$	$\frac{1}{2}q(1+q)(1+q^2)$	$q(1+q^2)$
A_{21}	$\frac{1}{2}(1-q)+q(1-q)\tilde{\varepsilon}$	$\frac{1}{2}(1+q)-q(1+q)\tilde{\varepsilon}$	$\frac{1}{2}(1-q)-q(1+q)\tilde{\varepsilon}$	$\frac{1}{2}(1+q)^2+q\tilde{\varepsilon}'$	$\frac{1}{2}q(1+q)+q^2\tilde{\varepsilon}'$	q
A_{31}	$\frac{1}{2}(1-q)$	$\frac{1}{2}(1+3q)$	$\frac{1}{2}(1-q)$	$\frac{1}{2}(1+q)$	q	q
A_{32}	$\frac{1}{2}(1-3q)$	$\frac{1}{2}(1+q)$	$\frac{1}{2}(1+q)$	$\frac{1}{2}(1+q)$		q
A_{41}	$-\tilde{\varepsilon}'$	$-\tilde{\varepsilon}$	$-\tilde{\varepsilon}$	$-\tilde{\varepsilon}$		
$B_1(i)$						
$B_2(i)$						
$B_3(i,j)$		$(-1)^j\alpha_{ik}+(-1)^i\alpha_{jk}$		$(-1)^i+(-1)^j$	$(-1)^i+(-1)^j$	$2(-1)^{i+j}$
$B_4(i,j)$	$(-1)^j\beta_{ik}+(-1)^i\beta_{jk}$					$2(-1)^{i+j+1}$
$B_5(i)$			$(-1)^i\alpha_{jk}$	$(-1)^j$	$(-1)^{j+1}$	
$B_6(i)$	$(-1)^i(1-q)\beta_{ik}$					q-1
$B_7(i)$	$(-1)^i\beta_{ik}$					-1
$B_8(i)$		$(-1)^i(1+q)\alpha_{ik}$		$(-1)^i(1+q)$	$(-1)^i(1+q)$	q+1
$B_9(i)$		$(-1)^i\alpha_{ik}$		$(-1)^i$	$(-1)^i$	1
$C_1(i)$	$(-1)^i(1-q)+$ $\frac{1}{2}(1-q)\beta_{ik}$		$(-1)^i(1+q)$	$\frac{1}{2}(1+q)$	$-\frac{1}{2}(1+q)$	$(-1)^i(q-1)$
$C_{21}(i)$	$(-1)^i-\beta_{ik}\tilde{\varepsilon}$		$(-1)^i$	$-\tilde{\varepsilon}'$	$\tilde{\varepsilon}'$	$(-1)^{i+1}$
$C_3(i)$		$(-1)^i(1+q)+$ $\frac{1}{2}(1+q)\alpha_{ik}$	$\frac{1}{2}(1-q)\alpha_{ik}$	$(-1)^i+\frac{1}{2}(1-q)$	$q(-1)^i+\frac{1}{2}(1-q)$	$(q+1)(-1)^i$
$C_{41}(i)$		$(-1)^i-\alpha_{ik}\tilde{\varepsilon}$	$-\alpha_{ik}\tilde{\varepsilon}$	$(-1)^i-\tilde{\varepsilon}'$	$-\tilde{\varepsilon}$	$(-1)^i$
D_1	$(1-q)^2$	$-(1+q)^2$	$(1-q)^2$	1-q	1-q	$-(q^2+1)$
D_{21}	$\frac{1}{2}(1-q)-(1-q)\tilde{\varepsilon}$	$-\frac{1}{2}(1+q)+(1+q)\tilde{\varepsilon}$	$\frac{1}{2}(1-q)-(1+q)\tilde{\varepsilon}$	$(1-q)-q\tilde{\varepsilon}$	$-\frac{1}{2}(1-q)-\tilde{\varepsilon}$	-1
D_{31}	$-2\tilde{\varepsilon}$	$2\tilde{\varepsilon}$	$-2\tilde{\varepsilon}$	1	$\tilde{\varepsilon}'-\tilde{\varepsilon}$	-1
D_{32}	$-\tilde{\varepsilon}-\tilde{\varepsilon}'$	$\tilde{\varepsilon}+\tilde{\varepsilon}'$	$-\tilde{\varepsilon}-\tilde{\varepsilon}'$	1		-1

Table 4-4 The Character table of PSp(4,q), q=4k+3

Character $\longrightarrow$ classes$\downarrow$	θ_1	θ_3	θ_9	θ_{10}	θ_{11}	θ_{12}	θ_{13}
A_1	$\frac{1}{2}q^2(1+q^2)$	$\frac{1}{2}(1+q^2)$	$\frac{1}{2}q(1+q)^2$	$\frac{1}{2}q(1-q)^2$	$\frac{1}{2}q(1+q^2)$	$\frac{1}{2}q(1+q^2)$	q^4
A_{21}	$-q^2\tilde{\varepsilon}$	$\frac{1}{2}(1+q)+q\tilde{\varepsilon}$	$\frac{1}{2}q(1+q)$	$\frac{1}{2}q(1-q)$	$\frac{1}{2}q(1-q)$	$\frac{1}{2}q(1+q)$	
A_{31}		$\frac{1}{2}(1+q)$	q		q		
A_{32}		$\frac{1}{2}(1-q)$		q		q	
A_{41}		$-\tilde{\varepsilon}'$					
$B_1(i)$			-1	1			1
$B_2(i)$	$(-1)^{i+1}$	$(-1)^i$			1	-1	-1
$B_3(i,j)$	$(-1)^{i+j}$	$(-1)^{i+j}$	2		1	1	1
$B_4(i,j)$	$(-1)^{i+j}$	$(-1)^{i+j}$		-2	-1	-1	1
$B_5(i,j)$					-1	1	-1
$B_6(i)$	-q	1		q-1	q	-1	-q
$B_7(i)$		1		-1		-1	
$B_8(i)$	q	1	1+q		1	q	q
$B_9(i)$		1	1		1		
$C_1(i)$	$\frac{1}{2}(-1)^i(1-q)$	$\frac{1}{2}(-1)^i(1-q)$		q-1	-1	q	-q
$C_{21}(i)$	$(-1)^{i+1}\tilde{\varepsilon}$	$(-1)^{i+1}\tilde{\varepsilon}$		-1	-1		
$C_3(i)$	$\frac{1}{2}(-1)^i(1+q)$	$\frac{1}{2}(-1)^i(1+q)$	1+q		q	1	q
$C_{41}(i)$	$(-1)^{i+1}\tilde{\varepsilon}$	$(-1)^{i+1}\tilde{\varepsilon}'$	1			1	
D_1	-q	-q	$\frac{1}{2}(1+q)^2$	$-\frac{1}{2}(1-q)^2$	$\frac{1}{2}(q^2-1)+q$	$\frac{1}{2}(1+2q-q^2)$	q^2
D_{21}	$q\tilde{\varepsilon}$	$-\tilde{\varepsilon}-\frac{1}{2}(1+q)$	$\frac{1}{2}(q+1)$	$\frac{1}{2}(q-1)$	$\frac{1}{2}(q-1)$	$\frac{1}{2}(1+q)$	
D_{31}		$\tilde{\varepsilon}'-\tilde{\varepsilon}$	$\frac{1}{2}(1+q)$	$\frac{1}{2}(q-1)$	$-\frac{1}{2}(1+q)$	$\frac{1}{2}(1-q)$	
D_{32}			$\frac{1}{2}(1-q)$	$-\frac{1}{2}(1+q)$	$\frac{1}{2}(q-1)$	$\frac{1}{2}(1+q)$	

References

[1] I.M.Isaacs,Character Theory of Finite Groups, Dover, (1994).
[2] A.Przygocki, Schur indices of symplectic group, Comm. inAlg., 10(3) (1982), 279-310.
[3] B.Srinivasan,The characters of the finite symplectic group Sp(4,q), Trans. Amer. Math. Soc., 131(1968), 488-525.

5 Appendix

This appendix contains the conjugacy classes and the character table of $\mathrm{Sp}(4,q)$ for odd q, given by B. Srinivasan [3]. In 1982 A. Przygocki found some small mistakes (See [2]), then F. Luebeck in CHEVIE made some corrections to Przygocki's version. Now by using these corrections and the corrections provided us by Professor B. Srinivasan , we state the conjugacy classes and the character table of $\mathrm{Sp}(4,q)$, in the tables 5 and 6 .

Table 5-1 The Conjugacy Classes of $Sp(4,q)$

Class Representation	*No. classes*	*Order of centralizer*	notation
$\begin{pmatrix}1&&&\\&1&&\\&&1&\\&&&1\end{pmatrix}, \begin{pmatrix}-1&&&\\&-1&&\\&&-1&\\&&&-1\end{pmatrix}$	$1, 1$	$q^4(q^2-1)(q^4-1)$	A_1, A_1'
$\begin{pmatrix}1&1&&\\&1&&\\&&1&\\&&&1\end{pmatrix}, \begin{pmatrix}-1&-1&&\\&-1&&\\&&-1&\\&&&-1\end{pmatrix}$	$1, 1$	$2q^4(q^2-1)$	A_{21}, A_{21}'
$\begin{pmatrix}1&\gamma&&\\&1&&\\&&1&\\&&&1\end{pmatrix}, \begin{pmatrix}-1&-\gamma&&\\&-1&&\\&&-1&\\&&&-1\end{pmatrix}$	$1, 1$	$2q^4(q^2-1)$	A_{22}, A_{22}'
$\begin{pmatrix}1&1&&\\&1&&\\&&1&-1\\&&&1\end{pmatrix}, \begin{pmatrix}-1&-1&&\\&-1&&\\&&-1&-1\\&&&-1\end{pmatrix}$	$1, 1$	$2q^3(q-1)$	A_{31}, A_{31}'
$\begin{pmatrix}1&1&&\\&1&&\\&&1&-\gamma\\&&&1\end{pmatrix}, \begin{pmatrix}-1&-1&&\\&-1&&\\&&-1&\gamma\\&&&-1\end{pmatrix}$	$1, 1$	$2q^3(q+1)$	A_{32}, A_{32}'
$\begin{pmatrix}1&1&&\\&1&&1\\-1&&1&\\&&&1\end{pmatrix}, \begin{pmatrix}-1&-1&&\\&-1&&-1\\1&&-1&\\&&&-1\end{pmatrix}$	$1, 1$	$2q^2$	A_{41}, A_{41}'
$\begin{pmatrix}1&\gamma&&\\&1&&1\\-1&&1&\\&&&1\end{pmatrix}, \begin{pmatrix}-1&-\gamma&&\\&-1&&-1\\1&&-1&\\&&&-1\end{pmatrix}$	$1, 1$	$2q^2$	A_{42}, A_{42}'
$\begin{pmatrix}\zeta^i&&&\\&\zeta^{-i}&&\\&&\zeta^{qi}&\\&&&\zeta^{-qi}\end{pmatrix}$	$\frac{1}{4}(q^2-1)$ $i \in R_1$	q^2+1	$B_1(i)$
$\begin{pmatrix}\theta^i&&&\\&\theta^{-i}&&\\&&\theta^{qi}&\\&&&\theta^{-qi}\end{pmatrix}$	$\frac{1}{4}(q-1)^2$ $i \in R_2$	q^2-1	$B_2(i)$

Table 5-2 The Conjugacy Classes of $Sp(4,q)$

Class Representation	*No. classes*	*Order of centralizer*	notation
$\begin{pmatrix} \gamma^i & & & \\ & \gamma^{-i} & & \\ & & \gamma^j & \\ & & & \gamma^{-j} \end{pmatrix}$	$\frac{(q-3)(q-5)}{8}$ $i,j \in T_1$, $i \neq j$	$(q-1)^2$	$B_3(i,j)$
$\begin{pmatrix} \eta^i & & & \\ & \eta^{-i} & & \\ & & \eta^j & \\ & & & \eta^{-j} \end{pmatrix}$	$\frac{(q-1)(q-3)}{8}$ $i,j \in T_2$, $i \neq j$	$(q+1)^2$	$B_4(i,j)$
$\begin{pmatrix} \eta^i & & & \\ & \eta^{-i} & & \\ & & \gamma^j & \\ & & & \gamma^{-j} \end{pmatrix}$	$\frac{(q-1)(q-3)}{4}$ $i \in T_2$, $j \in T_1$	q^2-1	$B_5(i,j)$
$\begin{pmatrix} \eta^i & & & \\ & \eta^{-i} & & \\ & & \eta^i & \\ & & & \eta^{-i} \end{pmatrix}$	$\frac{1}{2}(q-1)$ $i \in T_2$	$q(q+1)(q^2-1)$	$B_6(i)$
$\begin{pmatrix} \eta^i & & 1 & \\ & \eta^{-i} & & 1 \\ & & \eta^i & \\ & & & \eta^{-i} \end{pmatrix}$	$\frac{1}{2}(q-1)$ $i \in T_2$	$q(q+1)$	$B_7(i)$
$\begin{pmatrix} \gamma^i & & & \\ & \gamma^{-i} & & \\ & & \gamma^i & \\ & & & \gamma^{-i} \end{pmatrix}$	$\frac{1}{2}(q-3)$ $i \in T_1$	$q(q-1)(q^2-1)$	$B_8(i)$
$\begin{pmatrix} \gamma^i & & 1 & \\ & \gamma^{-i} & & 1 \\ & & \gamma^i & \\ & & & \gamma^{-i} \end{pmatrix}$	$\frac{1}{2}(q-3)$ $i \in T_1$	$q(q-1)$	$B_9(i)$
$\begin{pmatrix} \eta^i & & & \\ & \eta^{-i} & & \\ & & 1 & \\ & & & 1 \end{pmatrix}, \begin{pmatrix} \eta^i & & & \\ & \eta^{-i} & & \\ & & -1 & \\ & & & -1 \end{pmatrix}$	$\frac{1}{2}(q-1)$, $\frac{1}{2}(q-1)$ $i \in T_2$	$q(q+1)(q^2-1)$	$C_1(i)$, $C_i'(1)$
$\begin{pmatrix} \eta^i & & & \\ & \eta^{-i} & & \\ & & 1 & 1 \\ & & & 1 \end{pmatrix}, \begin{pmatrix} \eta^i & & & \\ & \eta^{-i} & & \\ & & -1 & -1 \\ & & & -1 \end{pmatrix}$	$\frac{1}{2}(q-1)$, $\frac{1}{2}(q-1)$ $i \in T_2$	$2q(q+1)$	$C_{21}(i)$, $C_{21}'(i)$

Table 5-3 The Conjugacy Classes of $Sp(4, q)$

Class Representation	*No. classes*	*Order of centralizer*	notation
$\begin{pmatrix} \eta^i & & & \\ & \eta^{-i} & & \\ & & 1 & \gamma \\ & & & 1 \end{pmatrix}, \begin{pmatrix} \eta^i & & & \\ & \eta^{-i} & & \\ & & -1 & -\gamma \\ & & & -1 \end{pmatrix}$	$\frac{1}{2}(q-1)$, $\frac{1}{2}(q-1)$ $i \in T_2$	$2q(q+1)$	$C_{22}(i)$, $C'_{22}(i)$
$\begin{pmatrix} \gamma^i & & & \\ & \gamma^{-i} & & \\ & & 1 & \\ & & & 1 \end{pmatrix}, \begin{pmatrix} \gamma^i & & & \\ & \gamma^{-i} & & \\ & & -1 & \\ & & & -1 \end{pmatrix}$	$\frac{1}{2}(q-3)$, $\frac{1}{2}(q-3)$ $i \in T_1$	$q(q-1)(q^2-1)$	$C_3(i)$, $C'_3(i)$
$\begin{pmatrix} \gamma^i & & & \\ & \gamma^{-i} & & \\ & & 1 & 1 \\ & & & 1 \end{pmatrix}, \begin{pmatrix} \gamma^i & & & \\ & \gamma^{-i} & & \\ & & -1 & -1 \\ & & & -1 \end{pmatrix}$	$\frac{1}{2}(q-3)$, $\frac{1}{2}(q-3)$ $i \in T_1$	$2q(q-1)$	$C_{41}(i)$, $C'_{41}(i)$
$\begin{pmatrix} \gamma^i & & & \\ & \gamma^{-i} & & \\ & & 1 & \gamma \\ & & & 1 \end{pmatrix}, \begin{pmatrix} \gamma^i & & & \\ & \gamma^{-i} & & \\ & & -1 & -\gamma \\ & & & -1 \end{pmatrix}$	$\frac{1}{2}(q-3)$, $\frac{1}{2}(q-3)$ $i \in T_1$	$2q(q-1)$	$C_{42}(i)$, $C'_{42}(i)$
$\begin{pmatrix} 1 & & & \\ & 1 & & \\ & & -1 & \\ & & & -1 \end{pmatrix}$	1	$q^2(q^2-1)^2$	D_1
$\begin{pmatrix} 1 & & & \\ & 1 & & \\ & & -1 & -1 \\ & & & -1 \end{pmatrix}, \begin{pmatrix} 1 & & & \\ & 1 & & \\ & & -1 & -\gamma \\ & & & -1 \end{pmatrix}$	1, 1	$2q^2(q^2-1)$	D_{21}, D_{22}
$\begin{pmatrix} 1 & 1 & & \\ & 1 & & \\ & & -1 & \\ & & & -1 \end{pmatrix}, \begin{pmatrix} 1 & \gamma & & \\ & 1 & & \\ & & -1 & \\ & & & -1 \end{pmatrix}$	1, 1	$2q^2(q^2-1)$	D_{23}, D_{24}
$\begin{pmatrix} 1 & 1 & & \\ & 1 & & \\ & & -1 & -1 \\ & & & -1 \end{pmatrix}, \begin{pmatrix} 1 & 1 & & \\ & 1 & & \\ & & -1 & -\gamma \\ & & & -1 \end{pmatrix}$	1, 1	$4q^2$	D_{31}, D_{32}
$\begin{pmatrix} 1 & \gamma & & \\ & 1 & & \\ & & -1 & -1 \\ & & & 1 \end{pmatrix}, \begin{pmatrix} 1 & \gamma & & \\ & 1 & & \\ & & -1 & -\gamma \\ & & & -1 \end{pmatrix}$	1, 1	$4q^2$	D_{33}, D_{34}

Table 6-1 The Character table of Sp(4,q)

Character $\longrightarrow$	$\chi_1(j)$ $j \in R_1$	$\chi_2(j)$ $j \in R_2$	$\chi_3(k,r)$ $k,r \in T_1,\ k \neq r$	$\chi_4(k,r)$ $k,r \in T_2,\ k \neq r$	$\chi_5(k,r)$ $k \in T_2,\ r \in T_1$	$\chi_6(k)$ $k \in T_2$	$\chi_7(k)$ $k \in T_2$	$\chi_8(k)$ $k \in T_1$
class $\downarrow$								
A_1	$(1-q^2)^2$	$1-q^4$	$(1+q)^2(1+q^2)$	$(1-q)^2(1+q^2)$	$1-q^4$	$(1-q)(1+q^2)$	$q(1-q)(1+q^2)$	$(1+q)(1+q^2)$
A_1'	$(-1)^j(1-q^2)^2$	$(-1)^j(1-q^4)$	$(-1)^{k+r}(1+q)^2 (1+q^2)$	$(-1)^{k+r}(1-q)^2 (1+q^2)$	$(-1)^{k+r}(1-q^4)$	$(1-q)(1+q^2)$	$q(1-q)(1+q^2)$	$(1+q)(1+q^2)$
A_{21}	$1-q^2$	$1-q^2$	$(1+q)^2$	$(1-q)^2$	$1+q^2$	1-q	q(1-q)	1+q
A_{31}	1-q	1+q	1+3q	1-q	1-q	1	-q	1+2q
A_{32}	1+q	1-q	1+q	1-3q	1+q	1-2q	-q	1
A_{41}	1	1	1	1	1	1		1
$B_1(i)$	$\tilde{\zeta}^{ij}+\tilde{\zeta}^{-ij}+ \tilde{\zeta}^{qij}+\tilde{\zeta}^{-qij}$							
$B_2(i)$		$\tilde{\theta}^{ij}+\tilde{\theta}^{-ij}+ \tilde{\theta}^{qij}+\tilde{\theta}^{-qij}$				β_{ik}	$-\beta_{ik}$	α_{ik}
$B_3(i,\ j)$			$\alpha_{ik}\alpha_{jr}+\alpha_{jk}\alpha_{ir}$					$\alpha_{ik}\alpha_{jk}$
$B_4(i\ j)$				$\beta_{ik}\beta_{jr}+\beta_{jk}\beta_{ir}$		$\beta_{ik}\beta_{jk}$	$\beta_{ik}\beta_{jk}$	
$B_5(i\ j)$					$\beta_{ik}\alpha_{jr}$			
$B_6(i)$		$(1+q)\beta_{ij}$		$(1-q)\beta_{ik}\beta_{ir}$		$\beta_{2ik}+1-q$	$-q\beta_{2ik}+1-q$	1+q
$B_7(i)$		β_{ij}		$\beta_{ik}\beta_{ir}$		$\beta_{2ik}+1$	1	1
$B_8(i)$		$(1-q)\alpha_{ij}$	$(1+q)\alpha_{ik}\alpha_{ir}$			1-q	-(1-q)	$\alpha_{2ik}+1+q$
$B_9(i)$		α_{ij}	$\alpha_{ik}\alpha_{ir}$			1	-1	$\alpha_{2ik}+1$
$C_1(i)$				$(1-q)(\beta_{ik}+\beta_{ir})$	$(1+q)\beta_{ik}$	$(1-q)\beta_{ik}$	$\beta_{ik}(1-q)$	
$C_{21}(i)$				$(\beta_{ik}+\beta_{ir})$	β_{ik}	β_{ik}	β_{ik}	
$C_3(i)$			$(1+q)(\alpha_{ik}+\alpha_{ir})$		$(1-q)\alpha_{ir}$			$(1+q)\alpha_{ik}$
$C_{41}(i)$			$(\alpha_{ik}+\alpha_{ir})$		α_{ir}			α_{ik}
D_1			$(1+q)^2s(k,r)$	$(1-q)^2s(k,r)$	$(1-q^2)s(k,r)$	$(-1)^k(1-q)^2$	$(-1)^k(1-q)^2$	$(-1)^k(1+q)^2$
D_{21}			(1+q)s(k,r)	(1-q)s(k,r)	$(-1)^r(1-q)+ (-1)^k(1+q)$	$(-1)^k(1-q)$	$(-1)^k(1-q)$	$(-1)^k(1+q)$
D_{31}			s(k,r)	s(k,r)	s(k,r)	$(-1)^k$	$(-1)^k$	$(-1)^k$
D_{32}			s(k,r)	s(k,r)	s(k,r)	$(-1)^k$	$(-1)^k$	$(-1)^k$

Table 6-2 The Character table of Sp(4,q)

Character $\longrightarrow$	$\chi_9(k)$ $k \in T_1$	$\xi_1(k)$ $k \in T_2$	$\xi_1'(k)$ $k \in T_2$	$\xi_3(k)$ $k \in T_1$	$\xi_3'(k)$ $k \in T_1$	$\xi_{21}(k)$ $k \in T_2$	$\xi_{21}'(k)$ $k \in T_2$
class $\downarrow$							
A_1	$q(q+1)(q^2+1)$	$(1-q)(q^2+1)$	$q(1-q)(q^2+1)$	$(q+1)(q^2+1)$	$q(q+1)(q^2+1)$	$\frac{1}{2}(1-q^4)$	$\frac{1}{2}(1-q)^2(q^2+1)$
A_1'	$q(q+1)(q^2+1)$	$(-1)^k(1-q)$ (q^2+1)	$(-1)^k q(1-q)$ (q^2+1)	$(-1)^k(q+1)(q^2+1)$	$(-1)^k q(q+1)$ (q^2+1)	$\frac{1}{2}(-1)^{k+t}(1-q^4)$	$\frac{1}{2}(-1)^{k+t+1}$ $(1-q)^2(q^2+1)$
A_{21}	q(q+1)	$1+q^2-q$	q	$1+q+q^2$	q	$\frac{1}{2}(1+q)+$ $q(1-q)\tilde{\varepsilon}$	$\frac{1}{2}(1-q)-$ $q(q-1)\tilde{\varepsilon}$
A_{31}	q	1-q		1+q	2q	$\frac{1}{2}(1-q)$	$\frac{1}{2}(1-q)$
A_{32}	q	1-q	2q	1+q		$\frac{1}{2}(1+q)$	$\frac{1}{2}(1-3q)$
A_{41}		1		1		$-\tilde{\varepsilon}'$	$-\tilde{\varepsilon}'$
$B_1(i)$							
$B_2(i)$	$-\alpha_{ik}$						
$B_3(i,j)$	$\alpha_{ik}\alpha_{jk}$			$\alpha_{ik}+\alpha_{jk}$	$\alpha_{ik}+\alpha_{jk}$		
$B_4(i,j)$		$\beta_{ik}+\beta_{jk}$	$-\beta_{ik}-\beta_{jk}$				$(-1)^j\beta_{ik}+(-1)^i\beta_{jk}$
$B_5(i,j)$		β_{ik}	β_{ik}	α_{jk}	$-\alpha_{jk}$	$(-1)^j\beta_{ik}$	
$B_6(i)$	-(1+q)	$(1-q)\beta_{ik}$	$-(1-q)\beta_{ik}$				$(-1)^i(1-q)\beta_{ik}$
$B_7(i)$	-1	β_{ik}	$-\beta_{ik}$				$(-1)^i\beta_{ik}$
$B_8(i)$	$q\alpha_{2ik}+1+q$			$(1+q)\alpha_{ik}$	$(1+q)\alpha_{ik}$		
$B_9(i)$	1			α_{ik}	α_{ik}		
$C_1(i)$		$1-q+\beta_{ik}$	$q-1+q\beta_{ik}$	1+q	-(1+q)	$\frac{1}{2}(1+q)\beta_{ik}$	$(-1)^i(1-q)+$ $\frac{1}{2}(1-q)\beta_{ik}$
$C_{21}(i)$		$1+\beta_{ik}$	-1	1	-1	$-\beta_{ik}\tilde{\varepsilon}$	$(-1)^i-\beta_{ik}\tilde{\varepsilon}$
$C_3(i)$	$(1+q)\alpha_{ik}$	1-q	1-q	$1+q+\alpha_{ik}$	$1+q+q\alpha_{ik}$	$(-1)^i(1-q)$	
$C_{41}(i)$	α_{ik}	1	1	$1+\alpha_{ik}$	1	$(-1)^i$	
D_1	$(-1)^k(1+q)^2$	$(1-q)s(2,k)$	q(1-q)s(2,k)	(1+q)s(2,k)	q(1+q)s(2,k)	$\frac{1}{2}(1-q^2)s(k,t)$	$\frac{1}{2}(1-q)^2 s(k,t+1)$
D_{21}	$(-1)^k(1+q)$	$s(2,k)-q$	$q(-1)^k$	$s(2,k)+q$	$q(-1)^k$	$\frac{1}{2}(-1)^k(1+q)+$ $(-1)^{t+1}(1-q)\tilde{\varepsilon}$	$\frac{1}{2}(-1)^k(1-q)+$ $(-1)^t(1-q)\tilde{\varepsilon}$
D_{31}	$(-1)^k$	$s(2,k)$		s(2,k)		$s(k+1,t+1)\tilde{\varepsilon}$	$s(k+1,t)\tilde{\varepsilon}$
D_{32}	$(-1)^k$	s(2,k)		s(2,k)		$(-1)^{k+1}\tilde{\varepsilon}+(-1)^{t+1}\tilde{\varepsilon}'$	$(-1)^{k+1}\tilde{\varepsilon}+(-1)^t\tilde{\varepsilon}'$

Table 6-3 The Character table of Sp(4,q)

Character $\longrightarrow$ / *classes* $\downarrow$	$\xi_{41}(k)$ $k \in T_1$	$\xi'_{41}(k)$ $k \in T_1$	Φ_1	Φ_3	Φ_5	Φ_7	Φ_9
A_1	$\frac{1}{2}(1+q)^2(1+q^2)$	$\frac{1}{2}(1-q^4)$	$\frac{1}{2}(1-q)(1+q^2)$	$\frac{1}{2}q(1-q)(1+q^2)$	$\frac{1}{2}(1+q)(1+q^2)$	$\frac{1}{2}q(1+q)(1+q^2)$	$q(1+q^2)$
A'_1	$\frac{1}{2}(-1)^{k+t}$ $(1+q)^2(1+q^2)$	$\frac{1}{2}(-1)^{k+t+1}(1-q^4)$	$\frac{1}{2}(-1)^{t+1}$ $(1-q)(1+q^2)$	$\frac{1}{2}(-1)^{t+1}q$ $(1-q)(1+q^2)$	$\frac{1}{2}(-1)^t$ $(1+q)(1+q^2)$	$\frac{1}{2}(-1)^t q$ $(1+q)(1+q^2)$	$q(1+q^2)$
A_{21}	$\frac{1}{2}(1+q)-q(1+q)\tilde{\varepsilon}$	$\frac{1}{2}(1-q)-q(1+q)\tilde{\varepsilon}$	$\frac{1}{2}(1-q)^2-q\tilde{\varepsilon}$	$\frac{1}{2}q(1-q)-q^2\tilde{\varepsilon}$	$\frac{1}{2}(1+q)^2+q\tilde{\varepsilon}'$	$\frac{1}{2}q(1+q)+q^2\tilde{\varepsilon}'$	q
A_{31}	$\frac{1}{2}(1+3q)$	$\frac{1}{2}(1-q)$	$\frac{1}{2}(1-q)$		$\frac{1}{2}(1+q)$	q	q
A_{32}	$\frac{1}{2}(1+q)$	$\frac{1}{2}(1+q)$	$\frac{1}{2}(1-q)$	q	$\frac{1}{2}(1+q)$		q
A_{41}	$-\tilde{\varepsilon}$	$-\tilde{\varepsilon}$	$-\tilde{\varepsilon}$		$-\tilde{\varepsilon}$		
$B_1(i)$							
$B_2(i)$							
$B_3(i,j)$	$(-1)^j\alpha_{ik}+(-1)^i\alpha_{jk}$				$(-1)^i+(-1)^j$	$(-1)^i+(-1)^j$	$2(-1)^{i+j}$
$B_4(i,j)$			$(-1)^i+(-1)^j$	$(-1)^{i+1}+(-1)^{j+1}$			$2(-1)^{i+j+1}$
$B_5(i)$		$(-1)^i\alpha_{jk}$	$(-1)^i$	$(-1)^i$	$(-1)^j$	$(-1)^{j+1}$	
$B_6(i)$			$(-1)^i(1-q)$	$(-1)^{i+1}(1-q)$			q-1
$B_7(i)$			$(-1)^i$	$(-1)^{i+1}$			-1
$B_8(i)$	$(-1)^i(1+q)\alpha_{ik}$				$(-1)^i(1+q)$	$(-1)^i(1+q)$	q+1
$B_9(i)$	$(-1)^i\alpha_{ik}$				$(-1)^i$	$(-1)^i$	1
$C_1(i)$		$(-1)^i(1+q)$	$(-1)^i+\frac{1}{2}(1-q)$	$q(-1)^i-\frac{1}{2}(1-q)$	$\frac{1}{2}(1+q)$	$-\frac{1}{2}(1+q)$	$(-1)^i(q-1)$
$C_{21}(i)$		$(-1)^i$	$(-1)^i-\tilde{\varepsilon}'$	$\tilde{\varepsilon}'$	$-\tilde{\varepsilon}'$	$\tilde{\varepsilon}'$	$(-1)^{i+1}$
$C_3(i)$	$(-1)^i(1+q)+$ $\frac{1}{2}(1+q)\alpha_{ik}$	$\frac{1}{2}(1-q)\alpha_{ik}$	$\frac{1}{2}(1-q)$	$\frac{1}{2}(1-q)$	$(-1)^i+\frac{1}{2}(1-q)$	$q(-1)^i+\frac{1}{2}(1-q)$	$(q+1)(-1)^i$
$C_{41}(i)$	$(-1)^i-\alpha_{ik}\tilde{\varepsilon}$	$-\alpha_{ik}\tilde{\varepsilon}$	$-\tilde{\varepsilon}$	$-\tilde{\varepsilon}$	$(-1)^i-\tilde{\varepsilon}'$	$-\tilde{\varepsilon}$	$(-1)^i$
D_1	$\frac{1}{2}s(k,t)(1+q)^2$	$\frac{1}{2}s(k,t+1)(1-q)^2$	$\frac{1}{2}s(2,t+1)(1-q)$	$\frac{1}{2}s(2,t+1)q(1-q)$	$\frac{1}{2}s(2,t)(1+q)$	$\frac{1}{2}s(2,t)q(1+q)$	$(q^2+1)(-1)^t$
D_{21}	$\frac{1}{2}(-1)^k(1+q)+$ $(-1)^{t+1}(1+q)\tilde{\varepsilon}$	$\frac{1}{2}(-1)^k(1-q)+$ $(-1)^t(1+q)\tilde{\varepsilon}$	$\frac{1}{2}s(2,t+1)(1-q)+$ $q(-1)^t\tilde{\varepsilon}$	$\frac{1}{2}(-1)^t(1-q)+$ $(-1)^t\tilde{\varepsilon}$	$\frac{1}{2}s(2,t)(1+q)+$ $q(-1)^t\tilde{\varepsilon}'$	$\frac{1}{2}(-1)^t(1+q)+$ $(-1)^t\tilde{\varepsilon}'$	$(-1)^t$
D_{31}	$s(k+1,t+1)\tilde{\varepsilon}$	$s(k+1,t)\tilde{\varepsilon}$	$\frac{1}{2}s(2,t+1)$	$\frac{1}{2}s(1,t)(\tilde{\varepsilon}-\tilde{\varepsilon}')$	$\frac{1}{2}s(2,t)$	$\frac{1}{2}s(2,t)(\tilde{\varepsilon}'-\tilde{\varepsilon})$	$(-1)^t$
D_{32}	$(-1)^{k+1}\tilde{\varepsilon}+(-1)^{t+1}\tilde{\varepsilon}'$	$(-1)^{k+1}\tilde{\varepsilon}+(-1)^t\tilde{\varepsilon}'$	$\frac{1}{2}s(2,t+1)$	$\frac{1}{2}s(2,t)(\tilde{\varepsilon}'-\tilde{\varepsilon})$	$\frac{1}{2}s(2,t)$	$\frac{1}{2}s(2,t+1)(\tilde{\varepsilon}'-\tilde{\varepsilon})$	$(-1)^t$

Table 6-4 The Character table of Sp(4,q)

Character $\longrightarrow$ classes$\downarrow$	θ_1	θ_3	θ_5	θ_7	θ_9	θ_{10}	θ_{11}	θ_{12}	θ_{13}
A_1	$\frac{1}{2}q^2(1+q^2)$	$\frac{1}{2}(1+q^2)$	$\frac{1}{2}q^2(1-q^2)$	$\frac{1}{2}(1-q^2)$	$\frac{1}{2}q(1+q)^2$	$\frac{1}{2}q(1-q)^2$	$\frac{1}{2}q(1+q^2)$	$\frac{1}{2}q(1+q^2)$	q^4
A_1'	$\frac{1}{2}q^2(1+q^2)$	$\frac{1}{2}(1+q^2)$	$-\frac{1}{2}q^2(1-q^2)$	$-\frac{1}{2}(1-q^2)$	$\frac{1}{2}q(1+q)^2$	$\frac{1}{2}q(1-q)^2$	$\frac{1}{2}q(1+q^2)$	$\frac{1}{2}q(1+q^2)$	q^4
A_{21}	$-q^2\tilde{\varepsilon}$	$\frac{1}{2}(1+q)+q\tilde{\varepsilon}$	$-q^2\tilde{\varepsilon}$	$\frac{1}{2}(1-q)-q\tilde{\varepsilon}$	$\frac{1}{2}q(1+q)$	$\frac{1}{2}q(1-q)$	$\frac{1}{2}q(1-q)$	$\frac{1}{2}q(1+q)$	
A_{31}		$\frac{1}{2}(1+q)$		$\frac{1}{2}(1-q)$	q		q		
A_{32}		$\frac{1}{2}(1-q)$		$\frac{1}{2}(1+q)$		q		q	
A_{41}		$-\tilde{\varepsilon}'$		$-\tilde{\varepsilon}$					
$B_1(i)$			$(-1)^{i+1}$	$(-1)^i$	-1	1			1
$B_2(i)$	$(-1)^{i+1}$	$(-1)^i$					1	-1	-1
$B_3(i,j)$	$(-1)^{i+j}$	$(-1)^{i+j}$			2		1	1	1
$B_4(i,j)$	$(-1)^{i+j}$	$(-1)^{i+j}$				-2	-1	-1	1
$B_5(i,j)$			$(-1)^{i+j}$	$(-1)^{i+j}$			-1	1	-1
$B_6(i)$	-q	1				q-1	q	-1	-q
$B_7(i)$		1				-1		-1	
$B_8(i)$	q	1			1+q		1	q	q
$B_9(i)$		1			1		1		
$C_1(i)$	$\frac{1}{2}(-1)^i(1-q)$	$\frac{1}{2}(-1)^i(1-q)$	$\frac{1}{2}(-1)^i(1+q)$	$\frac{1}{2}(-1)^i(1+q)$		q-1	-1	q	-q
$C_{21}(i)$	$(-1)^{i+1}\tilde{\varepsilon}$	$(-1)^{i+1}\tilde{\varepsilon}$	$(-1)^{i+1}\tilde{\varepsilon}$	$(-1)^{i+1}\tilde{\varepsilon}'$		-1	-1		
$C_3(i)$	$\frac{1}{2}(-1)^i(1+q)$	$\frac{1}{2}(-1)^i(1+q)$	$\frac{1}{2}(-1)^i(1-q)$	$\frac{1}{2}(-1)^i(1-q)$	1+q		q	1	q
$C_{41}(i)$	$(-1)^{i+1}\tilde{\varepsilon}$	$(-1)^{i+1}\tilde{\varepsilon}'$	$(-1)^{i+1}\tilde{\varepsilon}$	$(-1)^{i+1}\tilde{\varepsilon}$	1			1	
D_1	$(-1)^t q$	$(-1)^t q$			$\frac{1}{2}(1+q)^2$	$-\frac{1}{2}(1-q)^2$	$\frac{1}{2}(q^2-1)+q$	$\frac{1}{2}(1+2q-q^2)$	q^2
D_{21}	$(-1)^{t+1}q\tilde{\varepsilon}$	$(-1)^t(\tilde{\varepsilon}+$ $\frac{1}{2}(1+q))$	$(-1)^t q\tilde{\varepsilon}$	$(-1)^t(\tilde{\varepsilon}+$ $\frac{1}{2}(1-q))$	$\frac{1}{2}(1+q)$	$\frac{1}{2}(q-1)$	$\frac{1}{2}(q-1)$	$\frac{1}{2}(1+q)$	
D_{31}		$(-1)^{t+1}(\tilde{\varepsilon}'-\tilde{\varepsilon})$			$\frac{1}{2}(1+q)$	$\frac{1}{2}(1-q)$	$-\frac{1}{2}(1+q)$	$\frac{1}{2}(1-q)$	
D_{32}				$(-1)^t(\tilde{\varepsilon}'-\tilde{\varepsilon})$	$\frac{1}{2}(1-q)$	$-\frac{1}{2}(1+q)$	$\frac{1}{2}(q-1)$	$\frac{1}{2}(1+q)$	

EXPONENT OF FINITE GROUPS WITH AUTOMORPHISMS

PAVEL SHUMYATSKY[1]

Department of Mathematics, University of Brasilia-DF, 70910-900 Brazil
E-mail: pavel@ipe.mat.unb.br

Abstract

Let A be a finite group acting coprimely on a finite group G. It was recently discovered that the exponent of $C_G(A)$ may have strong impact over the exponent of G. In this paper we discuss results on the exponent of a group with coprime automorphisms, as well as some applications and open problems. No detailed proofs are given.

1 Introduction

Let A be a finite group acting coprimely on a finite group G. It is well-known that the structure of the centralizer $C_G(A)$ (the fixed-point subgroup) of A has strong influence over the structure of G. To exemplify this we mention the following results.

The celebrated theorem of Thompson [27] says that if A is of prime order and $C_G(A) = 1$, then G is nilpotent. On the other hand, any nilpotent group admitting a fixed-point-free automorphism of prime order q has nilpotency class bounded by some function $h(q)$ depending on q alone. This result is due to Higman [13]. The reader can find in [15] and [16] an account on the modern developments related to Higman's theorem. The next result is a consequence of the classification of finite simple groups [29]: If A is a group of automorphisms of G whose order is coprime to that of G and $C_G(A)$ is nilpotent or has odd order, then G is soluble.

Once the group G is known to be soluble, there is a wealth of results bounding the Fitting height of G in terms of the order of A and the Fitting height of $C_G(A)$. The proofs of those results mostly use representation theory in the spirit of the Hall-Higman work [8]. A general discussion of these methods and their use in numerous fixed-point theorems can be found in Turull [28].

It was recently discovered that the exponent of $C_G(A)$ may have strong impact over the exponent of G. In this paper we discuss the results on the exponent of a group with coprime automorphisms, as well as some applications and open problems. The proofs of results of this kind involve a number of deep ideas. In particular, Zel'manov's techniques that led to the solution of the restricted Burnside problem [35] are combined with the Lubotzky-Mann theory of powerful p-groups [20], Lazard's criterion for a pro-p group to be p-adic analytic [19], and a theorem of Bahturin and Zaicev on Lie algebras admitting an automorphism whose fixed-point subalgebra is PI [1]. In certain situations the classification of finite simple groups is required as well. No detailed proofs are presented in this survey. The interested

[1]Supported by FAPDF and CNPq-Brazil

reader can find the details in the original papers. Throughout the article we use the term "$\{a, b, c \dots\}$-bounded" to mean "bounded from above by some function depending only on the parameters $a, b, c \dots$".

2 The elementary abelian group of automorphisms

Let q be a prime, and let A be a non-cyclic group of order q^2 acting on a finite group G. It is well-known (see [6, Theorem 6.2.4, Theorem 5.3.16]) that if G is any group of order prime to q then $G = \langle C_G(a); a \in A^{\#} \rangle$, where $A^{\#} = A \setminus \{1\}$. If, moreover, G is a p-group then we even have $G = \prod_{a \in A^{\#}} C_G(a)$. Since A normalizes some Sylow p-subgroup of G for any p dividing $|G|$ ([6, Theorem 6.2.4]), it is immediate that if $|C_G(a)| \leq n$ for any $a \in A^{\#}$ then $|G| \leq n^{q+1}$ (we use that A has exactly $q+1$ cyclic subgroups). How profound is the connection between the structure of G and that of $C_G(a)$; $a \in A^{\#}$? One can show that if $C_G(a)$ has rank at most r for any $a \in A^{\#}$ then the rank of G is $\{q, r\}$-bounded (see for example [7]). It is known that if $C_G(a)$ is nilpotent for each $a \in A^{\#}$ then G is metanilpotent [30]. In some situations this result holds even if G is allowed to be infinite periodic [21]. The following theorem shows that the exponents of $C_G(a)$ have strong impact on that of G [17].

Theorem 2.1 *Suppose that A is a non-cyclic group of order q^2 acting on a finite group G of coprime order, and let m be such an integer that the exponents of the centralizers $C_G(a)$ of the non-trivial elements $a \in A^{\#}$ divide m. Then the exponent of G is $\{m, q\}$-bounded.*

The proof of this theorem runs as follows.

Since A normalizes some Sylow p-subgroup of G for any p dividing $|G|$, it is sufficient to deal with the case that G is p-group and m is a p-power. We may also assume G to be generated by d elements for some $d \leq q^2$, since every element $g \in G$ is contained in an A-invariant subgroup of G with at most q^2 generators. Let $L = L_p(G)$ be the Lie algebra over $\mathbb{F}_p$, the field with p elements, associated with the Lazard series of G (see [35]). The algebra L is generated by the subspace $L_1 = G/\Phi(G)$ of dimension d. We extend the ground field of L by a primitive qth root of unity ω, forming $\overline{L} = L \otimes \mathbb{F}_p[\omega]$. It is clear that $\overline{L}$ is generated by $\overline{L_1} = L_1 \otimes \mathbb{F}_p[\omega]$. The group A acts naturally on $\overline{L}$ and it is easy to see that $\overline{L_1}$ is invariant under the action. Since A is abelian and the ground field is now a splitting field for A, the space $\overline{L_1}$ decomposes into a direct sum of common eigenspaces for A. So $\overline{L_1}$ is spanned by d common eigenvectors for A. Hence $\overline{L}$ is generated by d common eigenvectors for A that belong to $\overline{L_1}$. With some work it can be shown that any Lie-product of the eigenvectors is ad-nilpotent in $\overline{L}$ of $\{q, m\}$-bounded index.

Choose $a \in A^{\#}$. Since the exponent of the centralizer $C_G(a)$ of a divides m, it follows that the identity

$$\sum_{\pi \in S_{m-1}} [x_0, x_{\pi(1)}, x_{\pi(2)}, \dots, x_{\pi(m-1)}] = 0$$

is satisfied in $C_L(a) \otimes \mathbb{F}_p[\omega] = C_{\overline{L}}(a)$. So the Bahturin-Zaicev theorem [1] now says that $\overline{L}$ satisfies some polynomial identity which depends only on m and q. Combined with the fact that any Lie-product of the eigenvectors is ad-nilpotent of $\{q,m\}$-bounded index in $\overline{L}$, this enables us to use a deep theorem of Zel'manov [35, III(0.4)] and to conclude that $\overline{L}$ (hence L) is nilpotent of $\{q,m\}$-bounded class.

Nilpotency of $L_p(G)$, with any given class, almost always has strong impact on the group G. In particular, in our situation it follows that G contains a characteristic powerful subgroup H of $\{q,m\}$-bounded index. Powerful p-groups were introduced by Lubotzky and Mann in [20]: a finite p-group G is powerful if and only if $G^p \geq [G,G]$ for $p \neq 2$ (or $G^4 \geq [G,G]$ for $p = 2$). Powerful p-groups have many nice linear properties, of which we need the following: if a powerful p-group G is generated by elements of order dividing m, then the exponent of G divides m too (see [5, Lemma 2.2.5]). Combining this with the fact that $H = \prod_{a \in A^{\#}} C_H(a)$, we conclude that H is of exponent dividing m. This completes the proof of Theorem 2.1.

Let us now assume that A is elementary abelian of order q^s with $s \geq 3$. There seems to hold a general principle: if a certain property of $C_G(a)$ for all $a \in A^{\#}$ implies a similar property for the whole group G then the property of $C_G(a)'$ for all $a \in A^{\#}$ implies a similar property for G' provided that the rank of A increases by 1. The following three results (obtained in [7]) provide a good illustration for this.

Theorem 2.2 *Let m be an integer, q a prime. Let G be a finite q'-group acted on by an elementary abelian group A of order q^3. Assume that $C_G(a)$ has derived group of order at most m for each $a \in A^{\#}$. Then the order of G' is $\{m,q\}$-bounded.*

Theorem 2.3 *Let m be an integer, q a prime. Let G be a finite q'-group acted on by an elementary abelian group A of order q^3. Assume that $C_G(a)$ has derived group of rank at most m for each $a \in A^{\#}$. Then the rank of G' is $\{m,q\}$-bounded.*

Theorem 2.4 *Let m be an integer, q a prime. Let G be a finite q'-group acted on by an elementary abelian group A of order q^3. Assume that $C_G(a)$ has derived group of exponent dividing m for each $a \in A^{\#}$. Then the exponent of G' is $\{m,q\}$-bounded.*

Each of the above theorems fails if $|A| = q^2$. Indeed, let G be a finite q'-group admitting a non-cyclic automorphism group A of order q^2 such that $C_G(a)$ is abelian for each $a \in A^{\#}$. Ward showed that G is necessarily soluble [32]. The author proved that if G has derived length k then G' is nilpotent of class bounded by some function of q and k [21]. However the derived length k can be arbitrarily large. For instance, for any odd prime p Khukhro constructed a p-group G of derived length bigger than $\log_2(p-1)$ acted on by a four-group A such that $C_G(a)$ is abelian for each $a \in A^{\#}$ [15, p. 149–150]. Thus Theorem 2.2 and Theorem 2.4 fail in the case $|A| = q^2$. Direct products of such groups show that in this case G' can have arbitrarily large rank. So the assumption that $|A| = q^3$ is essential in each of the above theorems.

Recall that the proof of Theorem 2.1 reduces very easily to the case that G is a p-group. This is no longer true for the three above theorems which reduce to the case of p-groups using the following result.

Theorem 2.5 *Let q be a prime. Let G be a finite q'-group acted on by an elementary abelian group A of order q^3. Let P be an A-invariant Sylow subgroup of G. Then $P \cap G' = \langle P \cap C_G(a)' | a \in A^{\#} \rangle$.*

The proof of Theorem 2.5 involves the classification of finite simple groups. The fact that is used is that any group of coprime automorphisms of a finite simple group is cyclic. Undoubtedly, Theorem 2.1 and Theorem 2.4 admit a common generalization. In particular, it is hoped that the following conjecture could be confirmed along the lines of the work [7] using an appropriate analogue of Theorem 2.5.

Conjecture 2.6 Let A be an elementary abelian group of order q^s with $s \geq 2$ acting on a finite q'-group G.

1. If $\gamma_{s-1}(C_G(a))$ has exponent dividing m for any $a \in A^{\#}$, then $\gamma_{s-1}(G)$ has $\{m, q, s\}$-bounded exponent;
2. If, for some integer d such that $2^d \leq s - 1$, the dth derived group of $C_G(a)$ has exponent dividing m for any $a \in A^{\#}$, then the dth derived group $G^{(d)}$ has $\{m, q, s\}$-bounded exponent.

Perhaps it is somehow possible to strengthen Theorem 2.4 without enlarging the rank of A. Let us see what can be expected in the case that A is of order q^3 and the hypothesis on $C_G(a)$ is weaker than that in Theorem 2.4. Ward showed that if $C_G(a)$ is nilpotent for any $a \in A^{\#}$, then the group G is nilpotent [31]. Recently the author of the present paper found that there exists a function $f = f(c, q)$ depending only on c and q such that if, under the assumptions of Ward's theorem, $C_G(a)$ is nilpotent of class at most c for any $a \in A^{\#}$, then the nilpotency class of G does not exceed f, that is the nilpotency class of G is $\{c, q\}$-bounded [24]. So we have certain grounds for suggesting the following, probably very risky, conjecture.

Conjecture 2.7 Let A be an elementary abelian group of order q^3, and let $f = f(c, q)$ be the same function as above. Given positive integers c, m, assume A acts on a finite q'-group G in such a way that $\gamma_{c+1}(C_G(a))$ has exponent dividing m for any $a \in A^{\#}$. Then $\gamma_{f+1}(G)$ has $\{c, m, q\}$-bounded exponent.

Theorem 2.1 has a number of interesting applications to profinite and locally finite groups. These will be discussed in the next two sections.

3 On the structure of periodic compact groups

Zel'manov proved that any periodic pro-p group is locally finite [34]. Combined with the results of Wilson [33], this implies that any periodic compact group is locally finite. A long-standing problem is whether any periodic compact group has finite exponent. Theorem 2.1 is an indispensable ingredient in the proof of the following theorem [23].

Theorem 3.1 *Assume that a periodic compact group G contains an abelian subgroup A whose centralizer in G is finite. Then G is of finite exponent.*

The following result of Hartley is used in the proof of Theorem 3.1 (see [18] for the proof of Hartley's result).

Theorem 3.2 *Assume that a finite group G is acted on by a nilpotent group A with $C_G(A) = 1$. Then G is soluble.*

An immediate corollary of the above theorem is that if, under the hypothesis of Theorem 3.2, N is any A-invariant normal subgroup of G then $C_{G/N}(A) = 1$. This is because A is a Carter subgroup of GA and so its image is Carter in any quotient of GA.

Now assume the hypothesis of Theorem 3.1. Since $C_G(A)$ is finite, we can choose a normal open subgroup H of G such that $H \cap C_G(A) = 1$. By Theorem 3.2 H is locally soluble. Of course, we use here that G is locally finite by the Wilson-Zel'manov theorem. By the remark made after Theorem 3.2 $C_Q(A) = 1$ for any A-invariant section Q of H. A well-known result of Dade now tells us that if Q is finite, then the Fitting height of Q is bounded by a function depending only on the order of A [4] (obviously the hypothesis of the theorem implies that A is finite). It follows that H has a finite series of characteristic closed subgroups, all of whose quotients are locally nilpotent. Thus, we can assume that H is locally nilpotent. Using a theorem of Herfort [12], the theorem can now be reduced to the case that H is a p-subgroup of G and A is a p'-group. Arguing by induction on the order of A we can assume that $C_H(B)$ has finite exponent for any proper subgroup B of A. Then Theorem 2.1 shows that if G is a counter-example then A is cyclic. This case has been handled in [23] by analyzing the structure of Lie algebras admitting a fixed-point-free automorphism of finite order.

It is worth mentioning that there is an interesting if implicit connection of results obtained in [23] with another well-known problem. Namely, suppose a finite p-group G admits a fixed-point-free automorphism a of order n. Is the derived length of G bounded by a function of n?

Slightly modifying some proofs in [23] one can obtain the following related result.

Theorem 3.3 *Let G be a d-generated finite p-group admitting a fixed-point-free automorphism a of order n. Let $b_1, \ldots, b_s$ be the elements of prime order in $\langle a\rangle$. Assume the exponent of $C_G(b_i)$ divides m for all $i \le s$. Then the derived length of G is $\{d, m, n\}$-bounded.*

Curiously, this somehow reminds the restricted Burnside problem.

4 On centralizers in locally finite groups

The study of centralizers is a well-established direction in the theory of locally finite groups [10]. The idea is that if a locally finite group G contains a finite subgroup A

(or, equivalently, admits a finite group of automorphisms A) with $C_G(A)$ satisfying certain conditions then the structure of G must be very special, usually close to that of a nilpotent or soluble group. The following theorem was obtained in [25].

Theorem 4.1 *Let G be a locally finite group containing a finite p-subgroup A such that $C_G(A)$ is finite and a non-cyclic subgroup B of order p^2 such that $C_G(b)$ has finite exponent for all $b \in B^{\#}$. Then G is almost locally soluble and has finite exponent.*

A group is said to almost have certain property if it contains a subgroup of finite index with that property. The proof of the above theorem uses Theorem 2.1 as well as the classification of finite simple groups and the classification of simple periodic linear groups [2, 3, 9, 26]. The next theorem on automorphisms of finite groups is especially helpful (its proof depends on the classification of finite simple groups).

Theorem 4.2 Let A be a finite p-group acting on a finite group G of exponent e. Assume $|C_G(A)| \leq m$. Then $|G : S(G)|$ is $\{|A|, e, m\}$-bounded.

Here $S(G)$ denotes the soluble radical of the group G. This is the product of all normal soluble subgroups or, equivalently, the maximal normal soluble subgroup of G. It is a deep result of Hartley that if, under the hypothesis of Theorem 4.2, A is cyclic, then the bound on $|G : S(G)|$ does not depend on e (but of course depends on $|A|$ and m) [11]. Examples of finite simple groups having a self-centralizing four-subgroup [6] show that if A is not cyclic the bound cannot be taken independent of e.

It would be interesting to see what happens to Theorem 4.1 if the assumption that $C_G(A)$ is finite is dropped. More precisely, what can be said about a locally finite group G containing a non-cyclic subgroup B of order p^2 such that $C_G(b)$ has finite exponent for all $b \in B^{\#}$? Perhaps the study of this question can be started along the lines of the following conjecture.

Conjecture 4.3 Suppose a locally finite group G contains a non-cyclic subgroup B of order p^2 such that $C_G(b)$ has finite exponent for all $b \in B^{\#}$. Then G possesses a finite series of normal subgroups $G = G_1 \geq G_2 \geq \ldots \geq G_s = 1$ all of whose quotients are either p-groups or groups of finite exponent.

One may wonder whether Theorem 2.4 is relevant to the study of centralizers in locally finite groups. In particular, the following conjecture looks plausible.

Conjecture 4.4 Let G be a locally finite group containing a finite p-subgroup A such that $C_G(A)$ is finite and an elementary abelian subgroup B of order p^3 such that $C_G(b)'$ has finite exponent for all $b \in B^{\#}$. Then G is almost locally soluble and G' has finite exponent.

5 Exponent of a group with a single automorphism

It is easy to see that the exponent of the centralizer of a single automorphism a of a finite group G has no impact over the exponent of G. Indeed, any abelian group of odd order admits a fixed-point-free automorphism of order two. Hence, we cannot bound the exponent of G solely in terms of the exponent of $C_G(a)$. Thus we have to make assumptions on orders of certain elements outside $C_G(a)$. It is well-known that if G is a group of odd order admitting an involutory automorphism a then G is generated by a-invariant cyclic subgroups. Hence, if G is abelian and the order of any a-invariant cyclic subgroup divides a given number m then the exponent of G is a divisor of m, too. The next theorem, proved in [22], shows that this can be generalized as follows.

Theorem 5.1 *Let m be a positive integer, G a finite group of odd order admitting an involutory automorphism a such that any a-invariant cyclic subgroup of G is of order dividing m. Then the exponent of G is bounded in terms of m.*

The general scheme of the proof of Theorem 5.1 is similar to that of Theorem 2.1.

Assume now that the finite group G admits a coprime automorphism a of order n. If $n \geq 3$ the group G need not be generated by a-invariant cyclic subgroups. So even if G is abelian the orders of a-invariant cyclic subgroups do not have much effect on the exponent of G. Hence, the next theorem is, in a sense, the best possible generalization of Theorem 5.1 to the case $n \geq 3$.

Theorem 5.2 *Let m be a positive integer, G a finite group admitting a coprime automorphism a of order n. Assume that any a-invariant $(n-1)$-generated subgroup of G is of exponent dividing m. Then the exponent of G is $\{m, n\}$-bounded.*

On the other hand, if we slightly modify the language of the discussion, we will see that Theorem 5.2 can be somewhat strengthened. Given a group G with an automorphism a, set $G_{-a} = \{x^{-1}x^a;\ x \in G\}$ and $G_a = C_G(a)$. Obviously, Theorem 5.1 is equivalent to the following theorem.

Theorem 5.3 *Let m be a positive integer, G a finite group of odd order admitting an involutory automorphism a such that any element of $G_a \cup G_{-a}$ is of order dividing m. Then the exponent of G is bounded in terms of m.*

It would be interesting to see whether this can be extended to the case of automorphisms of order greater than two.

Problem 5.4 Let m be a positive integer, G a finite group admitting a coprime automorphism a of order n. Assume that any element of $G_a \cup G_{-a}$ is of order dividing m. Is then the exponent of G bounded in terms of m and n alone?

The next two results give partial positive answers to the above problem. Note that Theorem 5.5 strengthens Theorem 5.2.

Theorem 5.5 *Let m be a positive integer, G a finite group admitting a coprime automorphism a of order n. Assume that any element of $G_a \cup G_{-a}$ is contained in an a-invariant subgroup of exponent dividing m. Then the exponent of G is $\{m, n\}$-bounded.*

Theorem 5.6 *Let m be a positive integer, G a finite group admitting a coprime automorphism a of order n and a fixed-point-free automorphism b of order k. Assume that any element of $G_a \cup G_{-a}$ is of order dividing m. If a and b commute, then the exponent of G is $\{k, m, n\}$-bounded.*

Hopefully, the proofs of Theorems 5.2, 5.5, 5.6 will be published in a separate paper.

References

[1] Yu. A. Bahturin, M. V. Zaicev, Identities of graded algebras, J. Algebra, **205**(1998), 1–12.

[2] V. V. Belyaev, Locally finite Chevalley groups (Russian), in "Studies in group theory" **150**, 39–50, Sverdlovsk, 1984.

[3] A. V. Borovik, Embeddings of finite Chevalley groups and periodic linear groups, Siberian Math. J. **24** (1983), 843–855.

[4] E. C. Dade, Carter subgroups and Fitting heights of finite solvable groups, Illinois J. Math., **13** (1969), 449-514.

[5] J. D. Dixon, M. P. F. du Sautoy, A. Mann, D. Segal, "Analytic pro-p groups" (London Math. Soc. Lecture Note Series **157**), Cambridge Univ. Press., 1991.

[6] D. Gorenstein, "Finite Groups", Harper and Row, New York, 1968.

[7] R. Guralnick and P. Shumyatsky, Derived subgroups of fixed points, Israel J. Math., to appear.

[8] P. Hall, G. Higman, The p-length of a p-soluble group and reduction theorems for Burnside's problem, Proc. London. Math. Soc., (3) **6** (1956), 1–42.

[9] B. Hartley and G. Shute, Monomorphisms and direct limits of finite groups of Lie type, Quart. J. Math. Oxford (2) **35** (1984), 49–71.

[10] B. Hartley, Centralizers in locally finite groups, in "Proc. 1st Bressanone Group Theory Conference 1986", Lecture Notes in Math. **1281**, Springer, 1987, pp. 36–51.

[11] B. Hartley, A general Brauer-Fowler theorem and centralizers in locally finite groups, Pacific J. Math. **152** (1992), 101–117.

[12] W. Herfort, Compact torsion groups and finite exponent, Arch. Math. **33** (1979), 404-410.

[13] G. Higman, Groups and Lie rings having automorphisms without non-trivial fixed points, J. London Math. Soc. **32** (1957), 321–334.

[14] G. Higman, Lie ring methods in the theory of finite nilpotent groups, in "Proc. Intern. Congr. Math. Edinburgh, 1958", Cambridge Univ. Press, 1960, 307–312.

[15] E. I. Khukhro, "Nilpotent Groups and their Automorphisms", de Gruyter–Verlag, Berlin, 1993.

[16] E. I. Khukhro, "p-Automorphisms of finite p-groups", London Math. Soc. Lecture Note Series **246**, Cambridge University Press, Cambridge, 1998.

[17] E. I. Khukhro and P. V. Shumyatsky, Bounding the exponent of a finite group with automorphisms, J. Algebra, **212** (1999), 363–374.

[18] M. Kuzucuoğlu and P. Shumyatsky, On local finiteness of periodic residually finite groups, preprint, 2001.

[19] M. Lazard, Groupes analytiques p-adiques, Publ. Math. Inst. Hautes Études Sci., **26** (1965), 389–603.

[20] A. Lubotzky, A. Mann, Powerful p-groups. I: finite groups, J. Algebra, **105** (1987), 484–505; II: p-adic analytic groups, *ibid.*, 506–515.

[21] P. Shumyatsky, On periodic solvable groups having automorphisms with nilpotent fixed-point groups, Israel J. Math., **87** (1994), 111–116.

[22] P. Shumyatsky, Exponent of finite groups admitting an involutory automorphism, J. Group Theory, **2** (1999), 367–372.
[23] P. Shumyatsky, On pro-p groups admitting a fixed-point-free automorphism, J. Algebra, **228** (2000), 357–366.
[24] P. Shumyatsky, Finite groups and the fixed points of coprime automorphisms, Proc. Amer. Math. Soc., to appear.
[25] P. Shumyatsky, On centralizers in locally finite groups, J. Algebra, to appear.
[26] S. Thomas, The classification of the simple periodic linear groups, Arch. Math. (Basel) **41** (1983), 103–116.
[27] J. G. Thompson, Finite groups with fixed-point-free automorphisms of prime order, Proc. Nat. Acad. Sci. USA, **45** (1959), 578–581.
[28] A. Turull, Character theory and length problems, in "Finite and Locally Finite Groups", Kluwer Academic Publ., NATO ASI Series **471**, Dordrecht-Boston-London, 1995, 377–400.
[29] Y. M. Wang and Z. M. Chen, Solubility of finite groups admitting a coprime order operator group, Boll. Un. Mat. Ital. A **7** (1993), 325–331.
[30] J. N. Ward, Automorphisms of finite groups and their fixed-point groups, J. Austral. Math. Soc., **9** (1969), 467–477.
[31] J. N. Ward, On finite groups admitting automorphisms with nilpotent fixed-point group, Bull. Austral. Math. Soc., **5**(1971), 281–282.
[32] J. N. Ward, On groups admitting a noncyclic abelian automorphism group, Bull. Austral. Math. Soc., **9**(1973), 363–366.
[33] J. S. Wilson, On the structure of compact torsion groups, Monatshefte für Mathematik, **96** (1983), 57-66.
[34] E. Zel'manov, On periodic compact groups, Israel J. Math. **77** (1992), 83-95.
[35] E. Zel'manov, *Nil Rings and Periodic Groups*, The Korean Math. Soc. Lecture Notes in Math., Seoul, 1992.

CLASSIFYING IRREDUCIBLE REPRESENTATIONS IN CHARACTERISTIC ZERO

ALEXANDRE TURULL[1]

Department of Mathematics, University of Florida, Gainesville, FL 32611, U.S.A.

Abstract

We discuss techniques which allow us to classify the representations of some finite groups over non-algebraically closed fields of characteristic zero. We propose the use of Clifford classes as a convenient and effective way to describe and calculate our answers. Furthermore, we argue for the use of global, rather than local, calculations for the classical groups.

1 Introduction

Given a finite group G and a field F in characteristic zero, a basic problem is to understand the representations of G as automorphisms of finite dimensional vector spaces over F. Maschke's Theorem allows us to concentrate on the *irreducible* representations of G. The use of characters is a convenient and useful way to work with isomorphism classes of representations. However, the computation of the isomorphism types of irreducible representations is often difficult. This problem amounts to the calculation of the Schur index $\mathrm{m}_F(\chi)$ for each complex irreducible character $\chi \in \mathrm{Irr}(G)$.

Richard Brauer contributed two important tools for the solution of these problems. First, instead of calculating the Schur indices of χ over F, for all F, one can now calculate an element of the Brauer group $[\chi]$ over a suitable field, and one deduces from it all the Schur indices of χ for all fields F. Second, his Characterization of Characters Theorem, and related ideas, show that one could calculate Schur indices by understanding the restriction of the character to certain solvable subgroups, and the Schur indices of the irreducible characters of these solvable subgroups.

Since the beginning of character theory, there has been a rightful emphasis on the characters of the classical groups. Thanks to work of Frobenius, Schur, and more recent results of Deligne and Lusztig and many others, we know quite a lot about the irreducible characters of many classical groups. We know far less about the Schur indices of these characters. In this paper, I want to discuss some new techniques that have allowed recent progress in the calculation of Schur indices in general, and in the calculation of Schur indices of specific families of classical groups in particular. These techniques should also prove useful to effectively investigate further classical groups.

When trying to calculate the Schur indices of the irreducible characters of groups such as the classical groups, one can go beyond the two fundamental contributions

[1] The author is supported by an NSA grant.

of Richard Brauer mentioned above in the following ways. The first way is suggested by the fact that many classical groups have normal subgroups which are also of interest. Classical groups often appear within a pair of groups G and N, where G is a finite group and N is a normal subgroup of G, and all the groups between G and N are of interest. Often N is quasi-simple, and G/N is uncomplicated, often cyclic. The irreducible characters of all these groups are related by Clifford theory, and such relation is used effectively when describing the characters systematically, see for example [1]. We show that we can define a generalization of the Brauer group of a field, which we call the set of Clifford classes $\mathrm{Clif}(G/N, F)$ of G/N over F. The elements of this set allow us to effectively work with all the characters of all these groups which are related by Clifford theory at once. Hence, in the same way that the calculation of $[\chi]$ allows one to know the Schur indices of χ over every field F, the calculation of $[[\chi]] \in \mathrm{Clif}(G/N, K)$ allows us to get all the Schur indices of χ and all the characters related to it by Clifford theory over every field.

If Brauer's first contribution mentioned above suggested the definition and use of the Clifford set $\mathrm{Clif}(G/N, F)$ to describe the Schur indices of large families of characters over all fields, new information about classical groups suggests that we need to supplement his second. Indeed, all indications are that Schur indices of quasi-simple groups are substantially less complicated that Schur indices of solvable groups. For example, all Schur indices of quasi-simple groups that have been calculated are at most 2, whereas the Schur indices of solvable groups can be any positive integer. Hence, a classification of the representations of the classical groups over fields of characteristic zero can not be expected to rely solely on restrictions of characters to solvable subgroups.

Indeed, the situation may be analogous to the classification of finite simple groups, where, Sylow's Theorem not withstanding, one wants to avoid relying of the detailed structure of the p-groups, because their structure is far more intricate than that of the simple groups. Here, we propose to try to work as much as possible with the whole group. We suggest some *global* and some *descent methods*, that allow us to calculate, starting from information over a larger field, information over smaller ones.

In this paper, we being with a review of the standard definitions and results that allows us to treat all F *at once*. We then discuss the more recent method of Clifford classes, that allows us to treat certain sets of related groups G together. We then discuss some methods to calculate the relevant invariants that rely on global techniques. Finally, we review briefly some results that can be obtained about the classical groups using these techniques.

* * *

Roughly speaking, given G, we can treat all the fields F at once using the following. (See Section 2, below for more details). First calculate the character table for G, in other words, solve the problem for $F = \mathbf{C}$. Next, for each $\chi \in \mathrm{Irr}(G)$, calculate its field of values $\mathbf{Q}(\chi)$, and an element $[\chi] \in \mathrm{Br}(\mathbf{Q}(\chi))$, in the Brauer group of $\mathbf{Q}(\chi)$. Then, given any field F contained in $\mathbf{C}$, the characters of the irreducible modules of G over F are exactly those obtained as follows. Take any

$\chi \in \mathrm{Irr}(G)$, we then obtain from χ a unique character $\widetilde{\chi}$ afforded by some irreducible module for G over F. Let $\chi = \chi_1, \ldots, \chi_n$ be the distinct Galois conjugates of χ over F. Let $[\chi]_F$ be the projection of $[\chi]$ to an element of $\mathrm{Br}(F(\chi))$, and let m be the order of $[\chi]_F$ ($m = m_F(\chi)$ is called the Schur index of χ over F). Then, $\widetilde{\chi} = m(\chi_1 + \cdots + \chi_n)$. The $\widetilde{\chi}$ depend only on the Galois conjugacy class of χ, different Galois conjugacy classes correspond to different irreducible representations of G over F, and all irreducible representations of G over F afford some $\widetilde{\chi}$ as character. Hence, using some Galois theory and some knowledge of Brauer groups, we can reduce our computations to calculating $\mathrm{Irr}(G)$, and, for each $\chi \in \mathrm{Irr}(G)$, to calculating $[\chi] \in \mathrm{Br}(\mathbf{Q}(\chi))$. However, neither calculation is easy in general.

When G is not simple, and it has some proper non-trivial normal subgroup N, one can use information about N and G/N to calculate the irreducible modules for G. Over algebraically closed fields, traditional Clifford Theory describes how to do this. Over non-algebraically closed fields F, there are additional difficulties, in part arising from the fact that an irreducible character of a subgroup that induces up to an irreducible character of G may have a larger field of values than the field of values of the character it induces to.

In the present paper, I want to argue that the problem of describing the irreducible representations of G over a field F of characteristic zero, in the presence of a normal subgroup N, can profitably be split into a number of distinct problems. I will give a brief overview of how the solution of all these solve the problem. Furthermore, I will describe some of the results I have obtained recently about the various components of the problem. In addition, I will discuss how this new method has already successfully helped solve the problem of finding all the representations of G over any field F of characteristic zero for some families of classical groups.

Suppose G has a non-trivial proper normal subgroup N. Let $\chi \in \mathrm{Irr}(G)$, and let $\psi \in \mathrm{Irr}(N)$ be contained in the restriction of χ to N. Clifford theory teaches us that there are a number of characters of G and its subgroups that contain N that are closely related to χ and should be studied together with χ. Our strategy is to define a set $\mathrm{Clif}(G/N, F)$, which has a natural definition analogous to that of the Brauer group, and whose elements encapsulate how to go from ψ to each of these characters. Thus, to each $\chi \in \mathrm{Irr}(G)$ is associated an element $[[\chi]] \in \mathrm{Clif}(G/N, F)$, which depends only on ψ, and which controls all these characters.

In this way, the problem of describing all the irreducible representations of G over a field F in characteristic zero, when N is a normal subgroup of G, and we have $\mathrm{Irr}(N)$, can be split into three distinct problems. The first two involving only the group G/N, and only the last one involving the embedding of N in G.

- First, we need to compute $\mathrm{Clif}(G/N, F)$.
- Second, we need to compute the Clifford theory of G corresponding to each element of $\mathrm{Clif}(G/N, F)$, i.e., given some element of $\mathrm{Clif}(G/N, F)$, this tells us exactly how to parametrize all the relevant characters, their Schur indices, etc., and we want to make this as explicit as possible, and as easy to use as possible.
- Third, and finally, for each $\psi \in \mathrm{Irr}(N)$, we need to compute the corresponding element of $\mathrm{Clif}(G/N, F)$ which, by our second step then solves our problem.

2 Many fields at once: Using the Brauer group

In this section, we review the definition of the Brauer group, and how it is used to solve our problem for fixed G and varying F. These classical results are the model for our definition of Clifford classes, and their use for our purposes. In fact, we can view this section as the development we are arguing for in the case when $G = N$. All definitions and results here are standard. See for example [9] for details and proofs. Let $F \subseteq \mathbf{C}$ be a field. In this section, we assume that $\mathrm{Irr}(G)$ is available to us.

2.1 First: Define and compute $\mathrm{Br}(F)$

Definition 2.1 An algebra A over F is said to be *simple* if it has exactly two two-sided ideals. It is said to be *central* if its center has dimension 1 over F. It is said to be a *full matrix algebra* if there exists some finite dimensional F-vector space $M \neq 0$ such that $A = \mathrm{End}_F(M)$.

Definition 2.2 (*Brauer group of F*). If A and B are central simple algebras over F, we say that A and B are equivalent if and only if there exist full matrix algebras E and E' such that

$$A \otimes E \simeq B \otimes E'.$$

The elements of $\mathrm{Br}(F)$ are the equivalence classes of central simple algebras over F. Multiplication in $\mathrm{Br}(F)$ is induced by the tensor product of algebras over F. So if A and B are central simple algebras over F, denoting by $[A]$ and $[B]$ their corresponding classes in $\mathrm{Br}(F)$, we have

$$[A][B] = [A \otimes B].$$

Remark 2.3 The number theory literature then discusses how to operate with $\mathrm{Br}(F)$. Namely, it describes the elements of $\mathrm{Br}(F)$, its multiplication, and how to compute the class $[A] \in \mathrm{Br}(F)$ of some central simple algebra A over F. This description is particularly effective for many F's of particular interest such as when F is a local or global field (e. g. $\mathbf{Q}$, $\mathbf{R}$, $\mathbf{Q}_p$ and their finite extensions).

2.2 Second: Given $[\chi] \in \mathrm{Br}(\mathbf{Q}(\chi))$, describe the representations

In this, the classical case, there is exactly one irreducible representation over F to be described. Its character is obtained as follows. Extend the scalars of $[\chi] \in \mathrm{Br}(\mathbf{Q}(\chi))$ to $F(\chi)$, and obtain an element $[\chi]_F \in \mathrm{Br}(F(\chi))$. Let m be the order of $[\chi]_F$ as an element of the Brauer group. Then, the unique character of G afforded by the irreducible representation over F arising from $\chi \in Irr(G)$ is

$$m\,(\chi_1 + \cdots + \chi_n)\,,$$

where the $\chi_1 = \chi, \ldots, \chi_n$ are the distinct Galois conjugates of χ over F.

2.3 Third: For each $\chi \in \mathrm{Irr}(G)$, calculate $[\chi] \in \mathrm{Br}(\mathbf{Q}(\chi))$

Let G be a finite group. Let $\chi \in \mathrm{Irr}(G)$. Then, we assign to χ the element $[\chi] \in \mathrm{Br}(\mathbf{Q}(\chi))$ as follows. We temporarily set $K = \mathbf{Q}(\chi)$.

Theorem 2.4 *There exist non-zero modules over K affording as character a multiple of χ. Let M be any one of them. Then, $\mathrm{End}_{KG}(M)$ is a central simple algebra over K. Furthermore, the class of $\mathrm{End}_{KG}(M)$ in $\mathrm{Br}(K)$ is independent of the choice of M.*

Definition 2.5 We define $[\chi]$ to be the class of $\mathrm{End}_{KG}(M)$ in $\mathrm{Br}(K)$, for any M as in Theorem 2.4.

Theorem 2.4 guarantees that $[\chi]$ is well defined for each $\chi \in \mathrm{Irr}(G)$. In this approach, the calculation of the *global* $[\chi]$ becomes the focus. We discuss some methods to explicitly calculate this invariant in Section 4.

3 Many groups at once: using Clifford classes

In this section, we define Clifford classes, and sketch their use in the solution of our classification of representations problem. Our basic set up is that we have a field F of characteristic zero, a finite group G, and a normal subgroup N of G. We assume, furthermore, that $\mathrm{Irr}(N)$ is available to us. For each $\psi \in \mathrm{Irr}(N)$, we classify all the representations for each subgroup over any field F of characteristic zero lying above ψ.

We set $\bar{G} = G/N$. We define the set $\mathrm{Clif}(\bar{G}, F)$ of Clifford classes, and we show how this set can be used to describe all the representations of G over F. This set was introduced in [9] for this purpose, but the name Clifford classes was introduced in [16].

3.1 First: Define and compute $\mathrm{Clif}(\bar{G}, F)$

If algebras over F are the basic ingredients for the definition of the Brauer group, it is $\bar{G}$-algebras over F that are the basic ingredients for the definition of the Clifford set of $\bar{G}$ over F.

Definition 3.1 A $\bar{G}$-algebra A over F is said to be *simple* if it has exactly two $\bar{G}$-invariant two-sided ideals. It is said to be *central* if the centralizer of $\bar{G}$ in the center of of the algebra A has dimension 1 over F. It is said to be a *trivial $\bar{G}$-algebra* if there exists some $F\bar{G}$-module $M \neq 0$ such that $A = \mathrm{End}_F(M)$, with the natural action of $\bar{G}$ on A.

Definition 3.2 (*Clifford Classes of $\bar{G}$ over F*). If A and B are central simple $\bar{G}$-algebras over F, we say that A and B are equivalent if and only if there exist trivial $\bar{G}$-algebras E and E' such that

$$A \otimes E \simeq B \otimes E',$$

as $\bar{G}$-algebras. The elements of $\mathrm{Clif}(\bar{G}, F)$ are the equivalence classes of central simple $\bar{G}$-algebras over F.

Remark 3.3 One difference with the ordinary Brauer group is that $\mathrm{Clif}(\bar{G}, F)$ does not have a natural operation on it, and so it is simply a set.

The problem of characterizing the elements of $\mathrm{Clif}(\bar{G}, F)$ becomes more difficult as $\bar{G}$ is more complicated.

Theorem 3.4 *For* $\bar{G} = 1$, $\mathrm{Clif}(\bar{G}, F)$ *is simply the set of elements of* $\mathrm{Br}(F)$.

This follows immediately from the definition. The case $|\bar{G}| = 2$ turns out to be classical too. The elements of $\mathrm{Clif}(\mathbf{Z}/2\mathbf{Z}, F)$ are naturally identified with the elements of the Brauer-Wall group of F, see for example [7] for a definition of the Brauer-Wall group and a characterization of its elements. Hence, the elements of $\mathrm{Clif}(\mathbf{Z}/2\mathbf{Z}, F)$ are in one-to-one correspondence with the elements of $\mathrm{Br}(F) \times \{0,1\} \times F^{\times}/(F^{\times})^2$. In [13], we obtain a more general result describing the elements of $\mathrm{Clif}(\mathbf{Z}/n\mathbf{Z}, F)$ as follows:

Theorem 3.5 *Let F be a field of characteristic zero and let G be a finite cyclic group, with preferred generator g_0. Given any triple $[Z, \alpha, b]$, where Z is a central simple commutative G-algebra, $\alpha \in F^{\times}$ and $b \in \mathrm{Br}(F)$, we may construct (see [13] Definition 2.6) a central simple G-algebra which we denote $[Z, \alpha, b]$. Then, for every central simple G-algebra A over F, there exists some $[Z, \alpha, b]$ which is equivalent to A. In addition, two G-algebras $[Z, \alpha, b]$ and $[Z_0, \alpha_0, b_0]$ are equivalent if and only if both of the following hold:*

1. *$Z \simeq Z_0$ as G-algebras.*
2. *Setting $I = C_G(Z) = C_G(Z_0)$ and $m = |I|$, we have that $\alpha\alpha_0^{-1} = \beta^m$ for some $\beta \in F^{\times}$ such that $b = b_0 \mathrm{br}(G/I, Z, \beta)$.*

A central simple commutative $\bar{G}$-algebra Z can be thought of as a Galois extension of F whose Galois group is a quotient of $\bar{G}$. Furthermore, here, $\mathrm{br}(G/I, Z, \beta) \in \mathrm{Br}(F)$ and $b_0\mathrm{br}(G/I, Z, \beta)$ represents the product in the Brauer group $\mathrm{Br}(F)$. The element $\mathrm{br}(G/I, Z, \beta)$ is constructed by a slight generalization of the usual crossed product. It is easy to see that Theorem 3.5 implies the classical [7] characterization of the elements of the Brauer-Wall group in the case $\bar{G} = \mathbf{Z}/2\mathbf{Z}$.

Further characterizations of the elements of $\mathrm{Clif}(\bar{G}, F)$ are obtained in [10, 11, 12], but the general problem remains open. The elements that correspond to characters whose inertia group is all of G, however, are easily classified as follows, see Corollary 5.4 in [11]. The relevant $\bar{G}$-algebras A are characterized by having a one-dimensional center, $Z(A) = F$.

Theorem 3.6 *Let $\bar{G}$ be any group, and let F be any field of characteristic zero. Then, the equivalence classes of central simple $\bar{G}$-algebras whose center is one dimensional are in natural one-to-one correspondence with the elements of*

$$H^2(\bar{G}, F^{\times}) \times \mathrm{Br}(F).$$

3.2 Second: Given $[[\chi]] \in \mathrm{Clif}(\bar{G}, F)$, describe the representations

If $\chi \in \mathrm{Irr}(G)$, we know that $[[\chi]]$ is sufficient to understand its Clifford theory. Indeed we have the following, see [9].

Theorem 3.7 *Let $\chi \in \mathrm{Irr}(G)$, and let $\psi \in \mathrm{Irr}(N)$ be contained in the restriction of χ to N. Suppose we know $[[\chi]] \in \mathrm{Clif}(\bar{G}, F)$ (which in fact depends only on ψ and the embedding of N in G, and not on χ itself). Then, we can use $[[\chi]]$ to parametrize all the irreducible characters ζ of any subgroup of G that contains N with the property that $(\zeta_N, \chi_N)_N > 0$. Furthermore, $[[\chi]]$ tells us, for each parametrized character ζ, its relative degree $\zeta(1)/\psi(1)$, its corresponding element $[\zeta] \in \mathrm{Br}(F(\zeta))$ (and hence the representations over all F), as well as the effects of restriction, induction, conjugation and even multiplication by any irreducible character of G/N.*

When $\bar{G}$ is cyclic, we have an explicit and convenient description of the elements of $\mathrm{Clif}(\bar{G}, F)$, see Theorem 3.5, above. This can be used to describe the relevant characters and their properties explicitly. See [13] for details.

Theorem 3.8 *Suppose $\bar{G}$ is cyclic, and $[[\chi]]$ is the class of the $\bar{G}$-algebra $[Z, \alpha, b]$ as in Theorem 3.5. Then, the parametrization and results of Theorem 3.7 can be expressed explicitly and conveniently.*

3.3 Third: For each $\chi \in \mathrm{Irr}(G)$, calculate $[[\chi]] \in \mathrm{Clif}(\mathbf{Q}(\chi_N))$

Let G be a finite group, and let N be a normal subgroup of G. Let $\chi \in \mathrm{Irr}(G)$, and let $\psi \in \mathrm{Irr}(N)$ be contained in the restriction of χ to N. It is easy to see that the field of values of each character ζ as in Theorem 3.7 contains $\mathbf{Q}(\chi_N)$, the field of values of the restriction of χ to N. We set $\bar{G} = G/N$ and $K = \mathbf{Q}(\chi_N)$. Then, we assign to χ the element $[[\chi]] \in \mathrm{Clif}(\bar{G}, K)$ as follows, see [9].

Theorem 3.9 *There exist non-zero modules over K affording as character a character whose restriction to N is a multiple of the sum of the distinct G-conjugates of ψ (or a rational multiple of χ_N). Let M be any one of them. Then, $\mathrm{End}_{KN}(M)$ is a central simple $\bar{G}$- algebra over K. Furthermore, the class of $\mathrm{End}_{KN}(M)$ in $\mathrm{Clif}(\bar{G}, K)$ is independent of the choice of M.*

Definition 3.10 We define $[[\chi]]$ to be the class of $\mathrm{End}_{KN}(M)$ in $\mathrm{Clif}(\bar{G}, K)$, for any M as in Theorem 3.9.

Theorem 3.9 guarantees that $[[\chi]]$ is well defined for each $\chi \in \mathrm{Irr}(G)$, and furthermore, it shows that $[[\chi]]$ depends only on the G-conjugacy class of ψ and not on χ itself. It is convenient that this definition has no requirement that M be irreducible. Indeed, if we were to assume that M is irreducible, then we would need to know the structure of some irreducible module before starting our computation, hence, we would need to know something about the Schur index of χ before calculating $[[\chi]]$, which is certainly undesirable. Furthermore, this way the definition can easily be extended to be functorial with respect to extensions of the ground field.

In this approach, the calculation of the *global* $[[\chi]]$ becomes the focus. We discuss some methods to explicitly calculate this invariant in Section 4.

4 Global methods

We have a multitude of solvable groups, but much fewer simple and almost simple groups. Hence, the Classification of Finite Simple group tries to avoid having to deal directly with p-groups as much as possible, trying instead to work with large subgroups of the unknown group instead. Likewise, I think that to effectively classify the representations of the classical groups over all fields of characteristic zero one needs to concentrate on large subgroups of the group in question. Brauer's Theorem implies that understanding all representations of the solvable subgroups of the group and their embedding in the group should be enough to calculate the Schur indices. However, the groups have a multitude of solvable subgroups, and their Schur indices can be extremely complicated, making such an approach impractical.

Indeed, my recent successes in effectively calculating the Schur indices of all the irreducible characters of some families of classical groups have ultimately been based on some *global* reduction. The next result refers to the reduction obtained in [8].

Theorem 4.1 *Let $\lambda_1 > \lambda_2 > \cdots > \lambda_r$ and $\mu_1 > \mu_2 > \cdots > \mu_s$ be strict partitions of n_1 and n_2 respectively, and suppose that the union of the two partitions is a strict partition of $n = n_1 + n_2$. Let $[[\lambda]], [[\mu]] \in \mathrm{Clif}(\mathbf{Z}/2\mathbf{Z}, \mathbf{Q})$ be the elements associated via Definition 3.10 to the corresponding characters of the double covers of S_{n_1} and S_{n_2}. Then, there is an effective formula to calculate the element $[[\lambda \cup \mu]] \in \mathrm{Clif}(\mathbf{Z}/2\mathbf{Z}, \mathbf{Q})$ corresponding to the character of the appropriate double cover of S_n.*

As another example, we offer the following from [14]. Further refinements of this result yield an even more explicit formula in certain cases, and this is the basis of the calculation of the Schur indices of all the irreducible characters of the special linear groups, see [15]. The next two results are examples of the *descent method*: they provide the needed information over a smaller field from properties of the representation of the group over a larger field.

Theorem 4.2 *Let G be a finite group, $\chi \in \mathrm{Irr}(G)$ and F be a field containing $\mathbf{Q}(\chi)$. Let U be a subgroup of G and $\theta \in \mathrm{Irr}(U)$ be such that $\theta(1) = 1$ and $(\mathrm{Res}^G_U(\chi), \theta) = 1$. Set $K = F(\theta)$ and set $n = |\mathrm{Gal}(K/F)|$, and assume that $\mathrm{Gal}(K/F) = \langle\sigma\rangle$ is cyclic. Assume that there exists a group H containing G as a normal subgroup, and a character $\tilde{\chi} \in \mathrm{Irr}(H)$ extending the character χ, such that $F(\tilde{\chi}) \cap K = F$. Assume, furthermore, that there is some element $x \in \mathrm{N}_H(U)$ such that, for all $u \in U$, $\theta(x^{-1}ux) = \sigma\theta(u)$, and $x^n \in U$. Then, there exists some h in the coset $x^{-1}G$ such that $\tilde{\chi}(h) \neq 0$ and for each such h, we have*

$$[\chi] = \mathrm{br}(K/F, \sigma, \tilde{\chi}(h)^n \theta(x^n)).$$

If we are interested not only on the characters of the special linear group, but on the characters of all the groups contained between the special linear groups and the general linear groups, then the above theorem is not enough, because it only calculates $[\chi]$ when what is needed is $[[\chi]]$. In [16], we obtain some explicit global results in view of obtaining what we need.

Theorem 4.3 *Let H be a finite group, and let J be a normal subgroup of H such that $G = H/J$ is cyclic with generator g_0. Let χ be an irreducible character of H, and assume that F is a field containing the values of $\mathrm{Res}_J^H(\chi)$. Then the Clifford class $[[\chi]] \in \mathrm{Clif}(F)$ is given by $[Z, \alpha, b]$ (see Theorem 3.5) as follows. Z and α_0 are easy to calculate, but we do not give the details here. Pick any $\beta \in F^\times$, we can then set $\alpha = \beta^{|I|}\alpha_0$, and the corresponding b is calculated as follows. Set $n = [H : I_0]$ and set γ to be an n-th root of $\beta\chi(j_0)$ so that $\gamma^n = \beta\chi(j_0)$. Take a finite Galois extension L of F such that $L \supseteq F(\gamma)$ and L is a splitting field for χ. Let M be a module for H over L affording χ. Then, the crossed product of L and S is a central simple algebra over F, and b is simply its class in $\mathrm{Br}(F)$.*

5 Applications to classical groups

There has been interest in the calculation of Schur indices for classical groups ever since the important contributions of Schur himself. Hence, we know the Schur indices of the symmetric groups, the alternating groups, more generally the Weyl groups. We also know the Schur indices of the sporadic simple groups, see [3, 4]. We also have a lot of partial information on the Schur indices of the groups of Lie type, see the work of Gow [5, 6] and others.

The techniques discussed above allow us to go beyond these results, and obtain *complete classifications* of the representations over every field in some new cases. In [8], we handle the case of the double covers of the symmetric and alternating groups completely. Similar techniques [2] handle all the irreducible characters of all the double covers of the Weyl groups of type B_n. In [15], we calculate the Schur index for each irreducible character of any special linear group. I intend to calculate explicitly the Schur index of every irreducible character of every group contained between the general linear group and the special linear group in a forthcoming paper.

We still do not have complete classifications for the irreducible characters of the other classical families. Hopefully, some extension of the ideas presented here should prove useful to solve these problems as well.

References

[1] J. H. Conway, R. T. Curtis, S. P. Norton, R. A. Parker, and R. A. Wilson, *Atlas of Finite Groups*, Clarendon Press, Oxford, 1985.

[2] Z. Du, *Schur indices of projective characters of the binary octahedral groups*, Ph. D. Thesis, University of Florida, Gainesville, 1998.

[3] W. Feit, *The computations of some Schur indices*, Israel J. Math. **46** (1983), 274–300.

[4] W. Feit, *Schur indices of characters of groups related to finite sporadic simple groups*, Israel J. Math. **93** (1996), 229–251.

[5] R. Gow, *Schur indices of some groups of Lie type*, J. Algebra **42** (1976), 102–120.

[6] R. Gow, *On the Schur indices of characters of finite classical groups*, J. London Math. Soc.(2) **24** (1981), 135-147.

[7] T. Y. Lam, *The algebraic theory of quadratic forms*, Benjamin, Reading, Mass., 1973.

[8] A. Turull, *The Schur indices of projective characters of symmetric and alternating groups*, Ann. Math. **135** (1992), 91–124.

[9] A. Turull, *Clifford Theory with Schur indices*, J. Algebra **170** (1994), 661–677.

[10] A. Turull, *Some invariants for equivalent G-algebras*, J. Pure Appl. Algebra **98** (1995), 209–222.

[11] A. Turull, *Equivalence of G-algebras with complemented centroid*, Comm. Algebra **22** (1994), 5037–5078.

[12] A. Turull, *Equivalence of G-algebras for abelian G*, Proc. Amer. Math. Soc. **123** (1995) 1655–1660.

[13] A. Turull, *Clifford Theory for cyclic quotient groups*, J. Algebra **227** (2000), 133–158.

[14] A. Turull, *A formula for calculating some Schur indices*, J. Algebra **227** (2000), 124–132.

[15] A. Turull, *The Schur indices of the irreducible characters of the special linear groups*, J. Algebra **235** (2001), 275–314.

[16] A. Turull, *Calculating Clifford Classes for Characters Containing a Linear Character Once*, to appear.

LIE METHODS IN GROUP THEORY

MICHAEL VAUGHAN-LEE

Christ Church, Oxford, OX1 1DP, England

Abstract

This article is based on four lectures I gave at "Groups St. Andrews" in Oxford in August 2001. The lectures were intended as a survey of some applications of Lie algebra techniques to group theory, and much of the material presented here has appeared elsewhere. But the application of the Baker-Campbell-Hausdorff formula to Burnside groups is new.

1 The Baker-Campbell-Hausdorff formula

The connection between groups and Lie algebras is most clearly illustrated by the Baker-Campbell-Hausdorff formula. We show how the formula gives an isomorphism between the category of nilpotent Lie algebras over $\mathbb{Q}$ and the category of torsion free divisible nilpotent groups. We also indicate how the formula might be used to investigate finite p-groups. Finally, we give a new application of the formula to the the computation of Burnside groups of prime exponent.

Let A be the free associative algebra with unity over the rationals $\mathbb{Q}$ which is freely generated by non-commuting indeterminates x, y. We extend A to the ring $\widehat{A}$ of formal power series consisting of the formal sums

$$\sum_{n=0}^{\infty} u_n,$$

where u_n is a homogeneous element of weight n in A. If $a \in \widehat{A}$, and if the homogeneous component of a of weight 0 is 0, then we define

$$\mathrm{e}^a = 1 + a + \frac{a^2}{2!} + \frac{a^3}{3!} + \ldots$$

in the usual way. The product $\mathrm{e}^x\mathrm{e}^y \in \widehat{A}$ can be expressed in the form $1+u$ for some $u \in \widehat{A}$, and

$$\mathrm{e}^x\mathrm{e}^y = \mathrm{e}^v$$

where

$$v = \sum_{n=1}^{\infty} (-1)^{n-1} \frac{u^n}{n}.$$

The Baker-Campbell-Hausdorff formula (see, for example, Jacobson [13]) enables us to compute the homogeneous components of v. The first few components are given below.

$$\begin{aligned} v \;=\; & x+y-\frac{1}{2}[y,x]+\frac{1}{12}[y,x,x]-\frac{1}{12}[y,x,y]+\frac{1}{24}[y,x,x,y] \\ & -\frac{1}{720}[y,x,x,x,x]-\frac{1}{180}[y,x,x,x,y] \\ & +\frac{1}{180}[y,x,x,y,y]+\frac{1}{720}[y,x,y,y,y] \\ & -\frac{1}{120}[y,x,x,[y,x]]-\frac{1}{360}[y,x,y,[y,x]] \\ & -\frac{1}{1440}[y,x,x,x,x,y]-\frac{1}{360}[y,x,x,x,y,y] \\ & -\frac{1}{1440}[y,x,x,y,y,y]-\frac{1}{240}[y,x,x,y,[y,x]] \\ & -\frac{1}{720}[y,x,y,y,[y,x]]+\frac{1}{240}[y,x,y,[y,x,x]]+\ldots \end{aligned}$$

One important (and surprising) feature of this expression is that the homogeneous components of v are all Lie elements of A (that is elements in the Lie subalgebra of A generated by x and y with respect to the Lie product $[a,b]=ab-ba$). A proof of this may be found in [32]. A similar formula holds for commutators:

$$[\mathrm{e}^y,\mathrm{e}^x]=\mathrm{e}^w,$$

where

$$w=[y,x]+\frac{1}{2}[y,x,x]+\frac{1}{2}[y,x,y]+\frac{1}{6}[y,x,x,x]+\frac{1}{4}[y,x,x,y]+\frac{1}{6}[y,x,y,y]+\ldots.$$

(Here $[\mathrm{e}^y,\mathrm{e}^x]$ is the group commutator $\mathrm{e}^{-y}\mathrm{e}^{-x}\mathrm{e}^y\mathrm{e}^x$, and w is an infinite sum of Lie elements in A.)

These formulae sometimes enable us to define a group structure on a Lie algebra. Perhaps the simplest situation where this applies is when L is a nilpotent Lie algebra over the rationals $\mathbb{Q}$. (See [1].) We write the element v above as

$$v=v_1+v_2+\ldots,$$

where v_i is a homogeneous Lie element of weight i, for $i=1,2,\ldots$, and we consider the truncated expression

$$\widetilde{v}(x,y)=v_1+v_2+\ldots+v_c,$$

where c is the class of L. If $a,b\in L$, then $\widetilde{v}(a,b)$ can be interpreted as an element of L. We define an operation "$\circ$" on L by setting

$$a\circ b=\widetilde{v}(a,b),$$

for $a,b\in L$. With this operation, $\langle L,\circ\rangle$ is a torsion free nilpotent group (of class c). Furthermore $\langle L,\circ\rangle$ is divisible, which is to say that for every $a\in L$ and every non-zero integer n there is an element $b\in L$ such that $b^n=a$. Every torsion free divisible nilpotent group arises in this way, and this connection provides an isomorphism between the category of nilpotent Lie algebras over $\mathbb{Q}$ and the category of torsion free divisible nilpotent groups. This isomorphism is known as the Mal'cev correspondence.

The Baker-Campbell-Hausdorff formula also provides a connection between finite p-groups and nilpotent Lie rings (over $\mathbb{Z}$) of prime power order, in the case when the groups and Lie rings are nilpotent of class at most $p-1$. This is because if $n < p$ then the denominators of the coefficients that occur in the homogeneous component v_n of v of weight n are coprime to p. We let $\widetilde{v}(x,y)$ be the truncated expression

$$\widetilde{v}(x,y) = v_1 + v_2 + \ldots + v_{p-1}.$$

If L is a Lie ring of order p^n (so that L has characteristic p^k for some k), and if L is nilpotent of class at most $p-1$, then we can define a group operation "$\circ$" on L by setting

$$a \circ b = \widetilde{v}(a,b) \text{ for } a, b \in L.$$

This turns L into a group of order p^n, and every finite p-group of class at most $p-1$ arises in this way from a finite Lie ring. This appears as an exercise in Bourbaki [2], Chapter 2, and is known as the Lazard correspondence.

One possible application of this is to the classification of groups of order p^n for small n. Groups of order p^n have class at most $n-1$, and so for $p \geq n$ we can classify groups of order p^n by classifying nilpotent Lie rings of order p^n.

Consider the case $n = 3$. The classification of groups of order p^3 is hardly a difficult problem, but I believe that it is even easier if we look at Lie rings. Clearly the problem of classifying abelian Lie rings of any given order is precisely the same as the problem of classifying abelian groups of that order. So consider a non-abelian nilpotent Lie ring L of order p^3. Clearly L will be nilpotent of class 2, with L/L^2 of order p^2 and L^2 of order p. So L will be generated by two elements a, b (say), L^2 will be spanned by $[b,a]$, and $[b,a]$ will be central. Clearly $p[b,a] = 0$, and $pa = \lambda[b,a]$, $pb = \mu[b,a]$ for some λ, μ. So L is completely determined by the values of λ, μ, which we may view as elements of the field $\mathbb{Z}_p$.

One possibility is $\lambda = \mu = 0$. On the other hand, if λ and μ are not both zero then (interchanging a and b if necessary) we may assume that $\lambda \neq 0$. Then, if we replace b by $\lambda b - \mu a$, we have $pa = [b,a]$, $pb = 0$. So there are two possibilities for L.

A slightly less trivial example is the case of groups of order p^4 and class 3. Let L be a Lie ring of order p^4 which is nilpotent of class 3. Then we have $|L/L^2| = p^2$, $|L^2/L^3| = |L^3| = p$. Clearly the centralizer of L^2 has index p in L, and we choose generators a, b for L such that b centralizes L^2. So L is spanned by a, b, $[b,a]$, $[b,a,a]$. And $[b,a,b] = 0$. Clearly $[b,a,a]$ has order p, and $pa, pb \in L^2$. So $p[b,a] = [b,pa] = 0$ since b centralizes L^2. Let $pa = \lambda[b,a] + \mu[b,a,a]$, where we can view λ, μ as elements of the field $\mathbb{Z}_p$. Then

$$0 = p[a,a] = [pa,a] = \lambda[b,a,a],$$

and so $\lambda = 0$, and $pa \in L^3$. Similarly, since

$$0 = p[b,a] = [pb,a],$$

we see that $pb \in L^3$. Let $pa = \mu[b,a,a]$, $pb = \nu[b,a,a]$.

One possibility is that $\mu = \nu = 0$.

If $\mu \neq 0$, but $\nu = 0$ we can replace b by μb, and then we have $pa = [b,a,a]$, $pb = 0$.

On the other hand if $\nu \neq 0$ then we can replace a by $a - (\mu/\nu)b$, and then we have $pa = 0$, $pb = \nu[b,a,a]$. Finally, replacing a by ka we have $pa = 0$, $pb = (\nu/k^2)[b,a,a]$. So the Lie rings with $pa = 0$, $pb = \nu[b,a,a]$ ($\nu \neq 0$) fall into two isomorphism classes, depending on whether ν is a square or not a square. (We are assuming p is odd.)

So there are four Lie rings of order p^4 which are nilpotent of class 3, and this implies that for $p \geq 5$ there are four groups of order p^4 and class 3.

The Baker-Campbell-Hausdorff formula has also been used to study the connection between Burnside groups of exponent p and their associated Lie rings. The r-generator Burnside group of exponent n, $B(r,n)$, is defined to be F_r/N, where F_r is the free group of rank r, and N is the (normal) subgroup of F_r generated by $\{g^n \mid g \in F_r\}$. Clearly any r-generator group of exponent n is a homomorphic image of $B(r,n)$. Zel'manov's solution of the restricted Burnside problem ([37],[38]) implies that $B(r,n)$ has a largest finite quotient $R(r,n)$. Any finite r-generator group of exponent n is a homomorphic image of $R(r,n)$. (The restricted Burnside problem for prime exponent p was solved by Kostrikin [18].)

Zel'manov's and Kostrikin's work both rely on the connection between a group and its associated Lie ring. The associated Lie ring of a group G is defined as follows. First we form the lower central series of G,

$$\gamma_1 \geq \gamma_2 \geq \ldots \geq \gamma_i \geq \ldots,$$

by setting $\gamma_1 = G$, and $\gamma_{i+1} = [\gamma_i, G]$ for $i = 1, 2, \ldots$. For $i = 1, 2, \ldots$ we let $L_i = \gamma_i/\gamma_{i+1}$ and we think of L_i as a $\mathbb{Z}$-module. Then we let

$$L(G) = L_1 \oplus L_2 \oplus \ldots \oplus L_i \oplus \ldots,$$

as a direct sum of $\mathbb{Z}$-modules. If $a \in L_i$ and $b \in L_j$ then we can write $a = g\gamma_{i+1}$, $b = h\gamma_{j+1}$ for some $g \in \gamma_i$, $h \in \gamma_j$. We define the Lie product $[a,b]$ to be $[g,h]\gamma_{i+j+1} \in L_{i+j}$, and we extend this Lie product to the whole of $L(G)$ by linearity. This turns $L(G)$ into a Lie ring, the associated Lie ring of G.

Now let p be prime, and let $L(r,p)$ denote the associated Lie ring of $B(r,p)$. The Lie ring $L(r,p)$ is an r-generator Lie ring of characteristic p, and can be viewed as a Lie algebra over $\mathbb{Z}_p$. Furthermore it is well known that $L(r,p)$ satisfies the $(p-1)$-Engel identity $[x, \underbrace{y, y, \ldots, y}_{p-1}] = 0$. (See Corollary 2.4.8 of [32] for a proof of this fact. We will be giving an alternative proof in Section 3 of this article.) Kostrikin proved that an r-generator Lie algebra over $\mathbb{Z}_p$ satisfying the $(p-1)$-Engel identity is nilpotent (of bounded class). It follows that $L(r,p)$ is nilpotent of class c (say), and hence that any finite quotient of $B(r,p)$ is nilpotent of class at most c. Thus

$$R(r,p) \cong B(r,p)/\gamma_{c+1}(B(r,p)).$$

Furthermore, the order of $R(r,p)$ is the same as the order of $L(r,p)$. Clearly $L(r,p)$ is a homomorphic image of $E(r,p)$, where $E(r,p)$ is the (relatively) free r-generator Lie algebra over $\mathbb{Z}_p$ satisfying the $(p-1)$-Engel identity.

Sanov [25] used the Baker-Campbell-Hausdorff formula to show that the class $2p-2$ quotients of $L(r,p)$ and $E(r,p)$ are isomorphic. For many years it was hoped that $L(r,p) \cong E(r,p)$, but in 1974 Wall [34] found a new identity of weight $2p-1$ which holds in $L(r,p)$, and he showed that it is not a consequence of the $(p-1)$-Engel identity, at least for $p=5$ and 7. Wall's result implies that the class $2p-1$ quotient of $L(3,p)$ is a proper homomorphic image of the class $2p-1$ quotient of $E(3,p)$ for $p=5,7$. It seems very likely that this is also the case for all $p>7$. In 1985 I found a complete basis $\{K_n \mid n \geq 1\}$ for all the multilinear identities which are satisfied by the associated Lie rings of groups of exponent p. (The $(p-1)$-Engel identity and Wall's identity are both consequences of these identities. See Chapter 2 of [32] for details.) If we let $W(r,p)$ be the largest r-generator Lie algebra over $\mathbb{Z}_p$ satisfying all these multilinear identities, then $L(r,p)$ is a homomorphic image of $W(r,p)$, which in turn is a homomorphic image of $E(r,p)$.

There remained the hope that $L(r,p) \cong W(r,p)$. It has been known since 1974 that $L(2,5) = W(2,5) = E(2,5)$ (see [11]). And in 1990 I showed that $L(3,5) = W(3,5)$ (see [30]). But Eamonn O'Brien and I ([24]) have recently shown that $L(2,7)$ is a proper homomorphic image of $W(2,7)$. Mike Newman and I ([22]) have computed $W(2,7)$, and have shown that it has class 29 and dimension 20418. But it turns out that $L(2,7)$ is the class 28 quotient of $W(2,7)$, and that $L(2,7)$ has dimension 20416.

The main tool for obtaining detailed information about groups of exponent p (and about groups of prime-power exponent p^k) has been the p-quotient algorithm [9], [21]. Subject to limitations of computer memory and computer time, the p-quotient algorithm can be used to compute a consistent power-commutator presentation (PCP) for the class c quotient of $B(r,p^k)$ for any c, r, p^k. In practice we are restricted to small values of these parameters, but consistent PCPs have been computed for $B(r,4)$ $(r=2,3,4,5)$, $R(2,5)$, $R(3,5)$ and $R(2,7)$. In addition consistent PCPs have been computed for various quotients of $R(2,8)$ and $R(2,9)$. (Groups of exponent 2 and 3 are well understood.)

A PCP for a finite p-group P consists of a set $\{a_1, a_2, \ldots, a_n\}$ of generators together with n power relations

$$a_i^p = a_{i+1}^{\alpha_{i\,i+1}} a_{i+2}^{\alpha_{i\,i+2}} \ldots a_n^{\alpha_{i\,n}}$$

$(1 \leq i \leq n)$ with $0 \leq \alpha_{ij} < p$ for $i+1 \leq j \leq n$, and $\binom{n}{2}$ commutator relations

$$[a_j, a_i] = a_{j+1}^{\alpha_{i\,j\,j+1}} a_{j+2}^{\alpha_{i\,j\,j+2}} \ldots a_n^{\alpha_{i\,j\,n}}$$

$(1 \leq i < j \leq n)$ with $0 \leq \alpha_{ijk} < p$ for $j+1 \leq k \leq n$. These relations can be used to express every element of P as a normal word

$$a_1^{\alpha_1} a_2^{\alpha_2} \ldots a_n^{\alpha_n}$$

$(0 \leq \alpha_1, \alpha_2, \ldots, \alpha_n < p)$, and the collection process can be used to express the product of two normal words as a normal word. The PCP is said to be consistent if different normal words represent different elements of P, so that P has order p^n. Once a consistent PCP for $R(r, p^k)$ has been computed then the order of the group can be read off from the PCP, as well as other useful information such as the nilpotency class and derived length.

Another approach has been to use the nilpotent quotient algorithm for graded Lie rings ([10]) to compute Lie rings associated with Burnside groups. For example, consistent product presentations for $E(2,5)$, $E(3,5)$, $E(2,7)$, $W(3,5)$, and $W(2,7)$ have been computed. (As mentioned above, $W(2,5) = E(2,5)$.) A consistent product presentation for a graded Lie algebra L of dimension n over a field K consists of a basis $\{a_1, a_2, \ldots, a_n\}$ for L over K together with $\binom{n}{2}$ Lie product relations

$$[a_j, a_i] = \alpha_{i\,j\,j+1}a_{j+1} + \alpha_{i\,j\,j+2}a_{j+2} + \ldots + \alpha_{i\,j\,n}a_n$$

$(1 \leq i < j \leq n)$ with $\alpha_{ijk} \in K$ for $j+1 \leq k \leq n$. (The coefficients α_{ijk} are the structure constants for the algebra L.) Since $L(r,p)$ is a homomorphic image of $W(r,p)$ and $W(r,p)$ is a homomorphic image of $E(r,p)$, information about these Lie algebras gives information about the corresponding groups $R(r,p)$.

One reason for working with product presentations for these Lie algebras rather than with PCPs for the groups is that the computing resources required to compute consistent product presentations for $E(r,p)$ or $W(r,p)$ are considerably less than the resources needed to compute a consistent PCP for $R(r,p)$. For example, the Burnside group $R(2,5)$ has order 5^{34} and nilpotency class 12, and nowadays it only takes a few seconds of computer time to verify these facts. But Kostrikin was able to show using hand calculation that these were upper bounds for the order and class of $R(2,5)$ long before computers were first applied to the problem. Kostrikin obtained his bounds for $R(2,5)$ by studying the Lie algebra $E(2,5)$. But it took another 20 years before Havas, Wall and Wamsley [11] were able to show that Kostrikin's bounds were sharp by computing a consistent PCP for $R(2,5)$.

Here we investigate the possibility of using the Baker-Campbell-Hausdorff formula to construct a consistent PCP for $R(r,p)$ from a consistent product presentation for $W(r,p)$. We show that a consistent PCP for the class $2p-2$ quotient of $R(r,p)$ can be computed from a consistent product presentation for the class $2p-2$ quotient of $E(r,p)$. We also show that, provided $L(r,p) = W(r,p)$, a *significant proportion* of a consistent PCP for $R(r,p)$ can be constructed from a consistent product presentation for $W(r,p)$. Eamonn O'Brien and I ([24]) exploited this technique in constructing a consistent PCP for $R(2,7)$. Note that although $L(2,7) \neq W(2,7)$, the class 28 quotient of $L(2,7)$ is isomorphic to the class 28 quotient of $W(2,7)$, and this was sufficient for our purposes.

1.1 An example

We give an example to show how a consistent product presentation for the associated Lie ring of a finite group of exponent p can be "read off" from a consistent PCP for the group. We also show how (in some circumstances) the group PCP can be reconstructed from the Lie ring presentation by using the Baker-Campbell-Hausdorff formula. The group we consider is the class 6 quotient of $B(2,5)$ — let us call it G. This group has order 5^{14} and has a consistent PCP with generators $a_1, a_2, \ldots, a_{14}$. The PCP has 14 power relations $a_i^5 = 1$, and 91 commutator relations. To save space we omit all "trivial" commutator relations $[a_i, a_j] = 1$. The non-trivial commutator relations are as follows.

$$
\begin{aligned}
[a_2, a_1] &= a_3 \\
[a_3, a_1] &= a_4 \\
[a_3, a_2] &= a_5 \\
[a_4, a_1] &= a_6 \\
[a_4, a_2] &= a_7 \\
[a_4, a_3] &= a_9 a_{11} a_{12}^4 \\
[a_5, a_1] &= a_7 a_9 a_{10}^2 a_{11} a_{12}^2 a_{13}^3 a_{14}^4 \\
[a_5, a_2] &= a_8 \\
[a_5, a_3] &= a_{10}^2 a_{12}^3 a_{13} a_{14}^4 \\
[a_5, a_4] &= a_{12}^2 a_{13} \\
[a_6, a_1] &= a_{11} \\
[a_6, a_2] &= a_9 \\
[a_6, a_3] &= a_{11} \\
[a_7, a_1] &= a_9^2 a_{11}^2 a_{12}^2 a_{13} \\
[a_7, a_2] &= a_{10} \\
[a_7, a_3] &= a_{12}^3 a_{13} \\
[a_8, a_1] &= a_{10}^3 a_{12}^2 a_{13}^2 a_{14}^3 \\
[a_8, a_2] &= a_{14}^3 \\
[a_8, a_3] &= a_{14}^2 \\
[a_9, a_1] &= a_{11} \\
[a_9, a_2] &= a_{12} \\
[a_{10}, a_1] &= a_{13} \\
[a_{10}, a_2] &= a_{14}
\end{aligned}
$$

The PCP generators $a_3, a_4, \ldots, a_{14}$ all have definitions in terms of a_1 and a_2. These are given by the following table.

$$
\begin{aligned}
a_3 &= [a_2, a_1] \\
a_4 &= [a_2, a_1, a_1] \\
a_5 &= [a_2, a_1, a_2] \\
a_6 &= [a_2, a_1, a_1, a_1] \\
a_7 &= [a_2, a_1, a_1, a_2] \\
a_8 &= [a_2, a_1, a_2, a_2] \\
a_9 &= [a_2, a_1, a_1, a_1, a_2] \\
a_{10} &= [a_2, a_1, a_1, a_2, a_2] \\
a_{11} &= [a_2, a_1, a_1, a_1, a_2, a_1] \\
a_{12} &= [a_2, a_1, a_1, a_1, a_2, a_2] \\
a_{13} &= [a_2, a_1, a_1, a_2, a_2, a_1] \\
a_{14} &= [a_2, a_1, a_1, a_2, a_2, a_2]
\end{aligned}
$$

(This is a standard feature of PCPs produced by the Canberra implementation of the p-quotient algorithm, though for a general group some of the definitions may involve p-th powers.) The PCP generators also all have weights, which can be read off from their definitions. Thus a_1 and a_2 have weight 1, a_3 has weight 2, a_4 and a_5 have weight 3, a_6, a_7, a_8 have weight 4, a_9 and a_{10} have weight 5, and $a_{11}, a_{12}, a_{13}, a_{14}$ have weight 6. For $i \leq 6$, the generators of weight i generate $\gamma_i(G)$ modulo $\gamma_{i+1}(G)$. If a_i has weight r and a_j has weight s, then $a_i \in \gamma_r(G)$ and $a_j \in \gamma_s(G)$ so that $[a_j, a_i] \in \gamma_{r+s}(G)$. This implies that the right hand side of the commutator relation $[a_j, a_i] = \ldots$ only involves generators with weight at least $r + s$. Let L be the associated Lie ring of G. We obtain a consistent product presentation for L on the same generating set $a_1, a_2, \ldots, a_{14}$ by writing all the commutator relations from the PCP for G additively, and discarding generators from the relation $[a_j, a_i] = \ldots$ which have weight greater than $wt(a_i)+wt(a_j)$. Thus the trivial commutator relations $[a_j, a_i] = 1$ from the PCP for G become trivial Lie product relations $[a_j, a_i] = 0$, and the 23 non-trivial commutator relations from the PCP for G become the Lie product relations

$$\begin{aligned}
[a_2, a_1] &= a_3 \\
[a_3, a_1] &= a_4 \\
[a_3, a_2] &= a_5 \\
[a_4, a_1] &= a_6 \\
[a_4, a_2] &= a_7 \\
[a_4, a_3] &= a_9 \\
[a_5, a_1] &= a_7 \\
[a_5, a_2] &= a_8 \\
[a_5, a_3] &= 2a_{10} \\
[a_5, a_4] &= 2a_{12} + a_{13} \\
[a_6, a_1] &= 0 \\
[a_6, a_2] &= a_9 \\
[a_6, a_3] &= a_{11} \\
[a_7, a_1] &= 2a_9 \\
[a_7, a_2] &= a_{10} \\
[a_7, a_3] &= 3a_{12} + a_{13} \\
[a_8, a_1] &= 3a_{10} \\
[a_8, a_2] &= 0 \\
[a_8, a_3] &= 2a_{14} \\
[a_9, a_1] &= a_{11} \\
[a_9, a_2] &= a_{12} \\
[a_{10}, a_1] &= a_{13} \\
[a_{10}, a_2] &= a_{14}
\end{aligned}$$

In principle a consistent product presentation for the associated Lie ring of any finite p-group can be read off from a consistent PCP for the group. But if the generators of the group do not have order p then its associated Lie ring need not have characteristic p, so that its structure is more complicated than the structure of a graded Lie algebra over $\mathbb{Z}_p$.

It turns out that it is possible to reverse this process and construct a consistent PCP for our example G from a consistent product presentation for L. The point of this is that it takes less computing resources to first compute L, and then apply the Baker-Campbell-Hausdorff formula to obtain a consistent PCP for G.

We know from Sanov's result that L is the class 6 quotient of $E(2,5)$. The nilpotent quotient algorithm for graded Lie rings can be used to compute a consistent product presentation for L, and the presentation computed is identical to that given above. It also gives definitions for the generators $a_3, a_4, \ldots, a_{14}$ identical to the definitions given by the p-quotient algorithm for the PCP generators of G, though of course the definitions of the generators of L are to be interpreted as Lie products, whereas the definitions of the PCP generators of G are to be interpreted as group commutators. We mentioned above that the commutator version of the Baker-Campbell-Hausdorff formula gives

$$[e^y, e^x] = e^w,$$

where

$$w = [y,x] + \frac{1}{2}[y,x,x] + \frac{1}{2}[y,x,y] + \frac{1}{6}[y,x,x,x] + \frac{1}{4}[y,x,x,y] + \frac{1}{6}[y,x,y,y] + \ldots.$$

Let us see how this formula enables us to obtain the group commutator relation

$$[a_7, a_1] = a_9^2 a_{11}^2 a_{12}^2 a_{13}$$

(to take one example). We note that $a_7 = [a_2, a_1, a_1, a_2]$, so that $[a_7, a_1] = [a_2, a_1, a_1, a_2, a_1]$. The first thing to do is to use the Baker-Campbell-Hausdorff formula to find an expression for $[e^y, e^x, e^x, e^y, e^x]$. This expression can be truncated at weight 6 (since G has class 6). Using the fact that the Lie element $[y,x,y,x]$ equals $[y,x,x,y]$ we first obtain

$$[e^y, e^x, e^x] = e^{[y,x,x]+[y,x,x,x]+\frac{1}{2}[y,x,x,y]+\ldots}.$$

Then we calculate

$$[e^y, e^x, e^x, e^y] = e^{[y,x,x,y]+[y,x,x,x,y]+[y,x,x,y,y]+\ldots},$$

and

$$[e^y, e^x, e^x, e^y, e^x] = e^{[y,x,x,y,x]+[y,x,x,x,y,x]+[y,x,x,y,y,x]+\frac{1}{2}[y,x,x,y,x,x]+\ldots}.$$

We truncate this expression to

$$[e^y, e^x, e^x, e^y, e^x] \sim e^{[y,x,x,y,x]+[y,x,x,x,y,x]+[y,x,x,y,y,x]+\frac{1}{2}[y,x,x,y,x,x]}.$$

If we substitute a_1 for x and a_2 for y in this truncated expression, and evaluate the right hand side of the expression as e^a for some $a \in L$ then we obtain

$$[e^{a_2}, e^{a_1}, e^{a_1}, e^{a_2}, e^{a_1}] \sim e^{2a_9+2a_{11}+a_{13}}.$$

(It is not at all clear what meaning this expression might have!) The definition of a_9 is $a_9 = [a_2, a_1, a_1, a_1, a_2]$, and we similarly obtain

$$[e^{a_2}, e^{a_1}, e^{a_1}, e^{a_1}, e^{a_2}] \sim e^{a_9+4a_{12}}.$$

Finally, a_{11}, a_{12}, a_{13} have definitions

$$\begin{aligned} a_{11} &= [a_2, a_1, a_1, a_1, a_2, a_1] \\ a_{12} &= [a_2, a_1, a_1, a_1, a_2, a_2] \\ a_{13} &= [a_2, a_1, a_1, a_2, a_2, a_1] \end{aligned}$$

and since they all have weight 6 we immediately obtain

$$[e^{a_2}, e^{a_1}, e^{a_1}, e^{a_1}, e^{a_2}, e^{a_1}] \sim e^{a_{11}},$$

$$[e^{a_2}, e^{a_1}, e^{a_1}, e^{a_1}, e^{a_2}, e^{a_2}] \sim e^{a_{12}},$$

$$[e^{a_2}, e^{a_1}, e^{a_1}, e^{a_2}, e^{a_2}, e^{a_1}] \sim e^{a_{13}}.$$

Let us define $b_1 = e^{a_1}$, $b_2 = e^{a_2}$, $b_3 = [b_2, b_1]$, ..., $b_{14} = [b_2, b_1, b_1, b_2, b_2, b_2]$, by analogy with the definitions of $a_3, a_4, \ldots, a_{14}$. Then we have

$$[b_7, b_1] \sim e^{2a_9+2a_{11}+a_{13}},\ b_9 \sim e^{a_9+4a_{12}},\ b_{11} \sim e^{a_{11}},\ b_{12} \sim e^{a_{12}},\ b_{13} \sim e^{a_{13}}.$$

Since $a_9, a_{11}, a_{12}, a_{13}$ all commute we immediately obtain

$$[b_7, b_1] \sim b_9^2 b_{11}^2 b_{12}^2 b_{13},$$

which corresponds to the group commutator relation

$$[a_7, a_1] = a_9^2 a_{11}^2 a_{12}^2 a_{13}$$

from the PCP for G. The whole of the PCP for G can be obtained from the product presentation for L in this way. In the next subsection we will see why this procedure is valid.

1.2 Justification

We let G be the class c quotient of $R(r,p)$ (for some c less than or equal to the class of $R(r,p)$), and we let L be the class c quotient of $W(r,p)$. We make the assumption that L is the associated Lie ring of G. We suppose that we have a consistent product presentation for L on generators $a_1, a_2, \ldots, a_n$. We suppose that each of the generators a_i has a weight $wt(a_i)$ such that

$$1 = wt(a_1) = wt(a_2) = \ldots = wt(a_r) < wt(a_{r+1}) \leq wt(a_{r+2}) \leq \ldots \leq wt(a_n) = c,$$

and we suppose that if $i > r$ then a_i has a definition as a left-normed Lie product of weight $wt(a_i)$ in the generators $a_1, a_2, \ldots, a_r$. (The product presentations produced by the nilpotent quotient algorithm for graded Lie rings satisfy all these conditions.) Then G has a consistent PCP on the same generating set $a_1, a_2, \ldots, a_n$ with the same weights, and with the same definitions for $a_{r+1}, a_{r+2}, \ldots, a_n$. (Though as PCP generators for G the definitions of the generators have to be read as group commutators rather than as Lie products.) As we described above, if $wt(a_i) = r$ and $wt(a_j) = s$ then the weight $r+s$ part of the group commutator $[a_j, a_i] = \ldots$ can be read off from the product presentation for L. We show that if $r+s \geq c-p+2$ then the procedure outlined above can be used to compute the complete group commutator $[a_j, a_i] = \ldots$.

In the example above we had $c = 6$, $r = 2$, $p = 5$. So in this example we can compute the full group commutator $[a_j, a_i] = \ldots$ whenever $r + s \geq 3$. But the only commutator $[a_j, a_i]$ with $r + s < 3$ is $[a_2, a_1]$, and since the definition of a_3 is $a_3 = [a_2, a_1]$, the *full* group commutator relation in this case is $[a_2, a_1] = a_3$. Thus in this example we are able to compute the complete PCP for G from the product presentation for L. And we are able to do this even though the Lazard correspondence between Lie rings and finite p-groups is only valid for p-groups of class at most $p-1$, whereas G is a finite 5-group of class 6.

We will show that the complete PCP for G can always be computed from the product presentation for L provided $c \leq 2p-2$.

The first key observation is that since G is nilpotent of class c it follows that G is a homomorphic image of the free nilpotent of class c group of rank r, $F_{r,c}$. We let A be the free associative algebra with 1 over $\mathbb{Q}$ freely generated by non-commuting indeterminates $y_1, y_2, \ldots$. Then we let I be the ideal of A generated by all products of weight $c+1$ in the generators $y_1, y_2, \ldots$, and we let $A_c = A/I$. We let $x_i = y_i + I$ for $i = 1, 2, \ldots$. So A_c is $\mathbb{Q}$-algebra, and every element of A_c can be expressed as a polynomial in $x_1, x_2, \ldots$ of degree at most c, with coefficients in $\mathbb{Q}$. If $a \in A_c$ has zero constant term then we define

$$\mathrm{e}^a = 1 + a + \frac{a^2}{2!} + \ldots + \frac{a^c}{c!}.$$

Then $\mathrm{e}^{x_1}, \mathrm{e}^{x_2}, \ldots, \mathrm{e}^{x_r}$ generate a subgroup of the group of units of A_c which we can identify with $F_{r,c}$. We turn A_c into a Lie algebra over $\mathbb{Q}$ by defining the Lie product of two elements $a, b \in A_c$ to be $[a, b] = ab - ba$. The set of all $\mathbb{Z}$-linear combinations of Lie products of $x_1, x_2, \ldots, x_r$ is the free nilpotent of class c Lie ring of rank r — let us call this Lie ring $\Lambda_{r,c}$. The set of all $\mathbb{Q}$-linear combinations of Lie products of $x_1, x_2, \ldots, x_r$, $\mathbb{Q}\Lambda_{r,c}$, is the free nilpotent of class c Lie algebra over $\mathbb{Q}$ of rank r. And if we let $\mathbb{Q}_p$ be the subring of $\mathbb{Q}$ consisting of rationals of the form m/n with n coprime to p, then $\mathbb{Q}_p\Lambda_{r,c}$ is a free nilpotent of class c Lie algebra over $\mathbb{Q}_p$.

The Baker-Campbell-Hausdorff formula can be used to express products and commutators of $\mathrm{e}^{x_1}, \mathrm{e}^{x_2}, \ldots, \mathrm{e}^{x_r}$ in the form e^a with $a \in A_c$. Thus $[\mathrm{e}^{x_1}, \mathrm{e}^{x_2}] = \mathrm{e}^w$ where $w = w_2 + w_3 + \ldots + w_c \in A_c$, and where $w_i \in \mathbb{Q}\Lambda_{r,c}$ is homogeneous of weight i, for $i = 2, 3, \ldots, c$.

It is easy to see that if $i \leq p$ then $w_i \in \mathbb{Q}_p\Lambda_{r,c}$. More generally, any commutator of weight k in $\mathrm{e}^{x_1}, \mathrm{e}^{x_2}, \ldots, \mathrm{e}^{x_r}$ can be expressed in the form e^u where $u = u_k + u_{k+1} + \ldots + u_c$, with $u_i \in \mathbb{Q}\Lambda_{r,c}$ for $i = k, k+1, \ldots, c$. Again, it is easy to see that if $i \leq k+p-2$ then $u_i \in \mathbb{Q}_p\Lambda_{r,c}$. So if $k \geq c-p+2$, then $u \in \mathbb{Q}_p\Lambda_{r,c}$. The $(c-p+2)$-th term of the lower central series of $F_{r,c}$, $\gamma_{c-p+2}(F_{r,c})$, is generated by commutators of weight greater than or equal to $c-p+2$ in $\mathrm{e}^{x_1}, \mathrm{e}^{x_2}, \ldots, \mathrm{e}^{x_r}$. It follows that every element of $\gamma_{c-p+2}(F_{r,c})$ can be expressed in the form e^u, where u a linear combination of Lie products of weight at least $c-p+2$, and where $u \in \mathbb{Q}_p\Lambda_{r,c}$.

Recall that L is the class c quotient of $W(r,p)$, and that L is generated by $a_1, a_2, \ldots, a_r$. Also recall that G is the class c quotient of $R(r,p)$. Since $\mathbb{Q}_p\Lambda_{r,c}$ is free nilpotent of class c, there is a homomorphism $\pi : \mathbb{Q}_p\Lambda_{r,c} \to L$ mapping x_i to a_i for $i = 1, 2, \ldots, r$. We let N be the subgroup of $\gamma_{c-p+2}(F_{r,c})$ consisting of elements of the form e^u with $u \in \ker \pi$. We show that $N = \gamma_{c-p+2}(F_{r,c}) \cap F_{r,c}^p$, which implies that $\gamma_{c-p+2}(F_{r,c})/N \cong \gamma_{c-p+2}(G)$.

We assume for the moment that $N = \gamma_{c-p+2}(F_{r,c}) \cap F_{r,c}^p$. There is a homomorphism θ from $F_{r,c}$ to G mapping e^{x_i} to the PCP generator a_i for $i = 1, 2, \ldots, r$. We let $b_i = e^{x_i}$ for $i = 1, 2, \ldots r$, and if $r < i \leq n$ and a_i is defined to be $[a_{i_1}, a_{i_2}, \ldots, a_{i_k}]$ then we define $b_i = [b_{i_1}, b_{i_2}, \ldots, b_{i_k}]$. So $b_i\theta = a_i$ for $1 \leq i \leq n$. If k is the smallest integer such that $wt(a_k) = c - p + 2$ then every element of $\gamma_{c-p+2}(G)$ can be expressed uniquely as a normal word

$$a_k^{\alpha_k} a_{k+1}^{\alpha_{k+1}} \ldots a_n^{\alpha_n},$$

and so if $b \in \gamma_{c-p+2}(F_{r,c})$ then there are unique integers $\alpha_k, \alpha_{k+1}, \ldots, \alpha_n$ with $0 \leq \alpha_i < p$ such that

$$bN = b_k^{\alpha_k} b_{k+1}^{\alpha_{k+1}} \ldots b_n^{\alpha_n} N.$$

The integers $\alpha_k, \alpha_{k+1}, \ldots, \alpha_n$ can be determined as follows. Write $b = e^u$, with $u \in \mathbb{Q}_p\Lambda_{r,c}$. Then $u\pi = \sum_{r=k}^{n} \beta_r a_r \in L$, for some $\beta_k, \beta_{k+1}, \ldots, \beta_n$. We let $\alpha_k = \beta_k$, and then it is not hard to show that $b_k^{-\alpha_k} b = e^v$ where $v\pi = \sum_{r=k+1}^{n} \gamma_r a_r \in L$, for some $\gamma_{k+1}, \gamma_{k+2}, \ldots, \gamma_n$. We let $\alpha_{k+1} = \gamma_{k+1}$, and then $b_{k+1}^{-\alpha_{k+1}} b_k^{-\alpha_k} b = e^w$, where $w\pi = \sum_{r=k+2}^{n} \delta_r a_r \in L$, for some $\delta_{k+2}, \delta_{k+3}, \ldots, \delta_n$. We let $\alpha_{k+2} = \delta_{k+2}$, and so on. In this way we eventually obtain

$$b_n^{-\alpha_n} \ldots b_{k+1}^{-\alpha_{k+1}} b_k^{-\alpha_k} b = e^z$$

for some $z \in \ker \pi$, as required.

In particular if $wt(a_i) + wt(a_j) \geq c - p + 2$ then there are unique integers $\alpha_k, \alpha_{k+1}, \ldots, \alpha_n$ with $0 \leq \alpha_i < p$ such that

$$[b_j, b_i]N = b_k^{\alpha_k} b_{k+1}^{\alpha_{k+1}} \ldots b_n^{\alpha_n} N,$$

and this gives the commutator relation $[a_j, a_i] = a_k^{\alpha_k} a_{k+1}^{\alpha_{k+1}} \ldots a_n^{\alpha_n}$ from the PCP for G.

It remains to show that $N = \gamma_{c-p+2}(F_{r,c}) \cap F_{r,c}^p$. We will prove that $N \leq F_{r,c}^p$ at the end of Section 3, as an application of the theory of Lie relators. Given this result it is quite easy to show that $N = \gamma_{c-p+2}(F_{r,c}) \cap F_{r,c}^p$.

Recall that N is the subgroup of $\gamma_{c-p+2}(F_{r,c})$ consisting of elements of the form e^u with $u \in \ker \pi$, where $\pi : \mathbb{Q}_p\Lambda_{r,c} \to L$ is the homomorphism mapping x_i to a_i for $i = 1, 2, \ldots, r$. We show that if e^u, $e^v \in \gamma_{c-p+2}(F_{r,c})$ then $e^uN = e^vN$ if $u\pi = v\pi$. (This is not immediately obvious since $e^{-u}e^v$ is not necessarily equal to e^{v-u}.) So suppose that $e^u, e^v \in \gamma_{c-p+2}(F_{r,c})$ and that $u\pi = v\pi$. Then we can write $v = u+w$, for some $w \in \ker\pi$. So $e^{-u}e^v = e^{-u}e^{u+w} = e^z$, where

$$z = -u + (u+w) + \frac{1}{2}[u+w,u] + \frac{1}{12}[u+w,u,u] + \frac{1}{12}[u+w,u,u+w] + \ldots.$$

Note that $z \in \ker\pi$. So $e^u N = e^v N$, as claimed. It follows that the index of N in $\gamma_{c-p+2}(F_{r,c})$ is at most $|L^{c-p+2}| = |\gamma_{c-p+2}(G)|$. But

$$\gamma_{c-p+2}(G) \cong \gamma_{c-p+2}(F_{r,c})/(\gamma_{c-p+2}(F_{r,c}) \cap F_{r,c}^p),$$

and $N \leq \gamma_{c-p+2}(F_{r,c}) \cap F_{r,c}^p$. So $N = \gamma_{c-p+2}(F_{r,c}) \cap F_{r,c}^p$.

1.3 Groups of class $2p-2$

Let $B(r,p;2p-2)$ denote the class $2p-2$ quotient of $B(r,p)$, and let $E(r,p;2p-2)$ denote the class $2p-2$ quotient of $E(r,p)$. Sanov's result implies that the associated Lie ring of $B(r,p;2p-2)$ is $E(r,p;2p-2)$. So we can use the method described above to construct part of a consistent PCP for $B(r,p;2p-2)$ from a consistent product presentation for $E(r,p;2p-2)$. However we show that in this case we can actually construct the whole of a consistent PCP for $B(r,p;2p-2)$.

As described above, a consistent product presentation for $E(r,p;2p-2)$ gives us a set of PCP generators $a_1, a_2, \ldots, a_n$ for $B(r,p;2p-2)$, together with their definitions and weights. It also enables us to compute the group commutator relations $[a_j, a_i] = \ldots$ whenever $wt(a_i) + wt(a_j) \geq p$. But the group commutator relations $[a_j, a_i] = \ldots$ where $wt(a_i) + wt(a_j) < p$ can be deduced from relations which hold in *any* group. This is because the number of PCP generators of $B(r,p;2p-2)$ of weight at most $p-1$ is the same as the number of basic commutators of weight at most $p-1$ in r generators. Essentially, all the commutator relations of the class $p-1$ quotient of $B(r,p)$ correspond to commutator relations which hold in free nilpotent groups of class $p-1$.

To illustrate this, let us look at the class 8 quotient of $R(2,5)$. Here $p=5$, and the group commutator relations of weight less than 5 are

$$\begin{aligned}
[a_2, a_1] &= a_3\\
[a_3, a_1] &= a_4\\
[a_3, a_2] &= a_5\\
[a_4, a_1] &= a_6\\
[a_4, a_2] &= a_7\\
[a_5, a_1] &= a_7 a_9 a_{10}^2 a_{11} a_{12}^2 a_{13}^3 a_{14}^4 a_{15}^3 a_{16}^3 a_{20}^3 a_{21}\\
[a_5, a_2] &= a_8
\end{aligned}$$

All of these relations, except for the relation $[a_5, a_1] = a_7 a_9 \ldots$come from the definitions of $a_3, a_4, \ldots, a_8$. Group commutator relations coming from definitions can be read off from the corresponding relations in the associated Lie ring (or read off from the definitions). So the only commutator relation of weight at most 4 which we need to compute is $[a_5, a_1] = \ldots$. Now the definition of a_5 is $[a_2, a_1, a_2]$, so $[a_5, a_1] = [a_2, a_1, a_2, a_1]$. The Baker-Campbell-Hausdorff formula gives

$$[e^y, e^x, e^y, e^x] = e^{[y,x,x,y]+[y,x,x,y,x]+[y,x,y,y,x]+\ldots}.$$

It also gives

$$[e^y, e^x, e^x, e^y] = e^{[y,x,x,y]+[y,x,x,x,y]+[y,x,x,y,y]+\cdots},$$

and so

$$[e^y, e^x, e^y, e^x] = [e^y, e^x, e^x, e^y]e^{[y,x,x,y,x]+[y,x,y,y,x]-[y,x,x,x,y]-[y,x,x,y,y]+\cdots}.$$

If we compute this expression as far as weight 8 ($8 = 2p-2$ in this case) then we can obtain an explicit formula for $[e^y, e^x, e^y, e^x]$ as $[e^y, e^x, e^x, e^y] \cdot b$, where b is a product of commutators of weight 5 or more in e^x and e^y. This formula is valid in *any* group which is nilpotent of class 8, and so gives an explicit expression for $[a_2, a_1, a_2, a_1]$ as a product of $[a_2, a_1, a_1, a_2]$ with a product of commutators of weight at least 5 in a_1, a_2. Note that $[a_2, a_1, a_1, a_2] = a_7$. The normal form for this product of commutators of weight at least 5 can be computed in the way described above.

In this way all the group commutator relations of $B(r, p; 2p-2)$ can be computed from the presentation for $E(r, p; 2p-2)$.

1.4 $R(2,7)$

Eamonn O'Brien and I used the method described above in [24] in the computation of $R(2,7)$. It turns out that $R(2,7)$ has class 28 and order 7^{20416}. In constructing a consistent PCP for $R(2,7)$ on generators $a_1, a_2, \ldots, a_{20416}$ we used the Baker-Campbell-Hausdorff formula to compute all the commutators $[a_i, a_j]$ with $wt(a_i) + wt(a_j) \geq 23$ from a presentation for $W(2,7)$.

2 Torsion in outer commutator varieties

The parallels between varieties of groups and varieties of Lie rings are particularly close in the case of outer commutator varieties. Often, the associated Lie rings of free groups in outer commutator varieties are free Lie rings in the corresponding outer commutator varieties of Lie rings. And by studying these free Lie rings we are sometimes able to show that the lower central factors of the free groups are torsion free. Perhaps rather more surprisingly, we are sometimes able to use information about these Lie rings to show that the groups are residually nilpotent.

Outer commutator varieties are varieties of groups defined by a single "multilinear" commutator identity. Some important examples are the variety of groups which are nilpotent of class at most c, given by the identity

$$[x_1, x_2, \ldots, x_{c+1}] = 1,$$

the variety of metabelian groups, given by the identity

$$[[x_1, x_2], [x_3, x_4]] = 1,$$

and the variety of groups which are soluble of derived length at most k, given by a single commutator identity

$$\delta(x_1, x_2, \ldots, x_{2^k}) = 1.$$

More generally, an outer commutator variety is defined by a single identity of the form

$$[x_1, x_2, \ldots, x_n] = 1$$

with some arrangement of brackets. The polynilpotent variety $\mathcal{N}_{c_1}.\mathcal{N}_{c_2}.\ldots.\mathcal{N}_{c_k}$ is an outer commutator variety. It consists of groups G with a chain of normal subgroups

$$\{1\} = A_0 \leq A_1 \leq A_2 \leq \ldots \leq A_k = G,$$

where A_i/A_{i-1} is nilpotent of class at most c_i for $i = 1, 2, \ldots, k$.

In this section we describe one way in which Lie methods have been used to investigate outer commutator varieties with the aim of showing that the free groups in certain outer commutator varieties are torsion free and residually nilpotent. But first we give a brief overview of some of the results in this area.

Gruenberg [7] showed that free polynilpotent groups are residually nilpotent in 1957. And in 1963 Smel'kin [26] proved that the lower central factors of free polynilpotent groups are torsion free. The variety of centre-by-metabelian groups is determined by the identity

$$[[x_1, x_2], [x_3, x_4], x_5] = 1,$$

and in 1973 C.K. Gupta [8] proved that free centre-by-metabelian groups are residually nilpotent, but that they have torsion. The free groups of the variety determined by the identity

$$[[x_1, x_2, \ldots, x_m], [x_{m+1}, x_{m_2}, \ldots, x_{m+n}]] = 1$$

can be expressed in the form $F/[\gamma_m(F), \gamma_n(F)]$, where F is (absolutely) free. In 1978 Kuzmin [19] showed that $F/[\gamma_m(F), \gamma_n(F)]$ is residually nilpotent, and in 1995 Tasic and Vaughan-Lee [27] showed that $F/[\gamma_m(F), \gamma_n(F)]$ is torsion free (for all m, n).

The situation is more complicated with the groups $F/[\gamma_j(F), \gamma_i(F), \gamma_k(F)]$ and $F/[\gamma_j(F), \gamma_i(F), \gamma_k(F), \gamma_l(F)]$ since many of these groups are residually nilpotent and torsion free, but others (such as the centre-by-metabelian groups) have torsion. Groves [5] has recently shown that $F/[\gamma_j(F), \gamma_i(F), \gamma_k(F)]$ is residually nilpotent and torsion free if $i \leq j \leq 2i$ and $i \leq k \leq 2i + 2j$. Groves [6] has also shown that $F/[\gamma_j(F), \gamma_i(F), \gamma_k(F), \gamma_l(F)]$ is residually nilpotent and torsion free if $i \leq j \leq 2i$ and $j \leq k \leq l \leq i + j$.

The key idea in the proofs of the results of Tasic and Vaughan-Lee and of Groves is to show that free Lie rings in the corresponding varieties of Lie rings are torsion free. We make use of a result of Kuzmin and Shapiro [20]. They showed that if the corresponding relatively free Lie rings are torsion free then these Lie rings are the associated Lie rings of the relatively free groups. And sometimes we can use this fact to prove that the groups are residually nilpotent. We now outline this method.

Let $w(x_1, x_2, \ldots, x_n)$ be an outer commutator word, and let $\widetilde{w}$ be the corresponding Lie word. (Typographically, $\widetilde{w}$ is identical to w but $\widetilde{w}$ has to be read as a Lie product rather than as a group commutator.) Let F be a free group, and let $G = F/w(F)$. Let

$$L = L_1 \oplus L_2 \oplus \ldots \oplus L_i \oplus \ldots$$

be the associated Lie ring of G, as described in Section 1. We show that L satisfies the Lie identity $\widetilde{w} = 0$. Using the fact that $\widetilde{w}$ is multilinear in $x_1, x_2, \ldots, x_n$ we need only show that $\widetilde{w}(a_1, a_2, \ldots, a_n) = 0$ whenever $a_1, a_2, \ldots, a_n$ are homogeneous elements of L. So let $a_i = g_i\gamma_{j_i+1} \in L_{j_i}$ (with $g_i \in \gamma_{j_i}$) for $i = 1, 2, \ldots, n$. Let $m = j_1 + j_2 + \ldots + j_n$. Then

$$\widetilde{w}(a_1, a_2, \ldots, a_n) = w(g_1, \ldots, g_n)\gamma_{m+1} = 0.$$

So L satisfies the identity $\widetilde{w} = 0$, and this implies that if Λ is a free Lie ring of the same rank as G then L is a homomorphic image of $\Lambda/\widetilde{w}(\Lambda)$.

Theorem 2.1 (Kusmin and Shapiro [20]) *If $\Lambda/\widetilde{w}(\Lambda)$ is torsion free then $L \cong \Lambda/\widetilde{w}(\Lambda)$.*

Proof Since $\widetilde{w}$ is multilinear $\Lambda/\widetilde{w}(\Lambda)$ is graded: letting $B = \Lambda/\widetilde{w}(\Lambda)$ we have

$$B = B_1 \oplus B_2 \oplus \ldots \oplus B_i \oplus \ldots$$

where $[B_i, B_j] \leq B_{i+j}$, and

$$B^m = B_m \oplus B_{m+1} \oplus \ldots$$

for all $m \geq 1$. The homomorphism from B onto L is a graded homomorphism so that $L_i = B_i/K_i$ for $i = 1, 2, \ldots$ (for some K_i). We need to show that $K_i = \{0\}$ for all i. Note that it is sufficient to prove this in the situation when F and Λ are free of finite rank. The advantage of this is that the summands B_i and L_i are then finitely generated abelian groups.

Now we let $C = B/B^{i+1}$. So C is nilpotent of class i, and $C^i = B_i$. We let $M = \mathbb{Q} \otimes_{\mathbb{Z}} C$, and view M as a Lie algebra over the rationals $\mathbb{Q}$. Clearly M is also nilpotent of class i. Note that as an additive group C is isomorphic to the direct summand $B_1 \oplus B_2 \oplus \ldots \oplus B_i$ of B, so that C is torsion free. This implies that $C \leq M$.

We use the Baker-Campbell-Hausdorff formula to turn M into a group under $\circ$, as described in Section 1, and it is then easy to see that M satisfies the group identity $w = 1$.

For example, suppose

$$w = [[x_1, x_2, \ldots, x_m], [x_{m+1}, \ldots, x_{m+n}]].$$

Let $a_1, a_2, \ldots, a_{m+n} \in M$. Then

$$\begin{aligned} a &= [a_1, a_2, \ldots, a_m]_\circ \\ &= [a_1, a_2, \ldots, a_m]_L + \text{higher terms.} \end{aligned}$$

(Here we are using $[,]_\circ$ to denote commutators in the group $\langle M, \circ \rangle$, and $[,]_L$ to denote Lie products in M.) Similarly

$$\begin{aligned} b &= [a_{m+1}, \ldots, a_{m+n}]_\circ \\ &= [a_{m+1}, \ldots, a_{m+n}]_L + \text{ higher terms.} \end{aligned}$$

Since M satisfies the Lie identity $\widetilde{w} = 0$, it follows that $[a, b]_L = 0$, and hence that $[a, b]_\circ = 0$. So M satisfies the group identity $w = 1$.

Let X be a free generating set for Λ, let Y be the image of X in C, and let H be the subgroup of $\langle M, \circ \rangle$ generated by Y. Since $\langle H, \circ \rangle \leq \langle M, \circ \rangle$, H satisfies the group identity $w = 1$, and so H is a homomorphic image of $G = F/w(F)$. This implies that $\gamma_i(H)/\gamma_{i+1}(H)$ is a homomorphic image of $\gamma_i(G)/\gamma_{i+1}(G) = L_i$. However Y generates C as a Lie ring, and C is nilpotent of class i. This implies that $\gamma_i(H)/\gamma_{i+1}(H) = C^i = B_i$. So B_i is a homomorphic image of $L_i = B_i/K_i$. Since B_i and L_i are finitely generated abelian groups this implies that $L_i = B_i$ and $K_i = \{0\}$, as required. □

This theorem implies that if $\Lambda/\widetilde{w}(\Lambda)$ is torsion free then the lower central factors of G are torsion free. But to deduce that G is torsion free we need to know that G is residually nilpotent. However sometimes it is possible to deduce that G is residually nilpotent from information about $\Lambda/\widetilde{w}(\Lambda)$. In particular, if $\widetilde{w}(\Lambda)$ is spanned by basic commutators then G is torsion free and residually nilpotent.

Before giving a proof of this result we recall the definition of basic commutators, and some of their properties.

Let Λ be a free Lie ring with free generators $x_1, x_2, \ldots$. Then the basic commutators are:

$$\begin{aligned} &x_i \ (i = 1, 2, \ldots), \\ &[x_i, x_j] \ (i > j), \\ &[x_i, x_j, x_k] \ (i > j \leq k), \\ &\ldots . \end{aligned}$$

They are totally ordered, so that commutators of higher weight follow those of lower weight. If c and d are basic then $[c, d]$ is basic if

1. $c > d$,
2. if $c = [e, f]$ then $f \leq d$.

The first important property of basic commutators is that the basic commutators form a basis for Λ as a free $\mathbb{Z}$-module. Furthermore the basic commutators of weight at least n form a basis for Λ^n as a free $\mathbb{Z}$-module (so that Λ^n is a direct summand of Λ. We can say even more than this: Λ^n is a free Lie ring with free generating set consisting of basic commutators c such that

1. weight$(c) \geq n$,
2. if $c = [d, e]$ then weight$(e) < n$.

Proofs of all these important facts can be found in Bahturin [1]. It is not hard to show that $(\Lambda^n)^m$ is spanned by basic commutators. However $((\Lambda^2)^2)^2$ (the third term of the derived series of Λ) is not spanned by basic commutators. Groves's results mentioned above are obtained by proving that $[\Lambda^j, \Lambda^i, \Lambda^k]$ and $[\Lambda^j, \Lambda^i, \Lambda^k, \Lambda^l]$ are spanned by basic commutators under certain restrictions on i, j, k, l. On the other hand $[\Lambda^2, \Lambda^2, \Lambda]$ is not spanned by basic commutators.

As an easy example of the sort of argument that can be used we prove the following lemma.

Lemma 2.2 *Let $i \leq j \leq 2i$. Then $[\Lambda^j, \Lambda^i]$ is spanned by basic commutators.*

Proof Clearly $[\Lambda^j, \Lambda^i]$ is spanned by elements $[a, b]$ where a and b are basic and weight$(a) \geq j$, weight$(b) \geq i$. If weight$(b) \geq j$ then $[a, b] \in [\Lambda^j, \Lambda^j] \leq [\Lambda^j, \Lambda^i]$, and we are done since $[\Lambda^j, \Lambda^j] = (\Lambda^j)^2$ is spanned by basic commutators. So we may assume that $a > b$. If $[a, b]$ is basic we are done. If not, $a = [c, d]$ where $c > d > b$. But then $[a, b] \in [\Lambda^i, \Lambda^i, \Lambda^i] \leq [\Lambda^j, \Lambda^i]$, and we are done since $[\Lambda^i, \Lambda^i, \Lambda^i] = (\Lambda^i)^3$ is spanned by basic commutators. □

It is fairly easy to extend this result and show that $[\Lambda^{2i+1}, \Lambda^i]$ is spanned by basic commutators. But, on the other hand, $[\Lambda^6, \Lambda^2]$ is *not* spanned by basic commutators. The commutator $[[y, x, y], [y, x, x], [y, x]]$ lies in $[\Lambda^6, \Lambda^2]$, but when it is expressed as a linear combination of basic commutators we obtain

$$\begin{aligned}
&[[y, x, y], [y, x, x], [y, x]] \\
&= [[y, x, y], [y, x], [y, x, x]] - [[y, x, x], [y, x], [y, x, y]],
\end{aligned}$$

and neither of $[[y, x, y], [y, x], [y, x, x]]$, $[[y, x, x], [y, x], [y, x, y]]$ lies in $[\Lambda^6, \Lambda^2]$.

Similarly $[\Lambda^2, \Lambda^2, \Lambda]$ is *not* spanned by basic commutators since

$$\begin{aligned}
&[[z, x], [y, x], x] \\
&= [[z, x, x], [y, x]] - [[y, x, x], [z, x]].
\end{aligned}$$

We also need to introduce Gorchakov's two sided collection process ([3]). And before doing that it helps to recall some of the key features of the Hall collection process. In the Hall collection process we start with a positive word

$$w = x_i x_j \ldots x_k$$

on the free generators $x_1, x_2, \ldots$ of a free group F. (It is possible to apply the Hall collection process to words involving negative exponents, but there are difficulties with this.) Initially, we scan w looking for the smallest index i such that x_i occurs in w. Suppose, for example, that this index is 1. Then we use the commutator identity $vu = uv[v, u]$ to move occurrences of x_1 towards the left. Suppose that w has a subword $x_n x_1$, with $n > 1$. Then we replace the subword $x_n x_1$ in w by the subword $x_1 x_n [x_n, x_1]$. This moves the occurrence of x_1 towards the left, though it lengthens w by introducing the basic commutator $[x_n, x_1]$. If we systematically move the leftmost occurrence of x_1 further to the left in this way, then we eventually obtain a word of the form

$$x_1 c_1 c_2 \ldots c_k,$$

where $c_1, c_2, \ldots, c_k$ are basic (group) commutators. Note that some of the c_i which occur might be further occurrences of x_1. The general step of the collection process proceeds as follows. Suppose we have a word

$$u = c_1 c_2 \ldots c_r c_{r+1} \ldots c_s$$

satisfying the following conditions.

1. $c_1, c_2, \ldots, c_s$ are basic commutators,
2. $c_1 \leq c_2 \leq \ldots \leq c_r$ in the ordering on basic commutators,
3. $c_{r+1}, c_{r+2}, \ldots, c_s \geq c_r$,
4. if $c_i = [d, e]$ for some i with $r < i \leq s$ then $e \leq c_r$.

We call $c_1 c_2 \ldots c_r$ the collected part of u, and $c_{r+1} \ldots c_s$ the uncollected part of u. (With the initial word w the collected part is empty, and the uncollected part is the whole of w. After collecting one occurrence of x_1 all the way to the left we had a word $x_1 c_1 c_2 \ldots c_k$ with collected part x_1 and uncollected part $c_1 c_2 \ldots c_k$.) If the uncollected part is empty then we are done, but this almost never happens.

So we assume that the uncollected part is non-empty. We scan the uncollected part of u, looking for the least basic commutator (in the ordering on basic commutators) in the sequence $c_{r+1}, \ldots, c_s$. Suppose that the least one is c. We look for the leftmost occurrence of c in the uncollected part of u. Suppose this is as indicated below.

$$u = c_1 c_2 \ldots c_r c_{r+1} \ldots c_i c \ldots c_s.$$

Then we "collect c", by replacing the subword $c_i c$ by the subword $c c_i [c_i, c]$. (Condition (4) above guarantees that $[c_i, c]$ is basic.) This gives us the word

$$c_1 c_2 \ldots c_r c_{r+1} \ldots c c_i [c_i, c] \ldots c_s$$

with collected part $c_1 c_2 \ldots c_r$ and uncollected part $c_{r+1} \ldots c c_i [c_i, c] \ldots c_s$. Continuing in this way, the basic commutator c will eventually reach the left hand end of the uncollected part:

$$c_1 c_2 \ldots c_r c d_1 d_2 \ldots d_t.$$

We then say that this occurrence of c is fully collected, and repartition the word into a collected part $c_1 c_2 \ldots c_r c$ and an uncollected part $d_1 d_2 \ldots d_t$. As mentioned above, the process usually continues indefinitely. But if you specify any particular term $F^{(k)}$ of the derived series of F, then eventually the uncollected part will lie in $F^{(k)}$, so that the original word w has been expressed in the form $c_1 c_2 \ldots c_r z$ where $c_1, c_2, \ldots, c_r$ are basic commutators with $c_1 \leq c_2 \leq \ldots \leq c_r$ and $z \in F^{(k)}$. Alternatively, you can specify a particular term $\gamma_k(F)$ of the lower central series of F, and obtain an expression for w of in the form $c_1 c_2 \ldots c_r z$ where $c_1, c_2, \ldots, c_r$ are basic commutators with $c_1 \leq c_2 \leq \ldots \leq c_r$ and $z \in \gamma_k(F)$. In this case we may assume that $wt(c_r) < k$, and the expression is then unique (for that value of k).

The two sided collection process handles negative exponents as well as positive exponents, and in the process basic commutators with positive exponents are moved towards the left, and basic commutators with negative exponents are moved towards the right. At an intermediate stage of the two sided process we have a word

$$u = c_1 c_2 \dots c_r c_{r+1}^{\pm 1} c_{r+2}^{\pm 1} \dots c_s^{\pm 1} c_{s+1}^{-1} c_{s+2}^{-1} \dots c_t^{-1}$$

satisfying the following conditions.

1. $c_1, c_2, \dots, c_t$ are basic commutators,
2. $c_1 \leq c_2 \leq \dots \leq c_r$,
3. $c_{s+1} \geq c_{s+2} \geq \dots \geq c_t$,
4. $c_{r+1}, c_{r+2}, \dots, c_s \geq c_r, c_{s+1}$,
5. if $c_i = [d, e]$ for some i with $r < i \leq s$ then $e \leq c_r$ and $e \leq c_{s+1}$.

Here $c_1 c_2 \dots c_r$ is called the positive collected part of u, $c_{r+1}^{\pm 1} c_{r+2}^{\pm 1} \dots c_s^{\pm 1}$ is called the uncollected part, and $c_{s+1}^{-1} c_{s+2}^{-1} \dots c_t^{-1}$ is called the negative collected part. As with the Hall collection process we scan the uncollected part looking for the least basic commutator, c say, occurring in the sequence $c_{r+1}, c_{r+2}, \dots, c_s$. If c occurs with positive exponent we move it to the left using one of the two commutator identities

$$\begin{aligned} c_i c &= c c_i [c_i, c], \\ c_i^{-1} c &= c [c_i, c]^{-1} c_i^{-1}. \end{aligned}$$

And if c occurs with negative exponent we move it to the right using one of

$$\begin{aligned} c^{-1} c_i &= c_i [c_i, c] c^{-1}, \\ c^{-1} c_i^{-1} &= [c_i, c]^{-1} c_i^{-1} c^{-1}. \end{aligned}$$

As with the Hall collection process, if we specify any particular term $F^{(k)}$ of the derived series, then we eventually obtain an expression for our original word w in the form

$$w = c_1 c_2 \dots c_m u c_{m+1}^{-1} c_{m+2}^{-1} \dots c_n^{-1}$$

for some $u \in F^{(k)}$, where $c_1, c_2, \dots, c_n$ are basic commutators with

$$c_1 \leq c_2 \leq \dots \leq c_m, \; c_{m+1} \geq c_{m+2} \geq \dots \geq c_n.$$

The main advantage of the two sided process is that it handles negative exponents effectively when you are working modulo some term of the derived series of F. The Hall process can be extended to handle negative exponents, but only modulo terms of the lower central series of F. The main disadvantage of two sided collection is that the expression

$$w = c_1 c_2 \dots c_m u c_{m+1}^{-1} c_{m+2}^{-1} \dots c_n^{-1}$$

is not unique.

We end this section by proving the theorem mentioned above. The argument that follows is due to Gorchakov, and was used in Groves [5], [6]. Recall that $G = F/w(F)$ where F is a free group and w is an outer commutator word. We let Λ be a free Lie ring of the same rank as F.

Theorem 2.3 *If $\widetilde{w}(\Lambda)$ is spanned by basic commutators then G is torsion free and residually nilpotent.*

Proof Since $\widetilde{w}(\Lambda)$ is spanned by basic commutators it is a direct summand of Λ, and $\Lambda/\widetilde{w}(\Lambda)$ is torsion free. So, by Theorem 2.1, $\Lambda/\widetilde{w}(\Lambda)$ is the associated Lie ring of G. If we can show that G is residually nilpotent then it immediately follows that G is torsion free.

We make use of the fact that a basic commutator c can be interpreted either as a basic commutator in the free Lie ring Λ, or as a basic commutator in the free group F. We make the assumption that $c \in \widetilde{w}(\Lambda)$ (when thought of an element of Λ) if and only if $c \in w(F)$ (when thought of as an element of F). There is a difficulty here, since it is not clear that this is generally true! However, in applications of this argument we determine whether or not $c \in \widetilde{w}(\Lambda)$ by looking at the "shape" of c (i.e. the arrangement of the brackets), and in all the applications of the argument that I am aware of it is clear that $c \in \widetilde{w}(\Lambda)$ if and only if $c \in w(F)$.

Now let $g \in \bigcap_{n=1}^{\infty} \gamma_n(G)$. We need to show that $g = 1$. Since G satisfies the outer commutator identity $w = 1$, G is soluble. We suppose that G has derived length k. We let π be the natural projection from F onto G, and we pick an element $w \in F$ such that $w\pi = g$. Then we apply the two sided collection process to w, working modulo $F^{(k)}$. We obtain an expression

$$w = c_1 c_2 \ldots c_m u c_{m+1}^{-1} c_{m+2}^{-1} \cdots c_n^{-1}$$

for w, where

1. $c_1, c_2, \ldots, c_n$ are basic commutators,
2. $c_1 \leq c_2 \leq \ldots \leq c_m$,
3. $c_{m+1} \geq c_{m+2} \geq \ldots \geq c_n$,
4. $u \in F^{(k)}$.

So

$$g = w\pi = d_1 d_2 \ldots d_m d_{m+1}^{-1} d_{m+2}^{-1} \cdots d_n^{-1}$$

where $d_i = c_i\pi$ for $i = 1, 2, \ldots, n$. (Note that $u\pi = 1$ since G is soluble of derived length k.) Now $c_i\pi = 1$ if and only if $c_i \in w(F)$. If we delete all d_i such that $c_i \in w(F)$ from this expression for g we obtain an expression

$$g = d_{i_1} d_{i_2} \ldots d_{i_r} d_{i_{r+1}}^{-1} \cdots d_{i_s}^{-1}$$

where $1 \leq i_1 < i_2 < \ldots < i_r \leq m < i_{r+1} < \ldots < i_s \leq n$, and where $c_{i_j} \notin w(F)$ for $1 \leq j \leq s$.

Let t be the minimum weight of the basic commutators $c_{i_1}, c_{i_2}, \ldots, c_{i_s}$, and let $c_{i_1}, c_{i_2}, \ldots, c_{i_p}$, and $c_{i_q}, c_{i_{q+1}}, \ldots, c_{i_s}$ have weight t, with the remaining c_{i_j} having weight at least $t+1$. Then

$$d_{i_1} d_{i_2} \ldots d_{i_p} d_{i_q}^{-1} \ldots d_{i_s}^{-1} \in \gamma_{t+1}(G).$$

However Theorem 2.1 implies that the images in $\gamma_t(G)/\gamma_{t+1}(G)$ of the basic commutators of weight t which do *not* lie in $w(F)$ forms a basis for $\gamma_t(G)/\gamma_{t+1}(G)$ as a free $\mathbb{Z}$-module. We have

$$d_{i_1}d_{i_2}\ldots d_{i_p}\gamma_{t+1}(G) = d_{i_s}\ldots d_{i_q}\gamma_{t+1}(G),$$

where $d_{i_1},\ldots d_{i_p}, d_{i_s},\ldots, d_{i_q}$ are members of a basis for $\gamma_t(G)/\gamma_{t+1}(G)$, and where $i_1 \leq i_2 \leq \ldots \leq i_p$ and $i_s \leq \ldots \leq i_q$. So $d_{i_1}d_{i_2}\ldots d_{i_p} = d_{i_s}\ldots d_{i_q} = u$, say.

But then

$$\begin{aligned} & g^u \\ = \; & d_{i_{p+1}}\ldots d_{i_r}d_{i_{r+1}}^{-1}\ldots d_{i_{q-1}}^{-1} \\ \in \; & \bigcap_{n=1}^{\infty}\gamma_n(G). \end{aligned}$$

Repeating the argument above, we eventually see that $g = 1$, as required. □

3 Lie relators in varieties of groups

By studying Lie relators in varieties of groups, we are able to recover useful information about the groups themselves. Perhaps the most famous examples of this are Kostrikin's solution of the restricted Burnside problem for groups of prime exponent and Zel'manov's solution of the restricted Burnside problem for groups of prime power exponent. In Section 2 of this article we saw an application of this method to outer commutator varieties.

Let $\mathcal{V}$ be a variety of groups, and let G be the free group of $\mathcal{V}$ with free generators $x_1, x_2, \ldots$. We form the associated Lie ring of G,

$$L = L_1 \oplus L_2 \oplus \ldots \oplus L_i \oplus \ldots$$

as described in Section 1. Note that L is generated as a Lie ring by the elements $a_1, a_2, \ldots \in L_1$, where $a_i = x_i\gamma_2(G)$.

Let Λ be the free Lie ring with free generators $y_1, y_2, \ldots$, and let

$$\pi : \Lambda \to L$$

be the unique homomorphism which maps y_i to a_i for $i = 1, 2, \ldots$. The elements of $\ker\pi$ are called *Lie relators* of $\mathcal{V}$. If $u \in \Lambda$ lies in the kernel of every homomorphism from Λ to L, then we say that u is an *identical Lie relator* of $\mathcal{V}$. Clearly, identical Lie relators are Lie relators, but it was an open question for many years whether or not all Lie relators are identical Lie relators. An alternative way of posing this question is to ask whether or not the associated Lie ring of a relatively free group is always relatively free. The associated Lie ring of a free group is a free Lie ring. In Section 2 we surveyed some results which implied that the associated Lie rings of relatively free groups in some outer commutator varieties are relatively free. However in 1999 Groves [4] showed that the variety $\mathcal{B}_4\mathcal{B}_2$ has Lie relators which are not identical Lie relators, so that the associated Lie rings of free groups in the variety $\mathcal{B}_4\mathcal{B}_2$ are not relatively free. (Here $\mathcal{B}_4\mathcal{B}_2$ is the variety of groups which are extensions of groups of exponent 4 by groups of exponent 2.)

Returning to our general description of Lie relators, note that Λ is a graded algebra:

$$\Lambda = \Lambda_1 \oplus \Lambda_2 \oplus \ldots \oplus \Lambda_i \oplus \ldots,$$

where Λ_i is spanned by Lie products of weight i in $y_1, y_2, \ldots$. The homomorphism $\pi : \Lambda \to L$ is a graded homomorphism, so that $L_i = \Lambda_i \pi$. And $K = \ker \pi$ is a graded ideal of Λ:

$$K = K_1 \oplus K_2 \oplus \ldots \oplus K_i \oplus \ldots,$$

where $K_i \leq \Lambda_i$. It is not hard to show that if $\theta : \Lambda \to \Lambda$ is a graded homomorphism (i.e. if $\Lambda_1 \theta \leq \Lambda_i$ for some i), then $K\theta \leq K$.

One way of characterizing the elements of K_i is as follows. Let F be the (absolutely) free group of countably infinite rank, so that the relatively free group G of the variety $\mathcal{V}$ is isomorphic to $F/\mathcal{V}(F)$. We can identify the free Lie ring Λ with the associated Lie ring of F, so that $\Lambda_i = \gamma_i(F)/\gamma_{i+1}(F)$. With this identification, K_i is the kernel of the natural map from $\gamma_i(F)/\gamma_{i+1}(F)$ to $\gamma_i(G)/\gamma_{i+1}(G)$, and so

$$K_i = (\gamma_i(F) \cap \mathcal{V}(F))\, \gamma_{i+1}(F)/\gamma_{i+1}(F).$$

3.1 Wall's theory of multilinear Lie relators

An element $a \in \Lambda$ is said to be multilinear in $y_1, y_2, \ldots, y_n$ if a can be expressed in the form

$$\sum_{\sigma} \alpha_\sigma [y_{1\sigma}, y_{2\sigma}, \ldots, y_{n\sigma}],$$

where the sum ranges over the symmetric group on n letters, and where $\alpha_\sigma \in \mathbb{Z}$. (We can similarly define multilinear elements in any finite subset of the free generators of Λ.) It is well known that multilinear Lie relators of $\mathcal{V}$ are identical Lie relators, and we let K_{mul} be the fully invariant ideal of Λ generated by the multilinear Lie relators of $\mathcal{V}$. Then $K_{mul} \leq K$. Furthermore, as an abelian group, K/K_{mul} is a torsion group. More precisely, suppose that a is a Lie relator of $\mathcal{V}$, and suppose that a lies in the subalgebra of Λ generated by $y_1, y_2, \ldots, y_m$. Suppose further that a has degree n_i in y_i for $i = 1, 2, \ldots, m$. Then

$$n_1! \cdot n_2! \cdot \ldots \cdot n_m! \cdot a \in K_{mul}.$$

To prove this, we substitute z_i for y_i in a for $i = 1, 2, \ldots, m$, where z_i is a sum of n_i free generators of Λ (with different free generators in each sum). We expand and pick out the multilinear components of the resulting expression. It is straightforward to show that these are all Lie relators of $\mathcal{V}$. But to recover the original relator a from these multilinear relators, we have to substitute y_i for all the free generators which occurred in the sum z_i (for $i = 1, 2, \ldots, m$) and this gives rise to the coefficient $n_1! \cdot n_2! \cdot \ldots \cdot n_m!$.

There is a beautiful theory of the multilinear Lie relators of $\mathcal{V}$, which is due to Wall [35]. Let $w = w(x_1, x_2, \ldots, x_m)$ be an element of the free group on $x_1, x_2, \ldots, x_m$, and suppose that $w = 1$ is an identical (group) relation in $\mathcal{V}$. We let A be the free associative $\mathbb{Q}$-algebra freely generated by the non-commuting indeterminates $x_1, x_2, \ldots$, and we let $\widehat{A}$ be the ring of formal power series over $\mathbb{Q}$ in these indeterminates. We can think of A as being embedded in $\widehat{A}$. Then e^{x_i} is a unit in $\widehat{A}$ for $i = 1, 2, \ldots$, and the subgroup F of the group of units of $\widehat{A}$ generated by e^{x_1}, $e^{x_2}, \ldots$ is a free group (with $e^{x_1}, e^{x_2}, \ldots$ as free generators). Let $(r_1, r_2, \ldots, r_m)$ be an m-tuple of non-negative integers, and let

$$\begin{aligned} z_1 &= e^{x_1} e^{x_2} \ldots e^{x_{r_1}}, \\ z_2 &= e^{x_{r_1+1}} e^{x_{r_1+2}} \ldots e^{x_{r_1+r_2}}, \\ &\ldots \\ z_m &= e^{x_{r_1+\ldots+r_{m-1}+1}} \ldots e^{x_{r_1+\ldots+r_m}}. \end{aligned}$$

Now consider the element

$$w(z_1, z_2, \ldots, z_m) \in \widehat{A}.$$

We let $T_w^{(r_1,\ldots,r_m)}$ be the multilinear component of degree 1 in each of the generators $x_1, x_2, \ldots, x_{r_1+\ldots+r_m}$ in the power series expansion of $w(z_1, z_2, \ldots, z_m)$. Note that this multilinear element lies in A. However, if we let B be the associative subring of A generated over $\mathbb{Z}$ by $x_1, x_2, \ldots$, it turns out that $T_w^{(r_1,\ldots,r_m)}$ lies in B. To see this note that if we are only concerned with the multilinear component of $w(z_1, z_2, \ldots, z_m)$, then we can substitute $1 + x_i$ for e^{x_i} and substitute $1 - x_i$ for e^{-x_i} in $w(z_1, z_2, \ldots, z_m)$ (for $i = 1, 2, \ldots$). Then we define a $\mathbb{Z}$-linear Dynkin bracket operator

$$\delta : B \to \Lambda$$

by setting

$$(x_i x_j \ldots x_k)\delta = \begin{cases} [y_i, y_j, \ldots, y_k] \text{ if } i = 1, \\ 0 \text{ otherwise.} \end{cases}$$

Finally we let

$$t_w^{(r_1,\ldots,r_m)} = (T_w^{(r_1,\ldots,r_m)})\delta.$$

Wall proves that $t_w^{(r_1,\ldots,r_m)}$ is an identical Lie relator of $\mathcal{V}$ for every identical group relation $w = 1$ of $\mathcal{V}$, and for every m-tuple $(r_1, r_2, \ldots, r_m)$ of non-negative integers. Furthermore, he shows that if W is a basis for the identical group relations of $\mathcal{V}$, then

$$\{t_w^{(r_1,\ldots,r_m)} \mid w \in W, \ r_1, r_2, \ldots, r_m \geq 0\}$$

is a basis for K_{mul}. (The word "basis" is somewhat misleading here, though it is conventional in this context. A basis for a set W of identical relations is a subset $U \subseteq W$ with the property that all the identical relations in W are consequences of those in U. There is no requirement that the elements of U be independent in any sense.)

Here is a proof of the fact that the elements $t_w^{(r_1,\ldots,r_m)}$ are Lie relators of $\mathcal{V}$. For each $i = 1, 2, \ldots$ we define a homomorphism $d_i : \widehat{A} \to \widehat{A}$ by

$$\begin{aligned} x_i d_i &= 0, \\ x_j d_i &= x_j \text{ for } j \neq i. \end{aligned}$$

Note that d_i induces an endomorphism sending e^{x_i} to 1 on the multiplicative group F generated by e^{x_1}, e^{x_2}, ... If g is an element of F we let

$$gD_i = (gd_i)^{-1} \cdot g.$$

So gD_i is in the fully invariant closure of g in F. Furthermore

$$gD_i d_i = (gd_i d_i)^{-1} \cdot gd_i = 1,$$

so gD_i is in the normal closure of e^{x_i}.

Now, recall our definition of $w(z_1, \ldots, z_m)$ above and let $r = r_1 + \ldots + r_m$. Let $w^* = w(z_1, \ldots, z_m)$. Then $u = w^* D_1 D_2 \ldots D_r$ lies in the intersection of the normal closures of e^{x_1}, e^{x_2}, ..., e^{x_r}. It follows that u can be expressed in the form

$$u = \prod_\sigma [e^{x_{1\sigma}}, e^{x_{2\sigma}}, \ldots, e^{x_{r\sigma}}]^{\alpha_\sigma} \bmod \gamma_{r+1}(F),$$

where the product ranges over the symmetric group on r elements, and where $\alpha_\sigma \in \mathbb{Z}$. Since $u = 1$ is an identity in $\mathcal{V}$ it follows immediately that

$$\sum_\sigma \alpha_\sigma [y_{1\sigma}, y_{2\sigma}, \ldots, y_{r\sigma}]$$

is a multilinear Lie relator of $\mathcal{V}$. Now if we expand u using the Baker-Campbell-Hausdorff formula we see that $u = e^v$ where

$$v = \sum_\sigma \alpha_\sigma [x_{1\sigma}, x_{2\sigma}, \ldots, x_{r\sigma}] + \text{ higher terms.}$$

It follows that the multilinear component of u is $\sum_\sigma \alpha_\sigma [x_{1\sigma}, x_{2\sigma}, \ldots, x_{r\sigma}]$. (In this expression $[,]$ denotes the ring commutator in A.) It is easy to see that

$$(\sum_\sigma \alpha_\sigma [x_{1\sigma}, x_{2\sigma}, \ldots, x_{r\sigma}])\delta = \sum_\sigma \alpha_\sigma [y_{1\sigma}, y_{2\sigma}, \ldots, y_{r\sigma}].$$

For example

$$([x_1, x_2])\delta = (x_1 x_2 - x_2 x_1)\delta = (x_1 x_2)\delta - (x_2 x_1)\delta = [y_1, y_2] - 0 = [y_1, y_2].$$

So the multilinear component of $w^* D_1 D_2 \ldots D_r$ is a Lie element $a \in B$, and $a\delta$ is a multilinear Lie relator of $\mathcal{V}$. But if we expand a as a linear combination of monomials, and if we let a' be the sum of the monomials in a which start with x_1 then $a\delta = a'\delta$. Furthermore it is not hard to see that a' is the sum of the multilinear terms of w^* which start with x_1. So $a\delta = a'\delta = t_w^{(r_1, \ldots, r_m)}$ is a Lie relator of $\mathcal{V}$.

3.2 Burnside varieties

As an illustration of Wall's theory, consider the Burnside variety $\mathcal{B}_p$ consisting of groups of exponent p. A basis for the identical relations of $\mathcal{B}_p$ is $\{x^p\}$.

The multilinear term of $(e^{x_1})^p$ is px_1, and $(px_1)\delta = py_1$. In Wall's notation, we have $t_{x^p}^{(1)} = py_1$. So we obtain the Lie relator py_1.

Now consider $(e^{x_1}e^{x_2}\ldots e^{x_p})^p$. Expanding $e^{x_1}e^{x_2}\ldots e^{x_p}$ we obtain $1+a$ where

$$a = x_1 + x_2 + \ldots + x_p + \text{ higher terms.}$$

So, working modulo p, we have

$$(e^{x_1}e^{x_2}\ldots e^{x_p})^p = 1 + a^p.$$

The multilinear component of a^p is

$$\sum_{\sigma \in \mathrm{Sym}(p)} x_{1\sigma}x_{2\sigma}\ldots x_{p\sigma},$$

and so

$$t_{x^p}^{(p)} = \left(\sum_{\sigma \in \mathrm{Sym}(p)} x_{1\sigma}x_{2\sigma}\ldots x_{p\sigma}\right)\delta = \sum_{\sigma \in \mathrm{Sym}\{2,3,\ldots,p\}} [y_1, y_{2\sigma}, \ldots, y_{p\sigma}].$$

Note that if we set $y_2 = y_3 = \ldots = y_p = z$ in $t_{x^p}^{(p)}$ then we obtain the relator $(p-1)![y_1, z, z, \ldots, z]$, which together with the Lie relator py_1 implies the $(p-1)$-Engel identity.

Wall's recipe also gives Lie relators $t_{x^p}^{(n)}$ for all $n \geq 2$. These are equivalent to the relators K_n mentioned in Section 1.

Next, consider the variety $\mathcal{E}_4 \cap \mathcal{B}_5$ of 4-Engel groups of exponent 5. This variety is determined by the two group identities $[x_2, x_1, x_1, x_1, x_1] = 1$, $x_1^5 = 1$. I proved that it is a locally finite variety in [33]. If we let $w = [x_2, x_1, x_1, x_1, x_1]$, then it is not hard to see that

$$t_w^{(1,4)} = \sum_{\sigma \in \mathrm{Sym}\{2,3,4,5\}} [y_1, y_{2\sigma}, y_{3\sigma}, y_{4\sigma}, y_{5\sigma}],$$

$$t_w^{(2,4)} = \sum_{\sigma \in \mathrm{Sym}\{3,4,5,6\}} [y_1, y_{3\sigma}, y_2, y_{4\sigma}, y_{5\sigma}, y_{6\sigma}],$$

$$t_w^{(3,4)} = \sum_{\sigma \in \mathrm{Sym}\{4,5,6,7\}} [y_1, y_{4\sigma}, y_2, y_3, y_{5\sigma}, y_{6\sigma}, y_{7\sigma}].$$

The group identity x_1^5 gives us the Lie relator $5y_1$, as above. Note also that $t_{x^5}^{(5)} = t_w^{(1,4)}$. It turns out that the associated Lie rings of 4-Engel groups of exponent 5 are free Lie rings in the variety $\mathcal{L}$ of Lie rings determined by these 4 identical Lie relators. It is convenient to think of $\mathcal{L}$ as a variety of Lie algebras over $\mathbb{Z}_5$.

We use a beautiful idea introduced by Graham Higman [12] in his solution of the restricted Burnside problem for exponent 5, and consider the Lie algebra of $\mathcal{L}$ with the following presentation (as an algebra in $\mathcal{L}$):

$$L = \langle x, a_1, a_2, \ldots \mid [a_i, a_j] = 0 \text{ for all } i, j \rangle.$$

It is an easy calculation to show that the ideal generated by x in L is nilpotent of class 4. In fact L itself is nilpotent of class 6. As Higman showed, this implies that if L is *any* Lie algebra in the variety $\mathcal{L}$, and if $x \in L$, then $\mathrm{Id}_L(x)$ is nilpotent of class 4. In other words, any product of elements of L is zero if it has degree 5 or more in any one element. This means that in considering Lie relators of $\mathcal{E}_4 \cap \mathcal{B}_5$, we need only consider relators which have degree 4 or less in each variable. Recall that if a is a Lie relator of degree n_i in y_i for $i = 1, 2, \ldots, m$. Then

$$n_1! \cdot n_2! \cdot \ldots \cdot n_m! \cdot a \in K_{mul}.$$

Using the relator $5y_1$, this implies that all Lie relators in 4-Engel groups of exponent 5 are consequences of multilinear relators.

So the Lie relators of $\mathcal{E}_4 \cap \mathcal{B}_5$ are all consequences of

$$t_{x^5}^{(n)} \ (n \geq 1)$$

and

$$t_w^{(m,n)} \ (m \geq 1, n \geq 4).$$

Another result of Wall's is that $t_w^{(m,n)}$ is symmetric in its first m variables modulo consequences of $\{t_w^{(k,n)} \mid k < m\}$, and also symmetric in its last n variables modulo consequences of $\{t_w^{(m,k)} \mid k < n\}$. Similarly $t_{x^5}^{(n)}$ is symmetric in its n variables modulo consequences of $\{t_{x^5}^{(k)} \mid k < n\}$. Using an extension of Higman's argument, it is possible to show that if L is a free Lie algebra in $\mathcal{L}$, and if $a \in L$ is multilinear in 9 or more variables, and if a is symmetric in 4 of those variables, then $a = 0$.

So the Lie relators in the variety of 4-Engel groups of exponent 5 are all consequences of

$$\{t_{x^5}^{(n)} \mid 1 \leq n < 9\} \cup \{t_w^{(m,n)} \mid m \geq 1,\, n \geq 4,\, m + n < 9)\}.$$

It is straightforward to show that these are all consequences of the defining relators of $\mathcal{L}$. So the associated Lie rings of free groups in the variety $\mathcal{E}_4 \cap \mathcal{B}_5$ are free Lie algebras in the variety $\mathcal{L}$.

Details of the above calculations may be found in [23], where we also show that the free rank m Lie algebra in $\mathcal{L}$ has dimension n where

$$n = m + \sum_{k=2}^{m} \binom{m}{k} (g_k + c_k),$$

where $g_k = (k-1)f_{2k} + (k+1)f_{2k-2}$, and where $c_k = 0$ for $k > 10$, and c_k has the value given in the following table for $2 \leq k \leq 10$.

c_2	c_3	c_4	c_5	c_6	c_7	c_8	c_9	c_{10}
3	87	595	1851	2996	2562	1094	224	35

(Here f_k is the k-th Fibonacci number.) It follows immediately that the free group of rank m in $\mathcal{E}_4 \cap \mathcal{B}_5$ has order 5^n.

We end this section by proving the lemma needed in Section 1. Recall that A is the free associative algebra with 1 over $\mathbb{Q}$ freely generated by $y_1, y_2, \ldots$, and that $A_c = A/I$ where I is the ideal of A generated by products of weight $c+1$ in the generators $y_1, y_2, \ldots$. We let $x_i = y_i + I$ for $i = 1, 2, \ldots$. The elements $\mathrm{e}^{x_1}, \ldots, \mathrm{e}^{x_r}$ generate a subgroup of the group of units of A_c, which we identify with $F_{r,c}$, the free nilpotent of class c group of rank r. We let $\Lambda_{r,c}$ be the Lie subring (over $\mathbb{Z}$) of A_c generated by $x_1, x_2, \ldots, x_r$ with respect to the Lie product $[a, b] = ab - ba$. So $\Lambda_{r,c}$ is the free nilpotent of class c Lie ring of rank r. Also $\mathbb{Q}\Lambda_{r,c}$ is a free nilpotent of class c Lie algebra over $\mathbb{Q}$, and if we let $\mathbb{Q}_p$ be the set of rationals of the form m/n where n is coprime to p, then $\mathbb{Q}_p\Lambda_{r,c}$ is a free nilpotent of class c Lie algebra over $\mathbb{Q}_p$.

Also recall that G is the class c quotient of $R(r,p)$ so that we can identify G with $F_{r,c}/F_{r,c}^p$. We are assuming that if L is the associated Lie ring of G then L is isomorphic to the class c quotient of $W(r,p)$. Since $W(r,p)$ is defined by multilinear Lie relators of $\mathcal{B}_p$, this implies that the Lie relators of G of weight at most c are all consequences of multilinear Lie relators of $\mathcal{B}_p$. In addition, we have a homomorphism $\pi : \mathbb{Q}_p\Lambda_{r,c} \to L$ mapping the generators $x_1, x_2, \ldots, x_r$ of $\mathbb{Q}_p\Lambda_{r,c}$ to the generators $a_1, a_2, \ldots, a_r$ of L. The elements of $\ker \pi \cap \Lambda_{r,c}$ are Lie relators of G (by definition).

Finally, recall that elements of $\gamma_{c-p+2}(F_{r,c})$ can be expressed in the form e^u where $u \in \mathbb{Q}_p\Lambda_{r,c}$, and that N is defined to be the subgroup of $\gamma_{c-p+2}(F_{r,c})$ consisting of elements e^u with $u \in \ker \pi$.

The lemma needed in Section 1 is the following.

Lemma 3.1 $N \leq F_{r,c}^p$.

Proof Let $b \in N$. We need to show that $b \in F_{r,c}^p$.

If $b = 1$ then there is nothing to prove, and so we suppose that $b \neq 1$. By definition $N \leq \gamma_{c-p+2}(F_{r,c})$, and so there is an integer k with $c - p + 2 \leq k \leq c$ such that $b \in \gamma_k(F_{r,c}) \backslash \gamma_{k+1}(F_{r,c})$. We may suppose by reverse induction on k that $\gamma_{k+1}(F_{r,c}) \cap N \leq F_{r,c}^p$.

Since $b \in \gamma_k(F_{r,c}) \backslash \gamma_{k+1}(F_{r,c})$ we have $b = \mathrm{e}^u$ for some $u = u_k + u_{k+1} + \ldots + u_c \in \mathbb{Q}_p\Lambda_{r,c}$, where u_j is homogeneous of weight j for $j = k, k+1, \ldots, c$. In addition we have $u_k \neq 0$, and $u_k \in \Lambda_{r,c}$. (For a proof of this well known fact see Chapter 2 of [32].) Since $b \in N$, $u \in \ker \pi$, which implies that $u_k \in \ker \pi$. So u_k is a Lie relator of G, which implies that there is an element $\mathrm{e}^v \in \gamma_k(F_{r,c}) \cap F_{r,c}^p$ with $v = u_k + v_{k+1} + \ldots + v_c \in \mathbb{Q}_p\Lambda_{r,c}$. The key to proving that $b \in F_{r,c}^p$ is to show that we can assume that $v \in \ker \pi$. For then $\mathrm{e}^v \in N \cap F_{r,c}^p$ and $\mathrm{e}^{-v}\mathrm{e}^u \in \gamma_{k+1}(F_{r,c}) \cap N$. By induction we have $\mathrm{e}^{-v}\mathrm{e}^u \in F_{r,c}^p$, and hence that $\mathrm{e}^u \in F_{r,c}^p$.

So we need to show that if u_k is a Lie relator of G of weight k (with $k \geq c-p+2$) then we can find $\mathrm{e}^v \in N \cap F_{r,c}^p$ with $v = u_k + v_{k+1} + \ldots + v_c \in \mathbb{Q}_p\Lambda_{r,c}$.

First note that this is straightforward if $k < p$. This is because all Lie relators of G of weight less than p are consequences of the Lie relator px_1. So if u_k is a Lie relator of G of weight k, and if $k < p$, then $u_k = pw_k$ for some $w_k \in \Lambda_{r,c}$ of weight k. There is some element $\mathrm{e}^w \in \gamma_k(F_{r,c})$ with $w = w_k + w_{k+1} + \ldots + w_c \in \mathbb{Q}_p\Lambda_{r,c}$.

Then

$$(e^w)^p = e^{pw} \in N \cap F^p_{r,c},$$

and we are done.

So from now on we assume that $k \geq p$. The advantage of this is that if $k \geq p$ then the restriction $k \geq c - p + 2$ implies that $2k > c$, so that if e^u, $e^v \in \gamma_k(F_{r,c})$ then u and v commute and $e^u \cdot e^v = e^{u+v}$.

It is at this point that we make use of the assumption that the Lie relators of G of weight at most c are all consequences of multilinear Lie relators of $\mathcal{B}_p$.

Since we are only concerned with Lie relators of $\mathcal{B}_p$ of weight at most c, we can identify them with elements of the Lie subring Λ_c of A_c generated by $x_1, x_2, \ldots$. Let R be the set of multilinear Lie relators of $\mathcal{B}_p$. (Their form does not concern us here.) We let S be the ideal generated by

$$\{a\theta \mid a \in R,\ \theta \text{ an endomorphism of } \Lambda_c\}.$$

First note that S is *spanned* by $\{a\theta \mid a \in R\}$, since if $a = a(x_1, x_2, \ldots, x_n) \in R$ then $[a, x_{n+1}] \in R$.

Next let $a = a(x_1, x_2, \ldots, x_n) \in R$, and let θ be an endomorphism of Λ_c. Then $a\theta$ is a linear combination of elements $a(b_1, b_2, \ldots, b_n)$, where $b_1, b_2, \ldots, b_n$ are Lie products of the generators $x_1, x_2, \ldots$. Consider such an element $a(b_1, b_2, \ldots, b_n)$, and choose elements $c_1, c_2, \ldots, c_n \in \Lambda_c$ where (for $i = 1, 2, \ldots, n$) c_i is a product of the generators with the same weight and bracketing as b_i, and where the generators occurring in c_i are disjoint from those occurring in c_j for $i \neq j$. Then $a(c_1, c_2, \ldots, c_n) \in R$, and

$$a(b_1, b_2, \ldots, b_n) = a(c_1, c_2, \ldots, c_n)\varphi$$

where $\varphi : \Lambda_c \to \Lambda_c$ is an endomorphism mapping generators of Λ_c to generators.

So S is spanned by elements $a\varphi$ where $a \in R$ and where $\varphi : \Lambda_c \to \Lambda_c$ is an endomorphism mapping generators of Λ_c to generators. Actually, if $a \in R$, then $na \in R$ for any $n \in \mathbb{Z}$, and so if $u \in \Lambda_{r,c}$ is a Lie relator of G of weight k for some $k \leq c$, then u is a *sum* of elements of the form

$$a(x_{i_1}, x_{i_2}, \ldots, x_{i_k}) \text{ with } a(x_1, x_2, \ldots, x_k) \in R,\ 1 \leq i_1, i_2, \ldots, i_k \leq r.$$

So consider a multilinear Lie relator $a(x_1, x_2, \ldots, x_k)$ in the variety $\mathcal{B}_p$ of groups of exponent p, with $p \leq k \leq c$. Let F_c be the subgroup of the group of units of A_c generated by e^{x_1}, e^{x_2}, ... Then F_c is the free nilpotent of class c group of countably infinite rank. Since $a(x_1, x_2, \ldots, x_k) \in R$, there is an element $e^w \in \gamma_k(F_c) \cap F_c^p$ where

$$w = a(x_1, x_2, \ldots, x_k) + a_{k+1} + \ldots + a_c \in \mathbb{Q}_p\Lambda_c.$$

Replacing e^w by $e^w D_1 D_2 \ldots D_k$ (as described earlier in this section), we may suppose that e^w lies in the intersection of the normal closures of e^{x_1}, e^{x_2}, ..., e^{x_k}. So we may suppose that $a_{k+1}, a_{k+2}, \ldots, a_c$ are all linear combinations of Lie products every one of which involves all of $x_1, x_2, \ldots, x_k$. Note that this, together with the restriction $k \geq c - p + 2$, implies that a_j has degree at most $p - 1$ in any of the variables $x_1, x_2, \ldots, x_k$ for $k + 1 \leq j \leq c$.

The main step in proving Lemma 3.1 lies in showing that there is an integer n coprime to p such that na_j is a Lie relator in $\mathcal{B}_p$ for $k + 1 \leq j \leq c$. Assuming this for the moment, we let θ be any endomorphism of A_c mapping x_j to x_{i_j} for $j = 1, 2, \ldots, k$. Then

$$\mathrm{e}^{w\theta} = (\mathrm{e}^w)\theta \in \gamma_k(F_{r,c}) \cap F_{r,c}^p,$$

and

$$w\theta = a(x_{i_1}, x_{i_2}, \ldots, x_{i_k}) + a_{k+1}\theta + \ldots + a_c\theta.$$

By our assumption, $na_j\theta$ is a Lie relator in $\mathcal{B}_p$ for $k + 1 \leq j \leq c$, and since $p \nmid n$ this implies that $a_j\theta \in \ker \pi$. So

$$\mathrm{e}^{w\theta} \in N \cap \gamma_k(F_{r,c}) \cap F_{r,c}^p.$$

Recall that we need to show that if u_k is a Lie relator of G of weight k (with $k \geq c - p + 2$) then we can find $\mathrm{e}^v \in N \cap F_{r,c}^p$ with $v = u_k + v_{k+1} + \ldots + v_c \in \mathbb{Q}_p\Lambda_{r,c}$. As we showed above, u_k is a sum of elements $a(x_{i_1}, x_{i_2}, \ldots, x_{i_k})$ where $a(x_1, x_2, \ldots, x_k) \in R$. For each $a(x_{i_1}, x_{i_2}, \ldots, x_{i_k})$ which occurs in this sum we obtain an element $\mathrm{e}^{w\theta}$, as above. The element e^v that we want is the product of the elements $\mathrm{e}^{w\theta}$.

So suppose that $\mathrm{e}^w \in \gamma_k(F_c) \cap F_c^p$ where

$$w = a(x_1, x_2, \ldots, x_k) + a_{k+1} + \ldots + a_c \in \mathbb{Q}_p\Lambda_c,$$

with $a(x_1, x_2, \ldots, x_k) \in R$. As we showed above, we may assume that $a_{k+1}, \ldots, a_c$ are all linear combinations of Lie products every one of which involves all of $x_1, \ldots, x_k$. We need to show that there is an integer n coprime to p such that na_j is a Lie relator in $\mathcal{B}_p$ for $k + 1 \leq j \leq c$.

As an illustration of the situation, consider the element $a_{k+1} \in \mathbb{Q}_p\Lambda_{r,c}$. It is homogeneous of degree $k + 1$, and is a linear combination of Lie products each of which involves all of $x_1, x_2, \ldots, x_k$. So we can express

$$a_{k+1} = a'_{k+1} + a''_{k+1},$$

where a'_{k+1} is homogeneous of degree 1 in x_1, and a''_{k+1} is homogeneous of degree 2 in x_1. Let φ be the endomorphism of A_c mapping x_1 to $2x_1$, and mapping x_i to x_i for $i > 1$. Then $\mathrm{e}^w\varphi = \mathrm{e}^u$ where

$$u = 2a + 2a'_{k+1} + 4a''_{k+1} + \text{ higher terms},$$

and $(\mathrm{e}^w)^2 = \mathrm{e}^{2w}$ where

$$2w = 2a + 2a'_{k+1} + 2a''_{k+1} + \text{ higher terms}.$$

So

$$\mathrm{e}^w\varphi \cdot \mathrm{e}^{-2w} = \mathrm{e}^v$$

where $v = 2a''_{k+1}+$ higher terms. So $\mathrm{e}^w\varphi{\cdot}\mathrm{e}^{-2w} \in \gamma_{k+1}(F_c) \cap F_c^p$, and $2a''_{k+1}$ is a Lie relator in $\mathcal{B}_p$.

We need a more general argument than this to deal with all the components of w. As we noted above, w has degree at most $p-1$ in x_1, and all the summands in w involve x_1 at least once. So we can express w in the form

$$w = w_1 + w_2 + \ldots + w_{p-1}$$

where w_i is homogeneous of degree i in x_1 (for $i = 1, 2, \ldots, p-1$). For each $r = 1, 2, \ldots, p-1$ we let $\varphi_r : A_c \to A_c$ be the endomorphism mapping x_1 to rx_1, and mapping x_i to x_i for $i > 1$. Then

$$w\varphi_r = rw_1 + r^2 w_2 + \ldots r^{p-1} w_{p-1}.$$

Now the $(p-1) \times (p-1)$ matrix with (i,j)-entry i^j is non-singular, and its determinant m is coprime p. Furthermore, for $i = 1, 2, \ldots, p-1$ the element mw_i is a $\mathbb{Z}$-linear combination of the elements $w\varphi_r$ $(1 \leq r \leq p-1)$. As we remarked above, our assumption that $k \geq p$ implies that the elements $w\varphi_r$ commute with each other. And so e^{mw_i} is in the subgroup generated by the elements $\mathrm{e}^{w\varphi_r}$. This implies that

$$\mathrm{e}^{mw_i} \in F_c^p \text{ for } i = 1, 2, \ldots, p-1.$$

We can now repeat this argument with the generator x_2. For each i $(1 \leq i \leq p-1)$ we write

$$mw_i = u_1 + u_2 + \ldots + u_{p-1}$$

where u_j is homogeneous of degree j in x_2. By considering endomorphisms of A_c mapping x_2 to rx_2 for $r = 1, 2, \ldots, p-1$ we show that there is an integer m' coprime to p such that

$$\mathrm{e}^{m'u_j} \in F_c^p \text{ for } j = 1, 2, \ldots, p-1.$$

We repeat this argument for the generators $x_3, x_4, \ldots, x_k$. The original element $w \in \mathbb{Q}_p\Lambda_{r,c}$ can be expressed in the form

$$w = \sum_i u_i$$

where each summand u_i is homogeneous in every generator $x_1, x_2, \ldots, x_k$. The argument given above shows that there is an integer n coprime to p such that $\mathrm{e}^{nu_i} \in F_c^p$ for all i. Since u_i is homogeneous in every generator this implies that $nu_i \in \Lambda_c$ and that nu_i is a Lie relator in $\mathcal{B}_p$.

Returning to our original expression

$$a(x_1, x_2, \ldots, x_k) + a_{k+1} + \ldots + a_c$$

for w, we see that each summand a_j is the sum of those u_i which have weight j. Since nu_i is a Lie relator of $\mathcal{B}_p$ for all i, we immediately see that na_j is a Lie relator of $\mathcal{B}_p$ for all j, as required. □

4 Superalgebras

Several authors have used the representation theory of the symmetric group and superalgebra techniques to prove that certain algebras are nilpotent. Zel'manov [36] used these techniques in his proof that Lie algebras of characteristic zero are nilpotent if they satisfy the n-Engel identity for any n. Zel'manov and Shestakov [39] used them in their proof that the radical of a free alternative algebra over a field of characteristic zero is nilpotent. I used them in [31] to prove that if A is an associative algebra of characteristic zero, and if $a^4 = 0$ for all $a \in A$, then $A^{10} = \{0\}$. And Traustason [28] used them in his proof that 4-Engel Lie algebras are nilpotent of class at most 7, provided their characteristic is not 2,3 or 5.

The first author to have systematically exploited these techniques is Kemer [15], [16]. In these two articles Kemer solved the Specht problem by proving that all varieties of associative algebras of characteristic zero are finitely based. His complete proof can be found in his book [17]. Here we give an illustration of this method, and outline an extension of the method to computing the dimensions of certain algebras.

A superalgebra is a $\mathbb{Z}_2$-graded algebra $A = A_0 \oplus A_1$, where $A_iA_j \leq A_{i+j}$ for all $i, j \in \mathbb{Z}_2$. (Here, by an algebra we mean a vector space equipped with a bilinear product.) An important example of a superalgebra is the Grassmann algebra, G. It is the associative algebra with unity generated by $e_1, e_2, \ldots$ subject to the relations $e_i^2 = 0$ $(i = 1, 2, \ldots)$, $e_ie_j = -e_je_i$ $(i < j)$. So G has a basis consisting of 1, together with the set of all products $e_ie_j \ldots e_k$ with $1 \leq i < j < \ldots < k$.

We can express

$$G = G_0 \oplus G_1,$$

where G_0 is spanned by 1 together with products $e_ie_j \ldots e_k$ of even length, and where G_1 is spanned by products $e_ie_j \ldots e_k$ of odd length. So G is a superalgebra.

If $g \in G_0$ and $h \in G$ then $gh = hg$. But if $g, h \in G_1$, then $gh = -hg$.

If X is any algebra then $X \otimes G = X_0 \oplus X_1$, where $X_0 = X \otimes G_0$ and $X_1 = X \otimes G_1$. So $X \otimes G$ is a superalgebra. We call X_0 the even part of $X \otimes G$, and X_1 the odd part.

If L is a Lie algebra then $L \otimes G$ is a Lie superalgebra. It is conventional to use $[,]$ to denote the product in Lie algebras and in Lie superalgebras. If $a \otimes g$ and $b \otimes h$ are elements of $L \otimes G$ then we define

$$[a \otimes g, b \otimes h] = [a, b] \otimes gh,$$

and we extend this Lie product to the whole of $L \otimes G$ by linearity. Note that

$$[b \otimes h, a \otimes g] = [b, a] \otimes hg = -[a, b] \otimes hg.$$

If either of g or h are even then $gh = hg$ and so

$$[a \otimes g, b \otimes h] = -[b \otimes h, a \otimes g].$$

But if g, h are both odd then $gh = -hg$, and so

$$[a \otimes g, b \otimes h] = [b \otimes h, a \otimes g].$$

The definition of a Lie superalgebra is as follows. Let $M = M_0 \oplus M_1$ be a superalgebra. Use $[,]$ to denote the product in M. Then M is a Lie superalgebra if

$$\begin{aligned} [y, x] &= -(-1)^{ij}[x, y] \\ [z, [x, y]] &= [z, x, y] - (-1)^{ij}[z, y, x] \end{aligned}$$

for $x \in M_i$, $y \in M_j$ and $z \in M$. Thus $[y, x] = -[x, y]$ if either of x or y is even, but if x and y are both odd then $[x, y] = [y, x]$, as in the algebra $L \otimes G$ above. In fact the defining identities for Lie superalgebras are obtained by considering the sort of identities satisfied by $L \otimes G$ when L is a Lie algebra — we make this more precise below. In characteristic 2 we also need the additional identity $[x, x] = 0$ when x is even. And if x is odd you can deduce the identity

$$3[x, x, x] = 0,$$

so in characteristic 3 you might want to add in the identity $[x, x, x] = 0$ for odd elements.

We want to define the notion of *varieties of superalgebras*, and so we need the notion of *free superalgebras*. If $x_1, x_2, \ldots$ are even generators and $y_1, y_2, \ldots$ are odd generators then we form the free superalgebra on $x_1, x_2, \ldots, y_1, y_2, \ldots$ as follows.

Let X be the free algebra with free generators $x_1, x_2, \ldots, y_1, y_2, \ldots$. As a vector space, X has a basis consisting of all possible products of the generators, with all possible bracketings. (Superalgebras need not be associative.) We let X_0 be the subspace of X spanned by all products of the free generators which have even degree in the odd generators $y_1, y_2, \ldots$, and we let X_1 be the subspace of X spanned by all products which have odd degree in the odd generators.

Then $X = X_0 \oplus X_1$ is a free superalgebra on $x_1, x_2, \ldots, y_1, y_2, \ldots$.

If $f(x_1, x_2, \ldots, x_m, y_1, y_2, \ldots, y_n) \in X$, and if $A = A_0 \oplus A_1$ is a superalgebra, then we say that A satisfies the *graded* identity

$$f(x_1, x_2, \ldots, x_m, y_1, y_2, \ldots, y_n) = 0$$

if $f(a_1, \ldots, a_m, b_1, \ldots, b_n) = 0$ for all $a_1, \ldots, a_m \in A_0$ and all $b_1, \ldots, b_n \in A_1$. In other words, A satisfies $f = 0$ if f is in the kernel of every *graded* homomorphism from X into A.

If S is a set of graded identities, then the variety of superalgebras determined by S is the class of all superalgebras which satisfy the graded identities in S. If $\mathcal{V}$ is a variety of ungraded algebras determined by a set of multilinear identities then $\mathcal{V}$ gives rise to a variety $\widetilde{\mathcal{V}}$ of superalgebras as follows. Suppose we have a multilinear element $f(x_1, x_2, \ldots, x_n)$ in a free algebra. (We can think of f as an element of the ungraded subalgebra of X generated by $x_1, x_2, \ldots$.) Let S be any subset of $\{1, 2, \ldots, n\}$ and define $z_1, z_2, \ldots, z_n \in X$ by setting $z_i = x_i$ if $i \notin S$, and setting $z_i = y_i$ if $i \in S$. Pick $g_1, g_2, \ldots, g_n$ in the Grassmann algebra G so that $g_1 g_2 \cdots g_n \neq 0$, and so that $g_i \in G_0$ if $i \notin S$ and $g_i \in G_1$ if $i \in S$. Then in $X \otimes G$

$$f(z_1 \otimes g_1, z_2 \otimes g_2, \ldots, z_n \otimes g_n)$$

$$= \widetilde{f}_S(z_1, z_2, \ldots, z_n) \otimes g_1 g_2 \ldots g_n,$$

where $\widetilde{f}_S$ is determined by f and the subset S. If we express $f(z_1, z_2, \ldots, z_n)$ and $\widetilde{f}_S(z_1, z_2, \ldots, z_n)$ as linear combinations of (multilinear) monomials, then the element $\widetilde{f}_S(z_1, z_2, \ldots, z_n)$ is identical to $f(z_1, z_2, \ldots, z_n)$, except for some sign changes.

We obtain 2^n graded identities $\widetilde{f}_S = 0$ in this way, one for each subset $S \subseteq \{1, 2, \ldots, n\}$. Note that $\widetilde{f}_\emptyset = f$. However the identity $\widetilde{f}_\emptyset = 0$ is thought of as a graded identity which only involves even generators, whereas the identity $f = 0$ is an ungraded identity.

If $\mathcal{V}$ is a variety of ungraded algebras determined by a set I of multilinear identities then $\mathcal{V}$ gives rise to the variety $\widetilde{\mathcal{V}}$ of superalgebras determined by the set of graded identities $\widetilde{f}_S = 0$ arising from identities $f = 0$ in I.

We now give an illustration of how the representation theory of the symmetric group and superalgebras can be used to proved nilpotency results in varieties of algebras. Let $\mathcal{V}$ be a variety of ungraded algebras over a field K determined by a set of multilinear identities, and suppose we want to show that algebras in $\mathcal{V}$ are nilpotent. In particular, suppose we want to show that any product of n elements in any algebra in $\mathcal{V}$ is zero. Let T be the free algebra of $\mathcal{V}$, with free generators $t_1, t_2, \ldots, t_n$, and let M be the multilinear component of T of degree 1 in $t_1, t_2, \ldots, t_n$. We want to show that $M = \{0\}$.

We turn M into an $K\mathrm{Sym}(n)$ module, letting permutations in $\mathrm{Sym}(n)$ permute $t_1, t_2, \ldots, t_n$. For example if $\mathcal{V}$ is a variety of Lie algebras then M is spanned by

$$[t_{1\sigma}, t_{2\sigma}, \ldots, t_{n\sigma}] \ (\sigma \in \mathrm{Sym}(n)),$$

and if $\tau \in \mathrm{Sym}(n)$ then we define

$$[t_{1\sigma}, t_{2\sigma}, \ldots, t_{n\sigma}]\tau = [t_{1\sigma\tau}, t_{2\sigma\tau}, \ldots, t_{n\sigma\tau}].$$

If we express the identity in $K\mathrm{Sym}(n)$ as a sum of orthogonal idempotents

$$1 = \sum f_i,$$

then

$$M = \bigoplus M f_i.$$

So $M = \{0\}$ provided $Mf_i = 0$ for all i.

Provided the characteristic of K is greater than n, we can use the classical theory of representations of the symmetric group to obtain a decomposition of 1 as a sum of primitive idempotents. The primitive idempotents of $\mathbb{Q}\mathrm{Sym}(n)$ are described in James and Kerber [14]: they correspond to Young tableaux. For each partition $(m_1, m_2, \ldots, m_s)$ of n with $m_1 + m_2 + \ldots + m_s = n$ and $m_1 \geq m_2 \geq \ldots \geq m_s > 0$ we associate a Young diagram, which is an array of m boxes arranged in s rows, with m_i boxes in the i-th row. The boxes are arranged so that the j-th column of the array consists of the j-th boxes out of the rows which have length j or more. We obtain a Young tableau from a Young diagram by filling in the n boxes of the diagram with $1, 2, \ldots, n$ in some order. We then let H be the subgroup of $\mathrm{Sym}(n)$ which permutes the entries within each row of the tableau, and we let V be the subgroup of Sym(n) which permutes the entries within each column of the tableau. We set

$$e = \sum_{\pi \in V, \rho \in H} sign(\pi)\pi\rho.$$

Then $(1/k)e$ is a primitive idempotent of $\mathbb{Q}\mathrm{Sym}(n)$, for some integer k dividing $n!$. Furthermore the identity element in $\mathbb{Q}\mathrm{Sym}(n)$ can be expressed as a sum of these primitive idempotents.

For example, let $n = 9$ and consider the partition (4,2,2,1). This gives the following Young diagram:

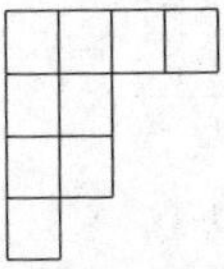

One possible Young tableau arising from this diagram is

1	3	5	7
2	4		
6	8		
9			

This is a standard tableau, since the entries are ascending in each row and column. By forming all possible standard tableaux for all possible Young diagrams we obtain a complete set of primitive idempotents which sum to 1.

We need to introduce a different notation to describe these idempotents. If S is a non-empty subset of $\{1, 2, \ldots, n\}$ then we let

$$\varphi_S^+ = \sum_{\sigma \in \mathrm{Sym}(S)} \sigma, \quad \varphi_S^- = \sum_{\sigma \in \mathrm{Sym}(S)} sign(\sigma) \cdot \sigma.$$

Then (for this tableau) $\sum_{\pi \in V, \rho \in H} sign(\pi)\pi\rho$ can be expressed in the form

$$\varphi^-_{\{1,2,6,9\}}\varphi^-_{\{3,4,8\}}\varphi^+_{\{1,3,5,7\}}\varphi^+_{\{2,4\}}\varphi^+_{\{6,8\}},$$

and

$$f = \frac{1}{1680}\varphi^-_{\{1,2,6,9\}}\varphi^-_{\{3,4,8\}}\varphi^+_{\{1,3,5,7\}}\varphi^+_{\{2,4\}}\varphi^+_{\{6,8\}}$$

is a primitive idempotent.

The relevance of all this to superalgebras is given by the following theorem from [29], which applies when M is spanned by the multilinear elements in n generators.

Theorem 4.1 *Let $S_1, S_2, \ldots, S_k$ be a partition of $\{1, 2, \ldots, n\}$ into disjoint non-empty subsets. Let $\varepsilon_1, \varepsilon_2, \ldots, \varepsilon_k = \pm$, and let $n_i = |S_i|$ for $i = 1, 2, \ldots, k$. Then*

$$\dim\left(M \cdot \varphi_{S_1}^{\varepsilon_1} \cdot \varphi_{S_2}^{\varepsilon_2} \cdot \ldots \cdot \varphi_{S_k}^{\varepsilon_k}\right) = \dim N,$$

where N is the multiweight $(n_1, n_2, \ldots, n_k)$ component of the free superalgebra Z of rank k in the variety $\tilde{\mathcal{V}}$, where for $i = 1, 2, \ldots, k$ the i-th generator of Z is even if $\varepsilon_i = +$ and is odd if $\varepsilon_i = -$.

We want to use this theorem to prove that $Mf = \{0\}$. That is, we want to show that

$$M\varphi^{-}_{\{1,2,6,9\}}\varphi^{-}_{\{3,4,8\}}\varphi^{+}_{\{1,3,5,7\}}\varphi^{+}_{\{2,4\}}\varphi^{+}_{\{6,8\}} = \{0\}.$$

Unfortunately the theorem does not apply directly since the subsets $\{1,2,6,9\}$, $\{3,4,8\}$, $\{1,3,5,7\}$, $\{2,4\}$, $\{6,8\}$ are not disjoint. We circumvent this problem by dividing the Young diagram into three disjoint strips, as shown below.

<table>
<tr><td>a</td><td>a</td><td>a</td><td>a</td></tr>
<tr><td>b</td><td>c</td><td></td><td></td></tr>
<tr><td>b</td><td>c</td><td></td><td></td></tr>
<tr><td>b</td><td></td><td></td><td></td></tr>
</table>

We can write

$$f = \frac{1}{1680}\psi\varphi^{-}_{\{2,6,9\}}\varphi^{-}_{\{4,8\}}\varphi^{+}_{\{1,3,5,7\}}\chi$$

where $\psi, \chi \in K\mathrm{Sym}(9)$, and where the subsets $\{1,3,5,7\}$, $\{2,6,9\}$, $\{4,8\}$, correspond to the entries in the strips indicated by a, b, c above. To show that $Mf = \{0\}$, it is sufficient to show that

$$M\psi\varphi^{-}_{\{2,6,9\}}\varphi^{-}_{\{4,8\}}\varphi^{+}_{\{1,3,5,7\}} = 0,$$

and since $M\psi \leq M$ it is sufficient to prove that

$$M\varphi^{-}_{\{2,6,9\}}\varphi^{-}_{\{4,8\}}\varphi^{+}_{\{1,3,5,7\}} = 0.$$

Conversely, if $M = \{0\}$ then $M\varphi^{-}_{\{2,6,9\}}\varphi^{-}_{\{4,8\}}\varphi^{+}_{\{1,3,5,7\}} = 0$. Theorem 4.1 implies that the dimension of $M\varphi^{-}_{\{2,6,9\}}\varphi^{-}_{\{4,8\}}\varphi^{+}_{\{1,3,5,7\}}$ is the same as the dimension of a certain subspace in a three generator superalgebra in the variety $\widetilde{\mathcal{V}}$. The essential idea is that the elements of $M\varphi^{-}_{\{2,6,9\}}\varphi^{-}_{\{4,8\}}\varphi^{+}_{\{1,3,5,7\}}$ are skew symmetric in t_2, t_6, t_9, skew symmetric in t_4, t_8, and symmetric in t_1, t_3, t_5, t_7. We replace t_2, t_6, t_9 by single odd element, replace t_4, t_8 by another odd element, and replace t_1, t_3, t_5, t_7 by a single even element. To show that $Mf = \{0\}$ we only need to show that products of weight 9 in three generator superalgebras in $\widetilde{\mathcal{V}}$ are zero.

More generally, any Young diagram with 9 cells can be partioned into a maximum of three horizontal and vertical strips. And so if products of weight 9 in three generator superalgebras in $\widetilde{\mathcal{V}}$ are zero, then $Mh = \{0\}$ for all primitive idempotents $h \in K\mathrm{Sym}(9)$. On the face of it, showing that $M = \{0\}$ involves calculating in 9 generator algebras in $\mathcal{V}$, but by using Theorem 4.1 we have reduced the problem to calculating in three generator algebras in $\widetilde{\mathcal{V}}$. This is usually a significant reduction.

The same ideas can be used to estimate the dimension of M when M is non-zero. We give an example of how a slightly different type of decomposition of the identity into a sum of orthogonal idempotents can be used to compute $\dim(M)$ even when the characteristic of K is less than n.

As described in Section 3, the associated Lie rings of free 4-Engel groups of exponent 5 are free Lie algebras in the variety $\mathcal{L}$ of Lie algebras over $\mathbb{Z}_5$ determined by the multilinear identities

$$\sum_{\sigma \in \mathrm{Sym}(4)} [x, y_{1\sigma}, y_{2\sigma}, y_{3\sigma}, y_{4\sigma}] = 0,$$

$$\sum_{\sigma \in \mathrm{Sym}(4)} [y_{1\sigma}, x_1, x_2, y_{2\sigma}, y_{3\sigma}, y_{4\sigma}] = 0,$$

$$\sum_{\sigma \in \mathrm{Sym}(4)} [y_{1\sigma}, x_1, x_2, x_3, y_{2\sigma}, y_{3\sigma}, y_{4\sigma}] = 0.$$

In [23], Mike Newman and I found a precise formula for the orders of free 4-Engel groups of exponent 5 by calculating the dimensions of the free Lie algebras in the variety $\mathcal{L}$. Let L be the free Lie algebra of $\mathcal{L}$ with free generators $x_1, x_2, \ldots$. Since $\mathcal{L}$ is determined by multilinear identities, L is multigraded. If we have a product a of the free generators of L, then we assign a multiweight $\underline{w} = (w_1, w_2, \ldots)$ to a if a has degree w_i in x_i for $i = 1, 2, \ldots$. Let $L_{\underline{w}}$ be the subspace of L spanned by all products of multiweight $\underline{w}$. Then L is the direct sum of its multiweight components $L_{\underline{w}}$ as $\underline{w}$ ranges over all possible multiweights.

We computed the dimension of $L_{\underline{w}}$ for all $\underline{w} = (w_1, w_2, \ldots)$ such that $\sum_i w_i \leq 10$. For higher multiweights $\dim L_{\underline{w}} = \dim C_{\underline{w}}$, where C is the free centre-by-(3-Engel) Lie algebra of countably infinite rank. The most difficult dimension to compute was $\dim L_{\underline{w}}$ where $\underline{w} = (1, 1, 1, 1, 1, 1, 1, 1, 1, 1)$. (Here we delete trailing zeros from all multiweights: for example let (2,2,2,2) represent $(2, 2, 2, 2, 0, 0, \ldots)$.)

So let M be the multiweight $(1, 1, 1, 1, 1, 1, 1, 1, 1, 1)$ component of L. Turn M into a $\mathbb{Z}_5\mathrm{Sym}(10)$ module, letting permutations in $\mathrm{Sym}(10)$ permute $x_1, x_2, \ldots, x_{10}$. The identity in $\mathbb{Z}_5\mathrm{Sym}(10)$ is the sum of the 32 orthogonal idempotents

$$3\varphi_{\{1,2\}}^{\varepsilon_1}\varphi_{\{3,4\}}^{\varepsilon_2}\varphi_{\{5,6\}}^{\varepsilon_3}\varphi_{\{7,8\}}^{\varepsilon_4}\varphi_{\{9,10\}}^{\varepsilon_5}$$

with $\varepsilon_i = \pm$ for $1 \leq i \leq 5$. If we let $f_1, f_2, \ldots, f_{32}$ denote these idempotents then

$$M = \bigoplus_{i=1}^{32} M f_i,$$

and

$$\dim M = \sum_{i=1}^{32} \dim(M f_i).$$

By symmetry, $\dim(M f_i)$ depends only on the number of minus signs in the definition of f_i. We let $d_0 = \dim(M f_i)$ where i is chosen so that there are no minus signs in the definition of f_i. And we let $d_1 = \dim(M f_i)$ where i is chosen so that there is one minus sign in the definition of f_i. We similarly define d_2, d_3, d_4, d_5. So

$$\dim M = d_0 + 5d_1 + 10d_2 + 10d_3 + 5d_4 + d_5.$$

Theorem 4.1 implies that d_i is the dimension of the multiweight $(2, 2, 2, 2, 2)$ component in the free superalgebra of rank 5 in the variety $\widetilde{\mathcal{L}}$, where i of the generators are odd. Note that $\widetilde{\mathcal{L}}$ is a variety of Lie superalgebras. Our computations showed that

$$d_0 = 4, d_1 = 5, d_2 = 7, d_3 = 12, d_4 = 18, d_5 = 32.$$

So $\dim M = 341$.

These 6 calculations in 5 generator Lie superalgebras also give us the dimensions of certain other multiweight components of L. Let $d(\underline{w}) = \dim L_{\underline{w}}$. Then

$$d(2,2,2,2,2) = d_0 = 4,$$

$$d(1,1,2,2,2,2) = d_0 + d_1 = 9,$$

$$d(1,1,1,1,2,2,2) = d_0 + 2d_1 + d_2 = 21,$$

$$d(1,1,1,1,1,1,2,2) = d_0 + 3d_1 + 3d_2 + d_3 = 52,$$

$$d(1,1,1,1,1,1,1,1,2) = d_0 + 4d_1 + 6d_2 + 4d_3 + d_4 = 132.$$

So 6 calculations in 5 generator Lie superalgebras give the dimensions of 6 different multiweight components in 5,6,7,8,9, and 10 generator Lie algebras.

References

[1] Yu.A. Bahturin, *Identical relations in Lie algebras*, VNU Science Press BV, 1987.

[2] N. Bourbaki, *Groupes et algebres de Lie*, Hermann, Paris, 1972.

[3] Y.M. Gorchakov, *Commutator subgroups*, Sibirsk. Mat. Zh. **10** (1969), 1023–1033.

[4] Daniel Groves, *A note on nonidentical Lie relators*, J. Algebra **211** (1999), 15–25.

[5] Daniel Groves, *Some properties of free groups of some soluble varieties of groups*, J. London Math. Soc. (2) **63** (2001), 592–606.

[6] Daniel Groves, *Torsion and residual nilpotence in some outer commutator varieties*, to appear.

[7] K.W. Gruenberg, *Residual properties of infinite soluble groups*, Proc. London Math. Soc. (3) **7** (1957), 29–62.

[8] C.K. Gupta, *The free centre-by-metabelian groups*, J. Austral. Math. Soc. **16** (1973), 294–295.

[9] G. Havas and M.F. Newman, *Applications of computers to questions like those of Burnside*, Lecture Notes in Mathematics, 806, Springer-Verlag, Berlin, 1980, pp. 211–230.

[10] G. Havas, M.F. Newman, and M.R. Vaughan-Lee, *A nilpotent quotient algorithm for graded Lie rings*, J. Symbolic Computation **9** (1990), 653–664.

[11] G. Havas, G.E. Wall, and J.W. Wamsley, *The two generator restricted Burnside group of exponent five*, Bull. Austral. Math. Soc. **10** (1974), 459–470.

[12] G. Higman, *On finite groups of exponent five*, Proc. Camb. Phil. Soc. **52** (1956), 381–390.

[13] N. Jacobson, *Lie algebras*, Wiley-Interscience, New York, 1962.

[14] G. James and A. Kerber, *The representation theory of the symmetric group*, Addison-Wesley, Reading, Massachusetts, 1981.

[15] A.R. Kemer, *Varieties and Z_2-graded algebras*, Izv. Akad. Nauk SSSR **48** (1985), 1042–1059.

[16] A.R. Kemer, *Finite bases of identities of associative algebras*, Algebra i Logika **26** (1987), 597–641; English transl. in Algebra and Logic **26** (1987).

[17] A.R. Kemer, *Ideals of identities of associative algebras*, Translations of Mathematical Monographs, 87, Amer. Math. Soc., Providence, Rhode Island, 1990.

[18] A.I. Kostrikin, *The Burnside problem*, Izv. Akad. Nauk SSSR, Ser. Mat. **23** (1959), 3–34.

[19] Yu. V. Kuz'min, *Some approximation properties of the variety AN_c*, Uspekhi Mat. Nauk **33** (1978), 217–218.

[20] Yu. V. Kuz'min and M.Z. Shapiro, *The connection between varieties of groups and varieites of Lie rings*, Sibirsk. Mat. Zh. **28** (1987).

[21] M.F. Newman and E.A. O'Brien, *Applications of computers to questions like those of Burnside, II*, Internat. J. Algebra and Computation **6** (1996), 593–605.

[22] M.F. Newman and Michael Vaughan-Lee, *Some Lie rings associated with Burnside groups*, ERA Amer. Math. Soc. **4** (1998), 1–3.

[23] M.F. Newman and Michael Vaughan-Lee, *Engel-4 groups of exponent 5. II. Orders*, Proc. London Math. Soc. (3) **79** (1999), 283–317.

[24] Eamonn O'Brien and Michael Vaughan-Lee, *Computing $R(2,7)$*, Internat. J. Algebra and Computation, to appear.

[25] I.N. Sanov, *Establishment of a connection between periodic groups with period a prime number and Lie rings*, Izv. Akad. Nauk SSSR, Ser. Mat. **16** (1952), 23–58.

[26] A.L. Smel'kin, *Free polynilpotent groups*, Dokl. Akad. Nauk SSSR **151** (1963), 73–75.

[27] Vladimir Tasic and Michael Vaughan-Lee, *Torsion in certain relatively free groups*, Bull. London Math. Soc. **27** (1995), 327–333.

[28] G. Traustason, *Engel Lie algebras*, D.Phil. thesis, Oxford University, 1993.

[29] Michael Vaughan-Lee, *Superalgebras and dimensions of algebras*, Internat. J. Algebra and Computation **8** (1998), 97–125.

[30] M.R. Vaughan-Lee, *Lie rings of groups of prime exponent*, J. Austral. Math. Soc. **49** (1990), 386–398.

[31] M.R. Vaughan-Lee, *An algorithm for computing graded algebras*, J. Symbolic Computation **16** (1993), 345–354.

[32] M.R. Vaughan-Lee, *The restricted Burnside problem*, second ed., Oxford University Press, 1993.

[33] M.R. Vaughan-Lee, *Engel-4 groups of exponent 5*, Proc. London Math. Soc. **74** (1997), 306–334.

[34] G.E. Wall, *On the Lie ring of a group of prime exponent*, Lecture Notes in Mathematics, 372, Springer-Verlag, Berlin, 1974, pp. 667–690.

[35] G.E. Wall, *Multilinear Lie relators for varieties of groups*, J. Algebra **157** (1993), 341–393.

[36] E.I. Zel'manov, *Engel Lie algebras*, Dokl. Akad. Nauk SSSR **292** (1987), 265–268.

[37] E.I. Zel'manov, *The solution of the restricted Burnside problem for groups of odd exponent*, Izv. Math. USSR **36** (1991), 41–60.

[38] E.I. Zel'manov, *The solution of the restricted Burnside problem for 2-groups*, Mat. Sb. **182** (1991), 568–592.

[39] E.I. Zel'manov and I.P. Shestakov, *Prime alternative superalgebras and nilpotence of the radical of a free alternative algebra*, Izv. Akad. Nauk SSSR **54** (1990), 676–693.

CHEVALLEY GROUPS OF TYPE G_2 AS AUTOMORPHISM GROUPS OF LOOPS

PETR VOJTĚCHOVSKÝ

Department of Mathematics, Iowa State University, Ames, Iowa, 50011, U.S.A.

Abstract

Let $M^*(q)$ be the unique nonassociative finite simple Moufang loop constructed over $GF(q)$. We prove that $\mathrm{Aut}(M^*(2))$ is the Chevalley group $G_2(2)$, by extending multiplicative automorphism of $M^*(2)$ into linear automorphisms of the unique split octonion algebra over $GF(2)$. Many of our auxiliary results apply in the general case. In the course of the proof we show that every element of a split octonion algebra can be written as a sum of two elements of norm one.

1 Composition algebras and Paige loops

Let C be a finite-dimensional vector space over a field k, equipped with a quadratic form $N : C \longrightarrow k$ and a multiplicative operation $\cdot$. Following [7], we say that $C = (C, N, +, \cdot)$ is a *composition algebra* if $(C, +, \cdot)$ is a *nonassociative* ring with identity element e, N is nondegenerate, and $N(u \cdot v) = N(u)N(v)$ is satisfied for every $u, v \in C$. The bilinear form associated with N will also be denoted by N. Recall /that $N : C \times C \longrightarrow k$ is defined by $N(u, v) = N(u + v) - N(u) - N(v)$. Write $u \perp v$ if $N(u, v) = 0$, and set $u^\perp = \{v \in C;\ u \perp v\}$.

The standard 8-dimensional real Cayley algebra $\mathbb{O}$ constructed by the *Cayley-Dickson process* (or *doubling* [7]) is the best known nonassociative composition algebra. There is a remarkably compact way of constructing $\mathbb{O}$ that avoids the iterative Cayley-Dickson process. As in [2], let $B = \{e = e_0, e_1, \dots, e_7\}$ be a basis whose vectors are multiplied according to

$$e_r^2 = -1, \quad e_{r+7} = e_r, \quad e_r e_s = -e_s e_r, \quad e_{r+1}e_{r+3} = e_{r+2}e_{r+6} = e_{r+4}e_{r+5} = e_r, \tag{1.1}$$

for $r, s \in \{1, \dots, 7\}$, $r \neq s$. (Alternatively, see [1, p. 122].) The norm $N(u)$ of a vector $u = \sum_{i=0}^7 a_i e_i \in \mathbb{O}$ is given by $\sum_{i=0}^7 a_i^2$.

Importantly, all the *structural constants* γ_{ijk}, defined by $e_i \cdot e_j = \sum_{k=0}^7 \gamma_{ijk} e_k$, are equal to ± 1, and therefore the construction can be imitated over any field k. For $k = GF(q)$ of odd characteristic, let us denote the ensuing algebra by $\mathbb{O}(q)$. When q is even, the above construction does not yield a composition algebra.

The following facts about composition algebras can be found in [7]. Every non-trivial composition algebra C has dimension 2, 4 or 8, and we speak of a *complex*, *quaternion* or *octonion* algebra, respectively. We say that C is a *division algebra* if it has no zero divisors, else C is called *split*. There can be many non-isomorphic octonion algebras over a given field. Exactly one of them is guaranteed to be split. Moreover, when k is finite, all octonion algebras over k are isomorphic (and thus split). Let $\mathbb{O}(q)$ be the unique octonion algebra constructed over $GF(q)$.

All composition algebras satisfy the so-called *Moufang identities*

$$(xy)(zx) = x((yz)x), \quad x(y(xz)) = ((xy)x)z, \quad x(y(zy)) = ((xy)z)y. \tag{1.2}$$

These identities are the essence of Moufang loops, undoubtedly the most investigated variety of nonassociative loops. More precisely, a quasigroup $(L, \cdot)$ is a *Moufang loop* if it possesses a neutral element e and satisfies one (and hence all) of the Moufang identities (1.2). We refer the reader to [6] for the basic properties of loops and Moufang loops in particular. Briefly, every element x of a Moufang loop L has a both-sided inverse x^{-1}, and a subloop $\langle x, y, z\rangle$ of L generated by x, y and z is a group if and only if x, y and z associate. Specifically, every two-generated subloop of L is a group.

Paige [5] constructed one nonassociative finite simple Moufang loop for every finite field $GF(q)$. Liebeck [3] used the classification of finite simple groups in order to prove that there are no other nonassociative finite simple Moufang loops. Reflecting the current trend in loop theory, we will call these loops *Paige loops*, and we denote the unique Paige loop constructed over $GF(q)$ by $M^*(q)$.

The relation between $\mathbb{O}(q)$ and $M^*(q)$ is as follows. Let $M(q)$ be the set of all elements of $\mathbb{O}(q)$ of norm one. Then $M(q)$ is a Moufang loop with center $Z(M(q)) = \{e, -e\}$, and $M(q)/Z(M(q))$ is isomorphic to $M^*(q)$. Note that $M(q) = M^*(q)$ in characteristic 2.

Historically, all split octonion algebras and Paige loops were constructed by Zorn [9] and Paige without reference to doubling. Given a field k, consider the *vector matrix algebra* consisting of all *vector matrices*

$$x = \begin{pmatrix} a & \alpha \\ \beta & b \end{pmatrix},$$

where a, $b \in k$, α, $\beta \in k^3$, addition is defined entry-wise, and multiplication by

$$\begin{pmatrix} a & \alpha \\ \beta & b \end{pmatrix}\begin{pmatrix} c & \gamma \\ \delta & d \end{pmatrix} = \begin{pmatrix} ac + \alpha \cdot \delta & a\gamma + d\alpha - \beta \times \delta \\ c\beta + b\delta + \alpha \times \gamma & \beta \cdot \gamma + bd \end{pmatrix}.$$

Here, $\alpha \cdot \beta$ (resp. $\alpha \times \beta$) is the standard dot product (resp. vector product) of α and β. Use $\det x = ab - \alpha \cdot \beta$ as a norm to obtain an octonion algebra. In fact, this is exactly the unique split octonion algebra over k. The identity element is

$$e = \begin{pmatrix} 1 & (0,0,0) \\ (0,0,0) & 1 \end{pmatrix},$$

and, when $N(x) = \det x \neq 0$, we have

$$x^{-1} = \begin{pmatrix} b & -\alpha \\ -\beta & a \end{pmatrix}.$$

The purpose of this paper is to initiate the investigation of automorphism groups of Paige loops. We will extend automorphisms of $M^*(2)$ into automorphisms of $\mathbb{O}(2)$ to prove that $\mathrm{Aut}(M^*(2))$ is the exceptional Chevalley group $G_2(2)$. We also present several results for the general case. Since $G_2(q) \leq \mathrm{Aut}(M(q))$, it is reasonable to expect that equality holds whenever q is prime. See Acknowledgement for more details. It is fun to watch how much information about the (boring) additive structure of a composition algebra can be obtained from the multiplication alone (cf. Lemma 2.4).

2 Multiplication versus addition

Perhaps the single most important feature of composition algebras is the existence of the minimal equation (cf. [7, Prop 1.2.3]). Namely, every element $x \in C$ satisfies

$$x^2 - N(x, e)x + N(x)e = 0. \tag{2.3}$$

Furthermore, (2.3) is the minimal equation for x when x is not a scalar multiple of e.

Lemma 2.1 *Let C be a composition algebra, $x, y \in C$. Then*

$$N(xy, y) = N(x, e)N(y). \tag{2.4}$$

When $N(y) \neq 0$, we have

$$(xy^{-1})^2 - N(x, y)N(y)^{-1}xy^{-1} + N(xy^{-1})e = 0. \tag{2.5}$$

In particular,

$$(xy^{-1})^2 - N(x, y)xy^{-1} + e = 0 \tag{2.6}$$

whenever $N(x) = N(y) = 1$. In such a case, $(xy^{-1})^2 = -e$ if and only if $x \perp y$.

Proof We have $N(xy, y) = N(xy + y) - N(x)N(y) - N(y)$, and $N(xy + y) = N(x+e)N(y) = (N(x, e)+N(x)+N(e))N(y) = N(x, e)N(y)+N(x)N(y)+N(y)$. Equation (2.4) follows.

Substitute xy^{-1} for x into (2.4) to obtain $N(x, y) = N(xy^{-1}, e)N(y)$. The minimal equation

$$(xy^{-1})^2 - N(xy^{-1}, e)xy^{-1} + N(xy^{-1})e = 0$$

for xy^{-1} can then be written as (2.5), provided $N(y) \neq 0$. The rest is easy. □

We leave the proof of the following Lemma to the reader.

Lemma 2.2 *Let*

$$x = \begin{pmatrix} a & \alpha \\ \beta & b \end{pmatrix}$$

be an element of a composition algebra C satisfying $N(x) = 1$. Then

(i) $x^2 = e$ *if and only if* $(\alpha, \beta) = (0, 0)$ *and* $a = b \in \{1, -1\}$,

(ii) $x^2 = -e$ *if and only if* $((\alpha, \beta) = (0, 0)$, $b = a^{-1}$, *and* $a^2 = -1)$ *or* $((\alpha, \beta) \neq (0, 0)$ *and* $b = -a)$,

(iii) $x^3 = e$ *if and only if* $((\alpha, \beta) = (0, 0), b = a^{-1}$, *and* $a^3 = 1)$ *or* $((\alpha, \beta) \neq (0, 0)$ *and* $b = -1 - a)$,

(iv) $x^3 = -e$ *if and only if* $((\alpha, \beta) = (0, 0), b = a^{-1}$, *and* $a^3 = -1)$ *or* $((\alpha, \beta) \neq (0, 0)$ *and* $b = 1 - a)$.

Let us denote the multiplicative order of x by $|x|$.

Lemma 2.3 *Let C be a composition algebra. Assume that $x, y \in C$ satisfy $N(x) = N(y) = 1$, $x \neq y$. The following conditions are equivalent:*

(i) $|xy^{-1}| = 3$,

(ii) $(xy^{-1})^2 + xy^{-1} + e = 0$,

(iii) $N(x, y) = -1$,

(iv) $N(x + y) = 1$.

Proof The equivalence of (ii) and (iii) follows from the uniqueness of the minimal equation (2.3). Condition (iii) is equivalent to (iv) since $N(x) = N(y) = 1$. It suffices to prove the equivalence of (i) and (ii).

As $(a^3 - e) = (a - e)(a^2 + a + e)$, there is nothing to prove when C has no zero divisors. The implication (ii) $\Rightarrow$ (i) is obviously true in any (composition) algebra. Let us prove (i) $\Rightarrow$ (ii) for a split octonion algebra C. Assume that $|xy^{-1}| = 3$, $x \neq y$, and

$$x = \begin{pmatrix} a & \alpha \\ \beta & b \end{pmatrix}, \quad y = \begin{pmatrix} c & \gamma \\ \delta & d \end{pmatrix}.$$

We prove that $N(x + y) = 1$. Direct computation yields $N(x + y) = 2 + r + s$, where $r = ad - \alpha \cdot \delta$, $s = bc - \beta \cdot \gamma$. Also,

$$xy^{-1} = \begin{pmatrix} r & \varepsilon \\ \varphi & s \end{pmatrix}$$

for some $\varepsilon, \varphi \in k^3$. Since $(xy^{-1})^3 = e$, we have either $((\varepsilon, \varphi) = (0, 0)$, $s = r^{-1}$, and $r^3 = 1)$, or $((\varepsilon, \varphi) \neq (0, 0)$, and $s = -1 - r)$, by Lemma 2.2. If the latter is true, we immediately get $N(x + y) = 1$. Assume the former is true. Then $r + s = r + r^{-1}$. Also, $r^3 = 1$ implies $r = 1$ or $r^2 + r + 1 = 0$. But $r = 1$ leads to $x = y$, a contradiction. Therefore $r^2 + r + 1 = 0$, i.e., $r + r^{-1} = -1$, and we get $N(x + y) = 1$ again. □

There is a strong relation between the additive and multiplicative structures in composition algebras.

Lemma 2.4 *Let C, x, and y be as in Lemma 2.3. Then $N(x + y) = 1$ if and only if $x + y = -xy^{-1}x$.*

Proof The indirect implication is trivial. Assume that $N(x + y) = 1$. Then $(xy^{-1})^2 + xy^{-1} + e = 0$, and $(xy^{-1})^3 = e$. Thus $yx^{-1} = (xy^{-1})^2 = -xy^{-1} - e$. Multiplying this equality on the right by x yields $y = -xy^{-1}x - x$. □

3 Doubling triples

Any composition algebra C can be constructed from the underlying field k in three steps. Proposition 1.5.1 and Lemma 1.6.1 of [7] tell us how to do it. Imitating these results, we say that a triple $(a, b, c) \in C^3$ is a *doubling triple* if $N(a) \neq 0$, $N(b) \neq 0$, $N(c) \neq 0$, $b \in e^\perp \cap a^\perp$, $c \in e^\perp \cap a^\perp \cap b^\perp \cap (ab)^\perp$, $a \in e^\perp$ (resp. $a \notin e^\perp$) and the characteristic of k is odd (resp. even). Then $B = \{e, a, b, ab, c, ac, bc, (ab)c\}$ is a basis for C.

Construction (1.1) immediately shows that there is a doubling triple with $N(a) = N(b) = N(c) = 1$ when k is of odd characteristic. Such a doubling triple exists for every split octonion algebra in even characteristic, too.

Lemma 3.1 *Let k be a field of even characteristic. Then*

$$\left(\begin{pmatrix} 0 & (1,0,0) \\ (1,0,0) & 1 \end{pmatrix}, \begin{pmatrix} 0 & (0,1,0) \\ (0,1,0) & 0 \end{pmatrix}, \begin{pmatrix} 0 & (0,0,1) \\ (0,0,1) & 0 \end{pmatrix} \right)$$

is a doubling triple consisting of elements of norm one.

Proof Straightforward computation. □

Doubling triples can be used to induce automorphisms. By an automorphism of a composition algebra C we mean a *linear* automorphism, i.e., a bijection $f : C \longrightarrow C$ satisfying $f(u + v) = f(u) + f(v)$, $f(\lambda u) = \lambda f(u)$, and $f(u \cdot v) = f(u) \cdot f(v)$ for every $u, v \in C$, $\lambda \in k$. By [7, Thm 1.7.1 and Cor 1.7.2], every such automorphism is an isometry (i.e., $N(f(u)) = N(u)$), and vice versa. Springer and Veldkamp [7, Ch. 2] use algebraic groups to show that $\mathrm{Aut}(\mathbb{O}(q))$ is the exceptional group $G_2(q)$, and more.

Proposition 3.2 *Let (a, b, c), (a', b', c') be two doubling triples of a composition algebra C. Then there is an automorphism of C mapping (a, b, c) onto (a', b', c') if and only if $N(x) = N(x')$, for $x = a, b, c$.*

Proof Let k be the underlying field. The necessity is obvious since every automorphism is an isometry. Now for the sufficiency. Let $A = ke \oplus ka$, $B = A \oplus Ab$, $C = B \oplus Bc$, and similarly for A', B', $C' = C$. Define $\psi_X : X \longrightarrow X' = \psi_X(X)$ (for $X = A, B, C$) by

$$\begin{aligned} \psi_A(x + ya) &= x + ya' \quad (x, y \in k), \\ \psi_B(x + yb) &= \psi_A(x) + \psi_A(y)b' \quad (x, y \in A), \\ \psi_C(x + yc) &= \psi_B(x) + \psi_B(y)c' \quad (x, y \in B). \end{aligned}$$

All maps ψ_X are clearly linear, and it is not hard to see that they are also multiplicative. (One has to use the assumption $N(x) = N(x')$.) Since a', b', c' generate $C' = C$, ψ_C is the automorphism we are looking for. □

4 Restrictions and Eextensions of automorphisms

The restriction of $h \in \mathrm{Aut}(\mathbb{O}(q))$ onto the loop $M(q)$ is a (multiplicative) automorphism. Moreover, two distinct automorphisms of $\mathbb{O}(q)$ differ on $M(q)$, because there is a basis for C consisting of unit vectors (cf. construction (1.1) and Lemma 3.1).

We would like to emphasize at this point how far are the metric properties of N from our intuitive understanding of (real) norms. Theorem 4.1 is not required for the rest of the paper, but is certainly of interest in its own right.

Theorem 4.1 *Every element of a split octonion algebra C is a sum of two elements of norm one.*

Proof We identify C with the vector matrix algebra over k, where the norm is given by the determinant. Let

$$x = \begin{pmatrix} a & \alpha \\ \beta & b \end{pmatrix}$$

be an element of C. First assume that $\beta \neq 0$. Note that for every $\lambda \in k$ there is $\gamma \in k^3$ such that $\gamma \cdot \beta = \lambda$. Pick $\gamma \in k^3$ so that $\gamma \cdot \beta = a + b - ab + \alpha \cdot \beta$. Then choose $\delta \in \gamma^\perp \cap \alpha^\perp \neq \emptyset$. (As usual, $\alpha \perp \delta$ if and only if $\alpha \cdot \delta = 0$.) This choice guarantees that $(a-1)(b-1) - (\alpha - \gamma) \cdot (\beta - \delta) = ab - a - b + 1 - \alpha \cdot \beta + \gamma \cdot \beta = 1$. Thus

$$\begin{pmatrix} a & \alpha \\ \beta & b \end{pmatrix} = \begin{pmatrix} 1 & \gamma \\ \delta & 1 \end{pmatrix} + \begin{pmatrix} a-1 & \alpha - \gamma \\ \beta - \delta & b-1 \end{pmatrix}$$

is the desired decomposition of x into a sum of two elements of norm 1. Note that the above procedure works even for $\alpha = 0$.

Now assume that $\beta = 0$. If $\alpha \neq 0$, we use a symmetrical argument as before to decompose x. It remains to discuss the case when $\alpha = \beta = 0$. Then the equality

$$\begin{pmatrix} a & 0 \\ 0 & b \end{pmatrix} = \begin{pmatrix} a & (1,0,0) \\ (-1,0,0) & 0 \end{pmatrix} + \begin{pmatrix} 0 & (-1,0,0) \\ (1,0,0) & b \end{pmatrix}$$

does the job. □

We now know that $G_2(q)$ is a subgroup of $\mathrm{Aut}(M(q))$. Let us consider the extension problem. Pick an automorphism g of the (not necessarily simple) Moufang loop $M(q)$. The ultimate goal is to construct $h \in \mathrm{Aut}(\mathbb{O}(q))$ such that $h \restriction M(q) = g$. If this can be done, we immediately conclude that $\mathrm{Aut}(M(q)) = G_2(q)$ for every q. We like to think of the problem as a notion "orthogonal" to Witt's lemma. Roughly speaking, Witt's lemma deals with extensions of partial isometries from subspaces onto finite-dimensional vector spaces, whereas we are attempting to extend a multiplicative, norm-preserving map from the first shell $M(q)$ into an automorphism (= isometry) of $\mathbb{O}(q)$. Naturally, g is not linear because $M(q)$ is not even closed under addition. However, the analogy with Witt's lemma will become more apparent once we prove that g *is*, in a sense, additive (cf. Proposition 7.1).

Lemma 4.2 *Let $g \in \mathrm{Aut}(M(q))$, and let (a, b, c) be a doubling triple for $\mathbb{O}(q)$ with $N(a) = N(b) = N(c) = 1$. Then $(g(a), g(b), g(c))$ is a doubling triple (with $N(g(a)) = N(g(b)) = N(g(c)) = 1$).*

Proof Since (a, b, c) is a doubling triple, we have $b \in e^\perp \cap a^\perp$, $c \in e^\perp \cap a^\perp \cap b^\perp$. Moreover, $a \in e^\perp$ (resp. $a \notin e^\perp$) if q is odd (resp. even). By Lemmas 2.1 and 2.3, this is equivalent to $b^2 = c^2 = (ab^{-1})^2 = (ac^{-1})^2 = (bc^{-1})^2 = ((ab)c^{-1})^2 = -e$, and $a^2 = -e$ (resp. $|a| = 3$). Because $g \in \mathrm{Aut}(M(q))$, we have $g(b)^2 = g(c)^2 = (g(a)g(b)^{-1})^2 = (g(a)g(c)^{-1})^2 = (g(b)g(c)^{-1})^2 = ((g(a)g(b))g(c)^{-1})^2 = -e$ and $g(a)^2 = -e$ (resp. $|g(a)| = 3$). Another application of Lemmas 2.1 and 2.3 shows that $(g(a), g(b), g(c))$ is a doubling triple. □

In particular, the mapping $h = \psi_{\mathbb{O}(q)}$ constructed from g and (a, b, c) by Proposition 3.2 is an automorphism of $\mathbb{O}(q)$ satisfying $\psi(x) = g(x)$, for $x = a, b, c$.

Remark 4.3 This extension h can be obtained in another way when the characteristic is odd. Namely, construct $\mathbb{O}(q)$ as in section 1, and define $h : \mathbb{O}(q) \longrightarrow \mathbb{O}(q)$ by

$$h(\sum_{i=0}^{7} a_i e_i) = \sum_{i=0}^{7} a_i g(e_i).$$

Obviously, h is linear. For fixed i, j, only one of the 8 structural constants γ_{ijk} is nonzero, and it is equal to ± 1. Using linearity of h, it is therefore easy to check that h is multiplicative.

By the construction, h coincides with g on a basis B. However, we do not know whether h is an extension of g. The fact that $h \restriction B = g \restriction B$ does not guarantee that $h \restriction M(q) = g$, since B does not need to generate $M(q)$ by multiplication. Interestingly enough, it seems to never be the case! The key to answering these questions is to look at the additive properties of g.

5 Automorphisms of finite octonion algebras

We have entered a more technical part of the paper. In this section, we construct a family of automorphisms of $\mathbb{O}(q)$.

Let $k = GF(q)$, and let $\mathrm{Lie}(q)$ be the three-dimensional Lie algebra k^3 with vector product $\times$ playing the role of a Lie bracket. A linear transformation $f : \mathrm{Lie}(q) \longrightarrow \mathrm{Lie}(q)$ belongs to $\mathrm{Aut}(\mathrm{Lie}(q))$ if and only if $f(\alpha \times \beta) = f(\alpha) \times f(\beta)$ is satisfied for every α, $\beta \in k^3$. We say that a linear transformation f is *orthogonal* if $f(\alpha) \cdot f(\beta) = \alpha \cdot \beta$ for every α, $\beta \in k^3$.

Lemma 5.1 *For a non-singular orthogonal linear transformation $f : k^3 \longrightarrow k^3$, let $\widehat{f} : \mathbb{O}(q) \longrightarrow \mathbb{O}(q)$ be the mapping*

$$\widehat{f}\begin{pmatrix} a & \alpha \\ \beta & b \end{pmatrix} = \begin{pmatrix} a & f(\alpha) \\ f(\beta) & b \end{pmatrix}.$$

Then $\widehat{f} \in \mathrm{Aut}(\mathbb{O}(q))$ if and only if $f \in \mathrm{Aut}(\mathrm{Lie}(q))$.

Proof The map $\widehat{f}$ is clearly linear and preserves the norm. Since f is one-to-one, so is $\widehat{f}$. We have

$$\widehat{f}\begin{pmatrix} a & \alpha \\ \beta & b \end{pmatrix}\widehat{f}\begin{pmatrix} c & \gamma \\ \delta & d \end{pmatrix}$$
$$= \begin{pmatrix} ac + f(\alpha)\cdot f(\delta) & af(\gamma) + df(\alpha) - f(\beta)\times f(\delta) \\ cf(\beta) + bf(\delta) + f(\alpha)\times f(\gamma) & f(\beta)\cdot f(\gamma) + bd \end{pmatrix}.$$

On the other hand,

$$\widehat{f}\left(\begin{pmatrix} a & \alpha \\ \beta & b \end{pmatrix}\begin{pmatrix} c & \gamma \\ \delta & d \end{pmatrix}\right) = \begin{pmatrix} ac + \alpha\cdot\delta & f(a\gamma + d\alpha - \beta\times\delta) \\ f(c\beta + b\delta + \alpha\times\gamma) & \beta\cdot\gamma + bd \end{pmatrix}.$$

Sufficiency is now obvious, and necessity follows by specializing the elements a, b, c, d, α, β, γ, δ. □

For a map f, let $-f$ be the map *opposite* to f, i.e., $(-f)(u) = -(f(u))$. Also, for a permutation $\pi \in S_3$, consider π as a linear transformation on k^3 defined by

$$\pi(\alpha_1, \alpha_2, \alpha_3) = (\alpha_{\pi(1)}, \alpha_{\pi(2)}, \alpha_{\pi(3)}).$$

Apparently, $-S_3 = \{-\pi;\ \pi \in S_3\}$ is a set of non-singular orthogonal linear transformations.

Lemma 5.2 $\widehat{-\pi} \in \mathrm{Aut}(\mathbb{O}(q))$ *for every* $\pi \in S_3$.

Proof Let $\pi \in S_3$ be the transposition interchanging 1 and 2, and let $\alpha, \beta \in k^3$. Then

$$\pi(\alpha\times\beta) = (\alpha_3\beta_1 - \alpha_1\beta_3,\ \alpha_2\beta_3 - \alpha_3\beta_2,\ \alpha_1\beta_2 - \alpha_2\beta_1),$$

and

$$\pi(\alpha)\times\pi(\beta) = (\alpha_1\beta_3 - \alpha_3\beta_1,\ \alpha_3\beta_2 - \alpha_2\beta_3,\ \alpha_2\beta_1 - \alpha_1\beta_2).$$

Hence $-\pi(\alpha\times\beta) = \pi(\alpha)\times\pi(\beta) = (-\pi)(\alpha)\times(-\pi)(\beta)$. Thanks to the symmetry of S_3, we have shown that $-\pi \in \mathrm{Aut}(\mathrm{Lie}(q))$ for every $\pi \in S_3$. The rest follows from Lemma 5.1. □

Observe there is another automorphism when q is even:

Lemma 5.3 *Define* $\partial : \mathbb{O}(q) \longrightarrow \mathbb{O}(q)$ *by*

$$\partial\begin{pmatrix} a & \alpha \\ \beta & b \end{pmatrix} = \begin{pmatrix} b & \beta \\ \alpha & a \end{pmatrix}.$$

Then $\partial \in \mathrm{Aut}(\mathbb{O}(q))$ *if and only if* $q = 2^n$.

Finally, we look at conjugations. Let L be a Moufang loop. For $x \in L$, define the conjugation $T_x : L \longrightarrow L$ by $T_x(y) = x^{-1}yx$, where $x^{-1}yx$ is unambiguous thanks to the properties of L. Not every conjugation of L is an automorphism. By [6, Thm IV.1.6], $T_x \in \mathrm{Aut}(L)$ if $x^3 = e$. (And it is not difficult to show that $x^3 = e$ is also a necessary condition, provided L is simple.)

6 Transitivity of the natural action

We take advantage of the automorphisms defined in section 5, and investigate the natural action of $\mathrm{Aut}(M^*(2))$ on $M^*(2) = M(2)$. The lattice of subloops of $M^*(2)$ was fully described in [8]. Here, we only focus on the action of $\mathrm{Aut}(M^*(2))$ on involutions and on subgroups of $M^*(2)$ isomorphic to V_4.

Once again, identify $\mathbb{O}(2)$ with the vector matrix algebra.

By Lemma 2.2, $M^*(2)$ contains only elements of order 1, 2 and 3. Furthermore, every involution $x \in M^*(2)$ is of the form

$$\begin{pmatrix} a & \alpha \\ \beta & a \end{pmatrix},$$

for some $a \in \{0, 1\}$ and $\alpha, \beta \in k^3$. In order to linearize our notation, we write $x = [\alpha, \beta]$ when the value of a is clear from α, β, or when it is not important; and $x = [\alpha, \beta]_a$ otherwise.

Similarly, every element $x \in M^*(2)$ of order 3 is of the form

$$\begin{pmatrix} a & \alpha \\ \beta & 1+a \end{pmatrix}.$$

To save space, we write $x = \{\alpha, \beta\}_a$.

We will sometimes leave out commas and parentheses. Thus, both 110 and (110) stand for $(1, 1, 0)$. The commutator of x and y will be denoted by $[x, y]$.

Proposition 6.1 *Let $x = [\alpha, \beta]_a$, $y = [\gamma, \delta]_b$ be two involutions of $M^*(2)$, $x \neq y$. Then:*

(i) *$[x, y] = e$ if and only if $|xy| = 2$ if and only if $\langle x, y\rangle \cong V_4$ if and only if $\alpha \cdot \delta = \beta \cdot \gamma$.*

(ii) *$[x, y] \neq e$ if and only if $|xy| = 3$ if and only if $\langle x, y\rangle \cong S_3$ if and only if $\alpha \cdot \delta \neq \beta \cdot \gamma$.*

(iii) *x is contained in a subgroup isomorphic to S_3,*

(iv) *every subgroup of $M^*(2)$ isomorphic to S_3 contains an involution of the form $[\ ,\ ]_0$.*

Proof The involution x commutes with y if and only if $|xy| = 2$. Since

$$xy = \begin{pmatrix} ab + \alpha \cdot \delta & - \\ - & ab + \beta \cdot \gamma \end{pmatrix},$$

parts (i) and (ii) follow.

Given $x = [\alpha, \beta]_a$, pick $\delta \in \alpha^\perp$, $\gamma \notin \beta^\perp$, and choose $b \in \{0, 1\}$ so that $y = [\gamma, \delta]_b \in M^*(2)$. Then $\langle x, y\rangle \cong S_3$, and (iii) is proved.

Let $G \leq M^*(2)$, $G \cong S_3$. Let $x = [\alpha, \beta]_1$, $y = [\gamma, \delta]_1 \in G$, $x \neq y$. Then

$$xy = \begin{pmatrix} 1 + \alpha \cdot \delta & \alpha + \gamma + \beta \times \delta \\ \beta + \delta + \alpha \times \gamma & 1 + \beta \cdot \gamma \end{pmatrix}.$$

Since $|xy| = 3$, we have $\alpha \cdot \delta \neq \beta \cdot \gamma$. In other words, $\alpha \cdot \delta + \beta \cdot \gamma = 1$. Then the third involution $xyx \in G$ equals

$$\begin{pmatrix} 1 + \alpha \cdot \delta + (\alpha + \gamma) \cdot \beta & - \\ - & - \end{pmatrix} = \begin{pmatrix} \alpha \cdot \beta & - \\ - & - \end{pmatrix}.$$

But $\alpha \cdot \beta = 0$, as $\det x = 1$. □

Lemma 6.2 *Let $x \in G \leq M$ be an element of order 3, $G \cong S_3$, and M a Moufang loop. Then $T_x \in \mathrm{Aut}(M)$ permutes the involutions of G.*

Proof Let $G = \langle x, y \rangle \cong S_3$ with $|y| = 2$. The remaining two involutions of G are xy and yx. Then $T_x(y) = x^{-1}yx = xy$, $T_x(xy) = yx$, and $T_x(yx) = y$. This can be seen from any presentation of G, or easily via the natural representation of S_3 with $x = (1, 2, 3)$, $y = (1, 2) \in S_3$. □

We are going to show that $\mathrm{Aut}(M^*(2))$ acts transitively on the subgroups of $M^*(2)$ isomorphic to C_2 — the *copies* of C_2 in $M^*(2)$. Let

$$x_0 = [111, 111]$$

be the canonical involution. For a vector α, let $\mathrm{w}(\alpha)$ be the *weight* of α, i.e., the number of nonzero coordinates of α.

Proposition 6.3 *The group* $\mathrm{Aut}(M^*(2))$ *acts transitively on the copies of C_2 in $M^*(2)$.*

Proof Let $x = [\alpha, \beta]_a$ be an involution. We transform x into x_0. By Proposition 6.1(iii), x is contained in some $G \cong S_3$. By Lemma 6.2 and Proposition 6.1(iv), we may assume that $a = 0$.

Let $r = \mathrm{w}(\alpha)$, $s = \mathrm{w}(\beta)$. Using the automorphism ∂ from Lemma 5.3 we can assume that $r \geq s$. We now transform x into x' so that $x' = x_0$ or $x' = x_1$ or $\langle x', x_0 \rangle \cong S_3$ or $\langle x', x_1 \rangle \cong S_3$, where $x_1 = [100, 100]$.

If $r \not\equiv s \pmod 2$, then $\langle x, x_0 \rangle \cong S_3$, by Proposition 6.1(ii). Suppose that $r \equiv s$. Every permutation of coordinates can be made into an automorphism of $M^*(2)$, by Lemma 5.2. The involution x_0 is invariant under all permutations. Since $a = 0$, we must have $s > 0$, and thus $(r, s) = (2, 2)$, $(1, 1)$, $(3, 1)$ or $(3, 3)$. If $(r, s) = (2, 2)$, transform x into $x' = [110, 011]$, and note that $\langle x', x_1 \rangle \cong S_3$. If $(r, s) = (1, 1)$, transform x into $x' = x_1$. If $(r, s) = (3, 1)$, transform x into $x' = [111, 001]$. Once again, $\langle x', x_1 \rangle \cong S_3$. Finally, if $(r, s) = (3, 3)$, we have $x = x' = x_0$.

Now, when $\langle x', x_i \rangle \cong S_3$ for some $i \in \{0, 1\}$, we can permute the involutions of $\langle x', x_i \rangle$ so that x' is mapped onto x_i, by Lemma 6.2.

It remains to show how to transform x_1 into x_0. For that matter, consider the element $y = \{001, 101\}_1$, and check that $x_0 = T_y(x_1)$. □

Let us now continue by showing that there are at most two orbits of transitivity for copies of V_4. (In fact, there are exactly two orbits but we will not need this fact.)

Put

$$u_0 = [000, 110], \quad u_1 = [001, 001], \quad u_2 = [100, 010].$$

Lemma 6.4 *Let $V_4 \cong \langle u, v \rangle \leq M^*(2)$. There is $f \in \mathrm{Aut}(M^*(2))$ such that $f(u) = x_0$ and $f(v)$ is one of the two elements u_1, u_2.*

Proof By Proposition 6.3, we may assume that $u = x_0$. Write $v = [\alpha, \beta]$, $r = \mathrm{w}(\alpha)$, $s = \mathrm{w}(\beta)$. We have $r \equiv s$, else $\langle u, v \rangle \cong S_3$. Thanks to the automorphism ∂, we may assume that $r \leq s$. If $(r, s) = (0, 2)$, transform v into u_0; if $(r, s) = (1, 1)$, into u_1 or u_2; if $(r, s) = (1, 3)$, into $u_3 = [001, 111]$; if $(r, s) = (2, 2)$, into $u_4 = [110, 110]$ or $u_5 = [011, 101]$.

Let $v_1 = \{010, 110\}_0$, $v_2 = \{001, 101\}_0$, and define $f_1 = T_{v_2^{-1}} \circ T_{v_1}$, $f_2 = T_{v_1^{-1}} \circ T_{v_2}$ (compose mappings from the right to the left). Then f_1, f_2 are automorphisms of $M^*(2)$, and one can check directly that $f_1(x_0) = f_2(x_0) = x_0$. Moreover, $f_1(u_4) = u_1$, $f_1(u_3) = u_2$, $f_1(u_5) = u_3$, and $f_2(u_5) = \partial(u_0)$. Thus u_4 can be transformed into u_1, and each of u_0, u_3, u_5 into u_2. □

7 Main result

We are now ready to demonstrate that the map $h : \mathbb{O}(2) \longrightarrow \mathbb{O}(2)$ constructed in section 4 is an extension of g.

Proposition 7.1 *Let C be a composition algebra, and let $M \subseteq C$ be the set of all elements of norm 1. Assume that $x, y \in M$ are such that $x + y \in M$. Then $g(x + y) = g(x) + g(y)$ for every $g \in \mathrm{Aut}(M)$.*

Proof If $x = y$, we have $1 = N(x + y) = N(2x) = 4N(x) = 4$. Therefore the characteristic is 3, and $g(x) + g(x) = -g(x) = g(-x) = g(x + x)$.

Assume that $x \neq y$. By Lemma 2.3, $|xy^{-1}| = 3$, and so $|g(x)g(y)^{-1}| = |g(xy^{-1})| = 3$ as well. Then $N(g(x) + g(y)) = 1$, again by Lemma 2.3. Consequently, we use Lemma 2.4 twice to obtain $g(x) + g(y) = -g(x)g(y)^{-1}g(x) = g(-xy^{-1}x) = g(x + y)$. □

We proceed to prove by induction on the number of summands that

$$g(\sum_{i=1}^{n} x_i) = \sum_{i=1}^{n} g(x_i)$$

for every $g \in \mathrm{Aut}(M^*(2))$ and $x_1, \ldots, x_n \in M^*(2)$ such that $x_1 + \cdots + x_n \in M^*(2)$.

Lemma 7.2 *Suppose that $x, y \in M^*(2)$, $x \neq y$, are such that none of $x + e$, $y + e$, $x + y$ belongs to $M^*(2)$. Then $\langle x, y \rangle \cong V_4$, and there are $a, b \in M^*(2)$ such that $a + b = e$, and $x + a$, $y + b \in M^*(2)$.*

Proof We have $N(x + e) = 0$, i.e., $N(x, e) = 0 - 1 - 1 = 0$. Then, by Lemma 2.1, $x^2 = (xe^{-1})^2 = -e = e$. Similarly, $y^2 = (xy^{-1})^2 = e$.

Since $\langle x, y \rangle \cong V_4$, we may assume that $(x, y) = (x_0, u_1)$ or $(x, y) = (x_0, u_2)$, where x_0, u_1, u_2 are as in Lemma 6.4. When $(x, y) = (x_0, u_1)$, let $a = \{011, 010\}_1$, else put $a = \{110, 100\}_1$. In both cases, let $b = e - a$, and verify that $x + a$, $y + b \in M^*(2)$. □

Proposition 7.3 *Let $x_1, \ldots, x_n \in M^*(2)$ be such that $x = \sum_{i=1}^n x_i$ belongs to $M^*(2)$. Then*

$$g(\sum_{i=1}^n x_i) = \sum_{i=1}^n g(x_i).$$

Proof The case $n = 1$ is trivial, and $n = 2$ is just Proposition 7.1. Assume that $n \geq 3$ and that the Proposition holds for all $m < n$. We can assume that at least two summands x_i are different, say $x_{n-2} \neq x_{n-1}$. Since $g(xx_n^{-1}) = g(x)g(x_n)^{-1}$, we can furthermore assume that $x_n = e$. When at least one of $x_{n-2} + e$, $x_{n-1} + e$, $x_{n-2} + x_{n-1}$ belongs to $M^*(2)$, we are done by the induction hypothesis. Otherwise, Lemma 7.2 applies, and there are $a, b \in M^*(2)$ such that $a + x_{n-2}$, $b + x_{n-1} \in M^*(2)$, and $a + b = e$. Therefore,

$$\begin{aligned} g(x) &= g(x_1 + \cdots + x_{n-3} + (x_{n-2} + a) + (x_{n-1} + b)) \\ &= g(x_1) + \cdots + g(x_{n-3}) + g(x_{n-2} + a) + g(x_{n-1} + b) \\ &= g(x_1) + \cdots + g(x_{n-1}) + g(a) + g(b) \\ &= g(x_1) + \cdots + g(x_{n-1}) + g(a + b), \end{aligned}$$

and we are through. □

Theorem 7.4 *Every automorphism of $M^*(2)$ can be uniquely extended into an automorphism of $\mathbb{O}(2)$. In particular, $\mathrm{Aut}(M^*(2))$ is isomorphic to $G_2(2)$.*

Proof Pick $g \in \mathrm{Aut}(M^*(2))$. Using the basic triple for $\mathbb{O}(2)$ from Lemma 3.1 construct an automorphism $h = \psi_{\mathbb{O}(2)}$ of $\mathbb{O}(2)$, as in Proposition 3.2. Then g, h coincide on a basis induced by the doubling triple. Every element of $M^*(2)$ is a sum of some of the basis elements. Hence, by Proposition 7.3, g and h coincide on $M^*(2)$.

This extension is unique. Thus $\mathrm{Aut}(M^*(2)) = \mathrm{Aut}(\mathbb{O}(2))$, and $\mathrm{Aut}(\mathbb{O}(2))$ is isomorphic to $G_2(2)$ by a theorem of Springer and Veldkamp. □

8 Acknowledgement

Most of this paper is extracted from the author's Ph. D. thesis [8]. I would like to acknowledge the support of the Department of Mathematics at Iowa State University. I also thank the Grant Agency of Charles University for partially supporting my visit to Oxford (grant number 269/2001/B-MAT/MFF).

Shortly before this paper was accepted for publication, Gábor P. Nagy and the author proved [4], using completely different methods, that $\mathrm{Aut}(M^*(k))$ is the semidirect product $G_2(k) \rtimes \mathrm{Aut}(k)$, for every perfect field k.

References

[1] J. H. Conway, N. J. A. Sloane, *Sphere packings, lattices and groups*, third edition, Springer Verlag (1999).

[2] H. S. M. Coxeter, *Integral Cayley Numbers*, Duke Mathematical Journal, Vol. **13**, No. 4 (1946). Reprinted in H. S. M. Coxeter, *Twelve Geometric Essays*, Southern Illinois University Press (1968).

[3] M. W. Liebeck, *The classification of finite simple Moufang loops*, Math. Proc. Cambridge Philos. Soc. **102** (1987), 33–47.

[4] G. P. Nagy, P. Vojtěchovský, *Automorphism Groups of Simple Moufang Loops over Perfect Fields*, submitted.

[5] L. Paige, *A class of simple Moufang loops*, Proc. Amer. Math. Soc. **7** (1956), 471–482.

[6] H. O. Pflugfelder, *Quasigroups and Loops: Introduction*, (Sigma series in pure mathematics; **7**), Heldermann Verlag Berlin (1990).

[7] T. A. Springer, F. D. Veldkamp, *Octonions, Jordan Algebras, and Exceptional Groups*, Springer Monographs in Mathematics, Springer Verlag (2000).

[8] P. Vojtěchovský, *Finite Simple Moufang Loops*, Ph. D. thesis, Iowa State University (2001). Available at `www.vojtechovsky.com`.

[9] M. Zorn, *Theorie der alternativen Ringe*, Abh. Math. Sem. Univ. Hamburg **8** (1931), 123–147.